SPACE

THE PHILIPS PAVILION

CALCULATED

LE CORBUSIER

IN SECONDS

EDGARD VARÈSE

MARC TREIB

SPACE
THE PHILIPS PAVILION
CALCULATED
LE CORBUSIER
IN SECONDS
EDGARD VARÈSE

MUSIC ANALYSIS:
RICHARD FELCIANO

Princeton University Press
Princeton, New Jersey

Published by Princeton University Press
41 William Street
Princeton, New Jersey 08540
In the United Kingdom:
Princeton University Press
Chichester, West Sussex

Book and Jacket Design: Marc Treib

This book has been composed in Sabon
and Gill Sans

CIP data appears on page 282

Princeton University Press books are printed
on acid-free paper and meet the guidelines
for permanence and durability of the Committee
on Production Guidelines for Book Longevity
of the Council on Library Resources
Printed in the United States of America

10 9 8 7 6 5 4 3 2 1

To / for Mary McLeod

CONTENTS

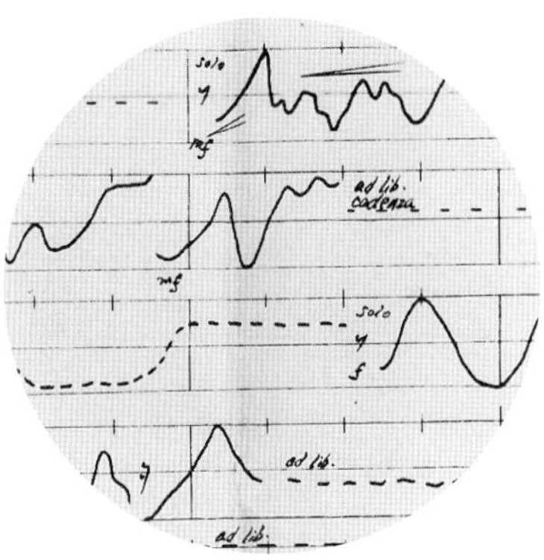

BRUXELLES 1958

PHILIPS

PREFACE

My renewed interest in the Philips Pavilion and the *Poème électronique* was stimulated by composers rather than architects. As a "resident outsider" on several Ph.D. oral examinations in musical composition at the University of California at Berkeley over several years, I was constantly reminded of the position Edgard Varèse occupies in twentieth-century music. Invariably it seemed, the discussion would reach a minor culmination with mention of Varèse's "spatialized" composition for the 1958 Philips Pavilion. While as a student I had known about the building, and had even owned a record of the piece before completing my architectural studies, I had never really grasped that the pavilion's music, or "organized sound," had been scored to be distributed from multiple points in space. Architecture was no longer restricted to playing the role of a reverberatory sounding board but became instead the source of sound: architecture and music were rendered coincident. Perhaps due to this initiating force outside my discipline, I have been less concerned in this volume with positioning the Philips Pavilion within the total output of Le Corbusier's architecture than as a story of collaboration between artists, architects, clients, and builders.

There is no denying that this pavilion was one eccentric building [Plate I]. Its raked structure and play of hyperbolic paraboloid shells had always struck me as curious, from the time I first encountered it in photographs while still a student (and had avidly sought the album of Varèse's music). Certainly, I thought, there must be a story behind a design like this, especially given its seeming variance from all other of Le Corbusier's works. There were other provocative aspects of the project, in particular the 480-second film—which fortunately remains in the Philips Company archives, and was available to me at the Fondation Le Corbusier on videocassette. Divorced from the architectural setting, denuded of its lighting, chroma and overprojections, the film is a linear skeleton for the multimedia project that constituted the *Poème*. And yet even in its "one track" form, the collage of images and its underlying humanism (and hubris) conveys a sense of what visitors to Brussels experienced in 1958.

As I began to research the building, I found little information within the standard publications on Le Corbusier's oeuvre, and even in the architect's self-compiled atlases, the project merits only two pages. Having been destroyed at the end of the exposition, the building was not available for investigation; like the film, this study drew upon a mosaic of available correspondence and text, film, photographs, and drawings. In spite of the plethora of architectural documentation, the Philips project possessed its own inherent interest, and I began to sense that the work on the pavilion constituted an intriguing model of postmodern collaboration *avant la lettre*.

I
The Philips Pavilion, World Exposition, Brussels, 1958. [Philips]

Architecture, color, voice, sound, and images were superimposed, without any full comprehension in advance of the nature of the resulting work. The completed *Poème électronique* would *emerge* as a conglomerate greater than any of its constituent parts, to some degree planned, to some degree the product of fortuitous accident.

If the collaboration and the work were not synchronized graphically they were rigidly structured in time. The piece would last exactly eight minutes—480 seconds dissected and planned to the instant. Here was Le Corbusier's greatest control over the work of the team.

In some ways, as I suggest in the text, the building is a less than beautiful result of a tormented process of design, development, and construction. The biography of the pavilion is, to my mind, not only revelatory in itself, but a mirrored microbiography of all its principal players. The calm and disciplined Louis Kalff, Philips Art Director and project liaison, and the experienced but intuitive Hoyte Duyster, the pavilion's engineer and builder, were two players whose contributions have been underplayed in the past recountings. And yet without either of them, the project would probably never have materialized. One can liken the *Poème électronique* to a Catholic requiem mass, to a Wagnerian operatic *gesamtkunstwerk*, more mundanely to a *son et lumière*, even to a 1960s psychedelic light show—or more drastically think of it as a prototype of virtual reality. Whatever we may think of the building and its spectacle in sound and light today, it was a wonder for its time—particularly because it eliminated the display of products that were the staple of expositions and because it was completely automated.

Notwithstanding the difficulties in realizing the structure and its presentation, the Philips Pavilion marked the end of one epoch of exposition thought, the mechanical, and the beginning of another, the electronic.

In spite of the relatively limited scope of this project, a number of people have been crucial to its realization. I should first cite Richard Felciano, Professor of Music Emeritus at the University of California at Berkeley, whose discussions of Varèse and his music provoked my interest in writing this book. I also thank him for his careful analysis of the *Poème électronique*, which I have cited at length in the text and in captions. Bart Lootsma, architectural historian, editor of the Dutch architectural journal *de Architect* and curator of an exhibition on the Philips project at the technical university in Eindhoven in 1983 (the twenty-fifth anniversary of the pavilion's rise and fall), was more than generous in sharing his extensive knowledge of the project and his sources for images and information. As the notes to the text will attest, in a number of places I have relied entirely on his knowledge, comments, and most of all his generosity, which far exceeded the norm. I am also grateful for a number of photographs and other images he provided. Harm Tillman helped in establishing contact with the various Dutch enterprises early in the project, and has remained a good tour guide, a rich source of information, and a good listener (a rare combination, and I thank him).

Holy Raveloarisoa at the Fondation Le Corbusier in Paris was the perfect archivist, with a singular command of the material and generous with her time and aid. Similarly, I. J. van Blanken and J. C. van den Biggelaar, archivists at the Philips Corporation Archives, and N.A.M. Verhoeckx, head of Philips Research Publicity in Eindhoven, were extremely helpful in making information and photographs available and providing answers to many trifling yet critical points. To both archives, I also give my appreciation for allowing me to reproduce photographs, especially since today these constitute the building's primary documentation. The two issues of the *Philips Technical Review*, the English editions of which were reprinted in 1958 due to popular demand, remain the most precise record of the pavilion's manifold aspects. That the material has always been open to use by others is a testament to the open-mindedness of the company. A. C. Pronk, public relations director for the Hollandsche Beton Groep in Rijswijk, amiably shared his time with me reviewing the biography of H. C. Duyster and Strabed's participation in the Philips project. HBG also provided a number of photographs and a copy of the film on constructing the building for which I am appreciative. The Arxiu de la Sagrada Família provided photographs of Gaudí's structural models, facilitated by Manuel Ribas y Piera—I thank them both. Other sources are cited in the captions. Thanks are also due to Edward Craanen, former Director of the Netherlands Institute in Paris and now again with the state department, who translated the majority of the Dutch source material with enviable skill and great enthusiasm, and to Wim de Wit, curator of special collections, the Getty Center for the History of Art and the Humanities.

I am told by native speakers that neither Le Corbusier nor Louis Kalff wrote particularly well in French, which created significant problems with translation into English. Should one render the material clause by clause, creating an English text difficult to follow and suggesting a bad, if literal, translation? Or should one adapt the writing, retaining the ideas and key words, while trying to reform the text into passable English? I have opted for the latter course. While I first struggled with the French correspondence and publications myself, it was Dorothée Imbert who continually pointed out my frequent detours from the true path, noted that translation is neither as literal nor as easy as I would have it, and helped shape the citations into reasonable renditions of their Gallic originals. Ms. Imbert also freely criticized a later draft of the book, a reading which every subsequent reader should greatly appreciate. I certainly do. Francesco Passanti first pointed out the connection between Le Corbusier and Emile Jaques-Dalcroze, which turned out to be a significant element of the story, and I am grateful for his insight. Mary McLeod has been a great source of sources, critical reader, instigator of depressing revisions and yet offering constant encouragement—all of which I appreciate greatly, considering that my wandering into Le Corbusier territory unannounced can be seen as a grave transgression in certain quarters. Jean-Louis Cohen and Robin Middleton reviewed the full manuscript and offered valuable suggestions for its improvement. Ellen Harrison helpfully critiqued the music chapter. To have addressed all the comments these readers raised would probably have doubled the length of the text and the time of writing—and one has to draw the line somewhere. Any errors in interpretation, however, can be assigned only to the author.

Although at times the funding was actually granted for other projects, this book was clandestinely written with some time taken from studies supported by the John S. Guggenheim Foundation, the Fulbright Western European Research program, and the Humanities Institute and Committee for Research of the University of California at Berkeley.

Finally, I wish to thank Elizabeth Powers for taking such a strong interest in the project and allowing for me to see the book through design and production. The collaboration has been a model one, and I am deeply grateful. Elizabeth Johnson has been an exemplary copy and production editor. Chuck Byrne provided valued computer aid in developing the book's design.

At this stage in any project, one becomes only too aware that any focused study is but a part of a larger arena. Like the concentric waves radiating outward from a stone dropped in a pond, each building suggests linkages with a limitless body of other architecture and ideas. Even if examined in terms of its architecture alone, the Philips Pavilion is but one part of the total work of Le Corbusier and Xenakis, which in turn is but a part of the work in France, Europe, and the world at mid-twentieth century, which in turn is but a part of the history of architecture. And so on. Where to stop? One stops either when exhaustion sets in (which it has), or the point at which it seems as if at least this chapter of the greater story appears to be reasonably complete (which I believe it does).

Marc Treib
Paris
December 1993

2
The expectant crowd before the Philips Pavilion, Brussels, 1958.
[*Philips Technical Review*]

INTRODUCTION

In the winter of 1956, when a building committee approached Le Corbusier about designing a pavilion for the Philips company at the upcoming world's fair, the architect was almost seventy years old with a long and distinguished career. He had built houses, housing, churches and exhibition pavilions and had conceived plans for parts of cities and cities in their entirety. He was then engaged in what would be his largest realized project, the governmental center for the new capital of Chandigarh for the Indian state of Punjab. To Le Corbusier, the design of an exhibition pavilion of 6,500 square feet must not have seemed like much of a challenge. He initially demurred.

One aspect of the proposal piqued his interest from the very outset, however, and ultimately led him to accept the commission: making the first electronic poem. Restricted in the past to the spaces and forms of architecture and painting, Le Corbusier now had the possibility of creating a truly synthetic artwork. As one of the world's premier manufacturers of electronic equipment for illumination and sound reproduction, Philips offered Le Corbusier access to some of the most advanced technological means of the day. Exploiting the aesthetic potential of that technology, the architect sought to create a synthetic work of art that integrated the properties of color, sound, light, and rhythm. That work, the *Poème électronique*, presented to over a million visitors to the Philips Pavilion during the six months of the fair, positioned Le Corbusier and his collaborators at the forefront of advanced visual media.

In early writings, such as *Towards a New Architecture* (1923), Le Corbusier focused on the problems of the dwelling and the city. The machine and rationalized production offered hope to a France still recovering from the devastation of the Great War. Le Corbusier's villa projects, which followed later in the decade, represented an architecture comprised of its primary elements, substituting the purity of architectonics for the iconography of historical style. Yet even within these seminal structures, a formal vocabulary related to cubism—or Le Corbusier's and Amédée Ozenfant's variant, purism—appeared in the free curves of walls that opposed the prismatic enclosing shell. Thus, even to an architecture drained of pictorial references, the presence of painting remained. Interrelated spaces replaced discrete rooms and joined diverse levels, while the distinction between interior and exterior was effaced in projects such as the noted Villa Savoye at Poissy-sur-Seine (1929).

Critics have held that Le Corbusier's paintings were important as investigations for the formal ideas that might later find realization in the architect's constructed works. Today, his paintings and sculptures are usually not regarded as first-rank, but lagging behind the more innovative and conceptually rich work of painters such as Fernand Léger and Joan Miró, not to mention Georges Braque, Juan Gris, and of course, Pablo Picasso. Le Corbusier himself, however, directed those interested in understanding his architecture to his paintings. The rooftop of the apartment block in Marseilles (1952), for example, comprised a sculptured landscape of forms recalling those appearing in his paintings for over two decades. In most cases, the painted shape preceded the sculptural and architectural form. Thus, the primary—but one assumes not exclusive—relationship between painting and architecture was sequential, the latter following from investigations in the former.

A second medium of inquiry was also consequential in the development of Le Corbusier's artistic ideas. From about the mid-1920s, the architect communicated architectural and social planning ideas with photomontage, creating a complex visual space layered on a two-dimensional plane. The cover for *Des canons, des munitions? Merci! Des logis . . . s.v.p.* (1938) positioned airplanes and howitzer shells over fragments of city plans, heightening the impact of the composition with radical dispositions of photographic images. The previous year, both Le Corbusier and his colleague Léger used montages of man, machine, painted shape, and architecture in the exhibits for the Pavillon des Temps Nouveaux at the 1937 Paris fair. Figures and polychromy articulated the surface of the mural, while their montage mixed abstraction with reality and detached each given element from its original context.

In his early houses in La Chaux-de-Fonds, Le Corbusier tinted the rooms a dominant shade in accord with tradition. As the definition of the room dissolved in his modernist villas, however, the wall planes assumed a more independent and assertive identity and instigated a new attitude toward color. Now, the wall became part of an overall composition in space, treated as a monochromatic plane. Fernand Léger had noted: "People are going to cover the walls. But as much as possible it should be done simply, through the application of color with no significance other than color itself (perhaps evocative objects can operate within it); color is itself a plastic reality. It is a new realism."[1] Using color's psycho-physiological properties, artists and architects attempted to create a sense of depth without using perspective. Only in the late 1930s did Le Corbusier truly attack the integrity and the purity of the wall as a plane in a series of murals for his own house at Cap Martin and another "villa on the Côte d'Azur." This transformation of the wall through color and drawing, with the architect's coeval spatial displacements using photomontage, prefigured the fracture and collision of the images of the electronic poem that filled the Philips Pavilion.

These investigations in two-dimensional representation and the four-dimensional (space-time) experience of architecture marked Le Corbusier's nascent interest in the integration of the arts that occupied his thinking after the second world war. He based his proposal for the 1950 "Synthesis of the Major Arts" pavilion at the Porte Maillot on the premise that "the major arts are empty, divided and isolated; the world is waiting: in fifty years architecture has left the stage of regeneration; a desire for synthesis is apparent, a desire for harmony." In response, Le Corbusier conceived the building "with the object of bringing painters and sculptors into contact with tasks of an architectural nature."[2] The essay "Ineffable Space," first published in 1946, revealed the distance between the machine infatuation of the 1920s and the more spiritual dimensions of architecture in the 1950s. Here, the architect declared that "the release of aesthetic emotion is a special function of space"; it was an emotion that devolved from the dissolution of spatial definition and perceived architectural limits.[3]

These were the years during which rough concrete walls superseded the flat planes of the 1920s, the years of the sculptural forms of Marseilles and the chapel of Notre-Dame-du-Haut at Ronchamp. This small chapel manifested the full range of Le Corbusier's talents, with space, form, color, painting, and spirituality announced by the sun as it traversed the skies. All was rendered by an ever changing light, creating an "ineffable space." Seen in this context, the Philips project was a true, if volatile, culmination of Le Corbusier's search for a synthesis of the arts. Ronchamp was an architectural entity whose form resonated with landscape and liturgy—what Le Corbusier termed "visual acoustics," whose apertures directed the passages of light and color. The Philips Pavilion, on the other hand, presented a collage liturgy for twentieth-century humankind, dependent on electricity instead of daylight and on virtual perspectives in place of terrestrial views.

In the architect's mind, the small structure to house this synthetic program of 480 seconds needed to evoke electronic space and employ mathematics. Le Corbusier's connection with the pavilion and spectacle was more as an impresario than as its project architect, and Iannis Xenakis developed the formal ideas for the pavilion and determined its geometries. Edgard Varèse, who composed the "organized sound" that augmented the projections of light and color, worked on the edge of musical development by mixing real sounds with those of the voice. His *Poème électronique* remains a key work in twentieth-century music. And there were other collaborators as well, each of whom played a significant part in the project. Philippe Agostini created the film montage from images selected by Le Corbusier but arranged by Jean Petit. The builder-engineer H. C. Duyster constructed the pavilion using calculations by C.G.J. Vreedenburgh, A. Bouma, and F. K. Ligtenberg; Willem Tak served as acoustical consultant; and S. L. de Bruin automated the production.[4] Considerable credit must also go to Louis Kalff, art director of Philips and project liaison, who kept the enterprise on track during a short but arduous two and a half years [figure 3]. Although the pavilion was the brainchild of Le Corbusier, its realization and its electronic poem were truly a col-

3
Louis Kalff in the unfinished pavilion. Early 1958. [Philips]

laborative effort, a complex undertaking completed in an incredibly condensed period of time.

The Philips Pavilion and its visual presentations are today almost completely forgotten; in the field of music, however, Varèse's *Poème électronique* retains a reputation as a masterwork—even if today one must hear the music detached from its original setting. And despite the shadow of oblivion that has fallen over the Philips Pavilion, the structure must be seen as a key work in Le Corbusier's career. For this small building in Brussels, which existed for barely eight months, came closer to materializing Le Corbusier's aspirations for a fusion of the arts than any other of his constructed works. Today there remain only a few photographs, a recorded score, a handful of drawings, and a film. As a result, this history of the Philips Pavilion is a reconstruction extracted from these documents and correspondence in the archives of the Fondation Le Corbusier, the Philips Company, the Hollandsche Beton Groep, and the Special Collections of the Getty Center for the History of Art and the Humanities.

Although I regard this study primarily as an architectural history of the pavilion, it will, I hope, function equally as a chronicle of artistic collaboration. It is not my purpose to advance Le Corbusier as the sole generating genius, under whose aegis the project was shaped from concept to detail. While the architect certainly conceived the building and determined the nature of the spectacle presented within, he benefited from dedicated and talented collaborators. Despite casting the book as an architectural history, I have also tried to examine the milieu and techniques for each of the media that comprised the electronic poem. For the most part, the story is told with only limited analysis, unfolding as a chronicle until the last chapters, which examine the project in greater critical detail.

The first chapters establish the context in which the project was initiated and trace the course of the pavilion's design. The development of the form through engineering and construction follows thereafter. Subsequent chapters examine the historical precedents for spatialized music and the evolution of Edgard Varèse's music, and the making of the Philips film and its relation to montage as conceived by Sergei Eisenstein. As light and color also figured prominently in the making of the multimedia presentation, they form the core of the fourth chapter, which investigates the ideas and techniques of the visual effects. The penultimate chapter recounts the reactions of journalists and professional writers to the exposition as a whole and the Philips project in particular. The final chapter offers a critical reappraisal of the project from our perspective at the end of the century and within the lives of each of its principal creators.

Although a little building of brief lifespan, the Philips Pavilion, with

4
Le Corbusier (left) and Edgard Varèse. Probably Eindhoven, February 1958. [Philips]

its spectacle of amplified sound and rhythmically orchestrated light and color, was a landmark in electronic media technology that concomitantly tested the limits of architecture, both concrete and virtual. When seen against the buildings and arts of its time, when seen as Le Corbusier's synthesis of the arts, the Philips project assumes justified importance [Figure 4]. While in some ways neither the architecture nor the spectacle fully realized its complete potential, in other ways all aspects of the project were prescient. If the Philips project did not locate the precise point at which all the arts—traditional and electronic—would intersect some time in the future, it did provide the unquestionable directional signs toward that point. For this reason, the Philips project warrants a careful examination.

5
The Philips Pavilion seen from above.
[Getty Center]

SPACE

THE PHILIPS PAVILION

CALCULATED

LE CORBUSIER

IN SECONDS

EDGARD VARÈSE

1. COMMISSION

On 16 January 1956, Mr. R. d'Aboville, an engineer and management "attaché" with the Philips corporation in Paris, wrote to Le Corbusier:

> We would be very pleased to discuss with you a project which we are planning to build at the next world's fair in Brussels, a project for which we would be quite pleased to have your participation.
>
> We would be very grateful if you would arrange a meeting with us to discuss the subject.[1]

It was the kind of letter that many architects shrug off, realizing that the vast majority of the time nothing of the inquiry ever comes to pass. On the other hand, architects rarely dismiss such requests without at least establishing the parameters of the project in greater detail. Penciled on Le Corbusier's copy of the letter is the note, "L.C. finds it interesting."

The meeting was confirmed for 24 February at the Le Corbusier studio, 35 rue de Sèvres, in Paris. Among those present were Camille Goemans, head of Philips public relations and information, and Louis C. Kalff, architect, engineer, and artistic director for Philips. Kalff would become the corporation's liaison for the pavilion project, and in many ways, he would be the real client. Both parties seem to have been pleased by the discussions, although Le Corbusier did not immediately accept the commission. Back in Eindhoven, Kalff issued an internal memo that recorded the proceedings and confirmed the idea for a pavilion without displays of consumer products:

> We enthusiastically put forward our ideas for a spatial-color-light-music production at the world's fair . . . and noted with pleasure the interest of Mr. Le Corbusier in our plans. Undoubtedly, he is the only great talent capable of dealing with such a work. Like many great men, he is very egocentric and works according to his own ideas, but we believe that in this case we should grant him a free hand if we truly wish to realize a unique, world-class demonstration in Brussels which will be one of the most interesting sights of the exposition.
>
> . . . Because this project was conceived as an "interior," he was indifferent to the external appearance of the building. He was, of course, willing to discuss this and proposed, for example, that the external construction could be made visible, which might be an interesting solution. In this regard, he would be willing to cooperate with the Dutch architects.
>
> In ten days, Mr. Le Corbusier will travel to the Punjab in India, where he is building an entire city. Within two months he will return. In the meantime, he will think about the project. In a second meeting, we will discuss more completely his plans and conditions. Mr. d'Aboville will see to it that [Le Corbusier] devotes appropriate attention to the project assigned to him.

> We are personally very optimistic about the uniqueness and the publicity value of the collaboration with Mr. Le Corbusier, who had already turned down two commissions for the fair, among them the Belgian Eternit Fabrieken.[2]

Although he was not without uncertainty—and perhaps even misgivings—about working with an architect of such an independent and uncompromising nature, Kalff could see the potential for producing a building and a spectacle that would establish a new standard for world exposition presentations many years to come. To realize this idea, he persuaded the Philips management to envision the potential for a spectacular design that would simultaneously test the limits of Philips technological resources while offering to the public an electronic display the likes of which had never been seen. The minutes of the directors' meeting reveal Kalff's influential role in conceiving the program:

> Mr. Kalff has had some talks with the architect Le Corbusier on an idea proposed by Messrs. Hartong, Numann, Kalff, and Van Lanschot. This idea consists in making a spectacular demonstration of a synthesis of light, music, space and color, all of this using the most advanced technical means. Within the pavilion, sessions of about eight minutes should take place for about 20,000 visitors a day, which means about 10% of all the visitors to the exposition. The entire project must be seen in the field of art and should be executed by internationally renowned artists: Le Corbusier for the pavilion, Benjamin Britten for the music, Zadkine for the decoration.
>
> In his talks with Engineer Kalff, Mr. Le Corbusier said that he had already formed his conditions but that he was willing to accept this idea. During the talks, Mr. Le Corbusier made known some wishes [demands] that should be fulfilled: that is, that he himself wanted to make the eight minutes of "scenario"; that he himself would be charged with the design of the interior (Mr. Rietveld for the exterior); for the music he wanted Mr. Varèse. . . . However, the difficulty is that Mr. Le Corbusier has no concrete ideas for the execution of the project—except some very raw ideas—and that he does not wish to give any guarantees concerning the costs and the time of delivery, though we raised these issues several times. An additional difficulty for us is that only after the scenario [is established] can we determine whether we will be capable of delivering the necessary equipment on time.
>
> Notwithstanding the high fee (according to Mr. Kalff, the building costs will work out better than expected), Mr. Kalff proposes to go with Mr. Le Corbusier because [he believes] that after solving the many vagaries, the result will be *the* success of the entire exposition.

The memo ended: "Mr. Tromp's proposal is accepted. Mr. Kalff will contact Mr. Le Corbusier."[3]

To diminish the disquiet remaining in Kalff's mind, a second meeting with the Philips group and the architect was set for 8 June. Some weeks after the conference, Kalff wrote to Le Corbusier, restating and clarifying the company's intentions:

> It goes without saying that the architectural form and the space to be created is of the greatest importance and will establish the artistic level of the production. For that reason we are soliciting your participation. However, the spectacle presented within your creation must in one way or another impress the crowd of visitors to the exposition. . . . Thus, it is necessary that the fundamental thought behind the presentation . . . will also be shared by the composer who creates the libretto. This libretto, however, must be developed not by the musician alone, but equally by light and color specialists, who must be artists as well. Our technicians anticipate collaborating on the last three aspects [of the presentation] and request several months to produce the music, make the tests for light and color, correct or modify the music during the trials. It is for that reason that we would prefer a composer who could stay and work not far from Eindhoven or from Paris.[4]

Kalff also discussed the matter of commissioning Varèse for the musical score, saying that due to their tight schedule, "his composition should be in our hands by the beginning of March 1957, at the latest, following the period of collaboration by several months."[5] Within the Philips organization, there was some discussion of relocating the pavilion to Eindhoven after the exposition ended, perhaps to serve as a corporate museum. Due to the additional problems and expenses such requirements would impose, however, the proposal was shelved indefinitely.[6] In the end, the building—and its electronic poem—stood for the six months of the fair, with only two more months of grace thereafter.

Giving Le Corbusier the summer of 1956—including his religiously observed August vacation in the south of France—Kalff requested that by September the conceptual design for the pavilion be established and the visual and musical presentations defined; the Philips technical staff could then proceed with their part of the operation. The architects for the neighboring Dutch pavilion were anxious to initiate construction in September (1956), and it was crucial for them to have a good idea of the surface area and height of the Philips project: in that way, "a harmony with the other buildings of the Dutch section" would result.[7] Could the architect present these materials by the end of August? Or, should he prefer, the Philips representatives could journey to the architect's summer house at Cap Martin on the French Riviera. Not missing the chance to pitch the importance of the commission, Kalff ended with a pleasant: "In any case, we hope that at the end of your vacation, we will find you refreshed and filled with ideas for our pavilion and the presentation."[8]

Given the course the design of the pavilion and the realization of its musical score and visual presentations would take—hampered by delays, revisions, differences in opinions and working methods, clashes with management, and internal divisions within the Le Corbusier studio itself—this schedule was optimistic, utopian, and almost comical. Thus began the Philips anxiety that would reach crisis proportions at the end of the following year.

By the end of summer 1956, Kalff had received nothing from Le Corbusier. The world's fair authorities, like the architects of the Dutch pavilion and the Philips team, were anxiously awaiting a definitive description of the pavilion. Three months had passed since Kalff had last written the architect; he now posed general questions and solicited specific answers:

> What will be the weight of the pavilion and how will the foundations be made? What electrical capacity will be necessary for the functioning of the building? What will be the assessed value for insurance purposes? Who will be the contractors and what guarantee will be given that the pavilion's construction will be completed by the opening day of the exposition?[9]

Given the state of the pavilion's design—i.e., virtually nonexistent as Le Corbusier had not really begun—these questions were rather ridiculous. Le Corbusier answered Kalff's letter with a flow of eyewash: he sent a program from a concert at the Munich Opera, which included the name of Iannis Xenakis, his project architect, and the catalog of his own exhibition at Lyons.[10]

If Kalff had become nervous by this time, he did his best to hide it from the Philips organization. At the beginning of October, he wrote J. Pugliesi-Conti at the Paris office informing him that the commission would be given to Le Corbusier:

> After our presentation, during which we expressed our concerns about working with Mr. Le Corbusier, we have nonetheless decided to go for it, that is to say, to ask Mr. Le Corbusier to design a preliminary project for the pavilion itself before he leaves [for India]. The sketches for this pavilion must be discussed with us, and must be sufficiently detailed before 1 November, so that we can continue the study of the pavilion with [Le Corbusier's] draftsman in Paris during his absence.[11]

Philips reserved the right to cancel the project if the cost estimates were too high or if the design or sound and light presentations were not to their liking, or if they believed they could not be realized in the given time.

The letter contract from Kalff dated 13 October confirmed a telegram he had sent Le Corbusier two days before. Kalff's letter was again both general and specific, discussing schedule while acknowledging the fee for designing the pavilion and the electronic presentation at 10 million francs. This sum was to be dispensed in four installments: 3.5 million at the commencement of the work; 3.5 million after the submission of the project and scenario (at the end of December); 2 million after the reception of the definitive scenario (1 April 1957); and the final million at the opening of the exposition.[12]

He again requested information on foundations, electrical load, and water services as they would affect the design of the Dutch pavilion. Given the very tentative state of the design, there was no way that these could be reasonably determined: the request was essentially futile. The original schedule called for work on the sound and light components of the presentation to begin at Eindhoven by December; at which time the completed scenario was to be in Kalff's hands. Philips reaffirmed their right to terminate the contract at that stage. The company agreed to Le Corbusier's nomination of Edgard Varèse as collaborator, and as requested, would pay him a sum of 3 million francs for his work, exclusive of travel expenses.[13]

At this moment, Varèse knew nothing about the project, with negotiations having taken place only between Kalff and Le Corbusier. The architect wrote to the composer on 12 June, for the first time soliciting his participation in the project. Le Corbusier described the building's program and his first thoughts on the sound and projected spectacle:

> *My idea is that music should have a part in this. In the darkness there will appear flashes of "black light," certain objects or atmospheres of violently different colors. The illumination (colored neons) will allow dynamic flashing drawings to be made and from time to time a realistic event, but occupying space with a striking presence.*
>
> *It is a scenario to be created wholly from relationships; light, plasticity, design, and music.*
>
> *Could you make the music? I can also tell you that it is Xenakis who will design the Pavilion and prepare the drawings to be used for the details and the synopsis of different sequences.*
>
> *. . . I hope that this will please you. It will be the first truly electric work and with symphonic power.*[14]

For Varèse, the commission was a boon, and he readily accepted: "As you have asked me [what I think about the proposal] I want to let you know immediately that I find your project superb and that I accept with great pleasure your offer of collaboration."[15] He would begin to ruminate at once on the musical ideas and their means of realization. Obviously excited at the prospects, Varèse added: "Like you [Le Corbusier] and the Philips company, the only thing in which I am interested is to give birth to 'the most extraordinary thing possible.'"

At this early stage Le Corbusier's ideas centered on the abstract properties of light, color, and sound, although the filmed episode of the *Poème électronique* became more photographic and illustrative as it developed during the ensuing year and a half. Abstraction would accompany figuration, but it would not dominate it.

In his reply to the proposed Philips letter contract, Le Corbusier questioned only one clause. Kalff had stipulated that Philips would retain the artistic rights to the pavilion and its contents after the exposition, and that the design and composition fees were, in essence, work for hire. Le Corbusier, in response, believed that the artistic rights should remain with the artist and asked to discuss the matter amicably at their next meeting. For the moment, it did not constitute a serious impediment.

Le Corbusier received the first payment from Philips on 19 October; the date of the first design review meeting was fixed for 24 October. Le Corbusier would leave for India shortly thereafter. Half a year had elapsed since the first meeting with the Philips representatives. With Iannis Xenakis assisting, Le Corbusier began to work.

2-1
Atomium, Brussels 1958.
A. and J. Polak with André
Waterkeyn, engineer.
[Marc Treib, 1994]

2. DESIGN CHRONICLE

In his response to the Philips commission, Le Corbusier deliberately avoided any discussion of the nature of an exhibition pavilion. Although early correspondence reveals that the scope of the project was restricted to the interior of the pavilion and its accompanying spectacle, Le Corbusier later stated that Philips had left him free to design the "facade" as he saw fit. But he did not intend to give them a facade for their products; instead, he would give them an "electronic poem":

> *I will not make a facade for Philips, but an electronic poem. Everything will happen inside: sound, light, color, rhythm. Perhaps, a scaffolding will be the pavilion's only exterior aspect.*[1]

The groundwork for the poem had been previously established by the Philips programming team, probably due to Louis Kalff's own interests in lighting and his personal intervention with the Philips management. The plan for realizing that idea and the building in which it would be housed would be formulated by Le Corbusier, however. Kalff and his associates had been struck by photographs they had seen of Le Corbusier's "small church in the Vosges" and thought that there might be some "elements also applicable to our pavilion."[2] The Brussels exposition would provide an ideal arena in which to verify Philips' emergence from setbacks during the war and by which to establish its multinational status. Technology in the service of the arts would be the medium.

The Brussels exposition was to be the first full-scale world's fair mounted since the war, and as such it was intended to celebrate the prosperity that followed the termination of hostilities in the European theater. The war had ushered in the Atomic Age, and in the 1950s the world was still guardedly optimistic about the peaceful applications of nuclear energy—in spite of the underlying anxiety created by the cold war and unnerving diplomatic incidents, such as those precipitating the Berlin Airlift of 1948–49. The Atomium, a colossal rendering of the steel molecule, served as the fair's theme-structure, observation tower, and constant reminder of technological potential [figure 2-1]. The pulse of the fair was thus caught between cautious confidence in growing economic prosperity and fission energy sources, and the lingering anxieties of the destruction that might descend, should tensions between the superpowers escalate.

Philips had survived the war years remarkably well in spite of heavy damage to their plants from aerial bombardment. On 10 May 1940 German troops invaded the Netherlands, and after four days of fighting, the government capitulated. Philips' top executives had made their way to England only a few days before the invasion; the company's fiftieth anniversary on 23 May 1941 was celebrated, somewhat sadly, in New York, where the directorship subsequently relocated.[3] During the war years the Philips headquarters was established—at least on paper—on the island of Curaçao in the Caribbean, where it remained until 1946. Factories in Britain and the United States, with other plants in Allied countries, took part in the effort against the Axis powers. The Eindhoven works continued to produce lightbulbs and other electrical products, although these became German war materiel rather than domestic products. On 6 December 1942, and again on 30 March 1943, British aircraft bombed Eindhoven, in attempts to destroy the Dutch industrial capability to deliver radio valves for German communication systems. As a result of the bombings, a quarter of the factory space and 70 percent of the company's offices were destroyed.

Spurred by widespread destruction, continuing economic recovery, and population growth in the 1950s, Philips began to diversify their consumer products. Even during the war years, management had been active in acquiring interests in other companies, for example, Philips' recording section operated under license from the British firm, Decca. With peace established, Philips developed new technologies in fluorescent lighting, loudspeakers, X-ray systems, telephones, among a host of other products, and augmented their interest in pharmaceuticals. More directly relevant to their pavilion for the Brussels fair, Philips researchers had been working on stereophonic sound reproduction since 1938, and in the postwar years they expanded their interests to include audio systems in wide-screen film projections such as Cinemascope and Todd-AO. These research and development efforts reverberated in architectural space. During the early 1950s, for example, Philips began producing sound systems that would invisibly ameliorate deficient natural sonics in spaces such as concert halls and churches.[4] In addition, the corporation was involved in the improvement of portable magnetic tape recorders and domestic stereo record players. Sales had been gratifying. By 1958, when the pavilion in Brussels was realized, Louis Kalff could write that now, sixty-seven years old, the multinational corporation employed 60,000 people in the Netherlands alone and 160,000 worldwide.[5]

The Philips corporation had participated in major trade and cultural exhibitions since the 1881 electrical fair in Paris, and had even erected their own pavilion at the Brussels exposition of 1935, and again in Brazilia in 1955.[6] At the 1937 Paris exposition, Kalff himself had executed a number of lighting assignments intended to showcase Philips products and know-how, enlivening the fair's nocturnal ambiance by using high-intensity light, color, and animation.[7] For their participation in Brussels in 1958, Philips sought a site that would maximize the corporation's presence in the public mind. There were three possibilities from which the management could choose. The Philips Pavilion could be located in the Belgian precinct, appropriate given that Belgium was the host country.[8] But Philips was not a Belgian company, and those products produced in Belgium did not represent the firm's full range of capabilities. Similarly, sharing space with the Dutch would also limit the potential to emphasize the firm's multinational operations. Clearly, the most attractive location was a site in the "neutral" zone, where the Philips Pavilion could stand free of any national associations.

The problem was timing. Given the late date on which the definitive decision about location was made, no site with sufficient area was still available. In the end, the best option for Philips was to share territory with the Netherlands on land ceded to them by the Dutch government.[9] The clock was moving quickly, with only two years remaining. The Netherlands pavilion, for example, was intended to be under construction by the end of 1956, a necessary timetable if construction were to be completed for the scheduled opening of the fair in April 1958. Time was critical for three additional reasons: the weather during the winter months seriously hampered construction, particularly fabrication in metal and concrete. Time was also necessary to install the displays, train the staff, and organize the logistics of handling millions of visitors for the some 200 days the fair would be open. And in the case of Philips, the company needed time to fine-tune the automation and electronic controls for the illumination, sound, and visual projections. Given these multiple concerns, the schedule was extremely tight, even if everything proceeded according to plan—which it did not. As it happened, the pavilion did not open on time and was closed for a month of reworking shortly after its official inauguration.

In spite of Philips' fiscal prosperity, a residual inferiority complex seems to have plagued corporate public relations. The minutes of the 17 January 1956 exhibition group meeting note that at least one committee member was convinced that the American pavilion—a palatial structure by Edward Durrell Stone, graced by gold poles and a decorative precast concrete screen—would feature a color television, rumored to be near or in production at IBM.[10] Although Philips products were highly regarded for their quality, color transmission was an arena in which they lagged behind their competition. Rather than face unfavorable product comparisons, Kalff offered a novel strategy: there would be no commercial goods on display within the pavilion. Instead, an international team comprising an artist, an architect, and a composer would create a pavilion utilizing electronic technology in the service of the arts—presenting Philips as a progressive, benevolent multinational corporation. The exhibition working group initially proposed the composer Benjamin Britten and the sculptor Ossip Zadkine. Although neither artist would ultimately figure in the making of the Philips spectacle, their nomination for the project reflects Kalff's initial ideas and taste.

Britten was a world figure in classical music, although his name may

2-2
Ossip Zadkine, *The City Destroyed*, Rotterdam, The Netherlands. 1953. [Marc Treib]

also have been advanced because of political motives: by involving an English composer Philips would seem to extend its artistic sphere and thus reflect its expanded trading domain. Britten's music had been extremely well received in the Netherlands, playing to "packed houses," and was more popular in the low countries than in England. His *Spring Symphony*, created on the theme of the seasons and rebirth, had its premier—and was acclaimed—in Amsterdam at the 1949 Holland Festival.[11] Of Britten's many works, his *Young Person's Guide to the Orchestra*, written in 1947, might have been the particular composition that attracted the attention of Philips' artistic management. The work, which could be performed with or without a narrator, was intended to introduce the various instruments of the orchestra to a young audience; it offered, however, a modern and yet rich collection of sonorities in a mode accessible to a broad audience. A symphonic composition such as this would have been a logical choice to showcase the achievements of Philips' reproduction technology.[12]

The company might also have selected Britten because he had no qualms about working under assignment. In 1940, even as war clouds were accumulating in the East, he had accepted a proposal to write a *Sinfonia de Requiem* solicited by the Japanese government.[13] The restrictions imposed by commissions presented no problems to Britten:

> For technical reasons, he is always prepared to work to order. He does not believe in allowing his talents to rust. As an artist he wants to serve the community and has shown himself ready to accept commissions of every kind. He finds virtue in serving all sorts of different persons and believes that even "hackwork" will not hurt an artist's integrity provided he does his best with every commission.[14]

The consideration of Zadkine as sculptor, on the other hand, is more difficult to understand from today's perspective, as he hardly shared the same celebrity as Britten. The Russian-born, French-resident sculptor was known primarily in a circle that extended only as far as Belgium and Holland, despite his residence in New York during the war years.[15] Zadkine's sculpture erratically moved between the expressive facture of Rodin and Matisse, the more purely abstract ideals of Brancusi, and a somewhat late flirtation with cubism. His most celebrated work, *The City Destroyed* (1953), relied on a dated cubist idiom, more convincingly evidenced decades earlier in the work of Jacques Lipchitz, for example [figure 2-2]. But in spite of its *retardataire* vocabulary, the sculpture's figurative expressionism powerfully conveys the horror of aerial bombardment while probing the formal relation of solid to void. Given a prominent site beside the maritime museum in the Rotterdam harbor the sculpture had been well received and had, in fact, become a symbol of the

Netherlands' postwar resurrection. Thus, Zadkine could have been of interest to the Philips management primarily because he was a known quantity working in a progressive but not avant-garde artistic idiom. In addition, his style was familiar to the general public, being well represented in the collection of the Kröller-Müller Museum near Otterlo.

All thoughts of collaboration with these artists ended with the selection of Le Corbusier as architect. In fact, as Le Corbusier's interest in the electronic production grew, he assumed even greater responsibility, easing out Gerritt Rietveld as the associate architect responsible for the exterior of the building. Le Corbusier would do it all. Kalff, Rietveld, and Le Corbusier were scheduled to confer on 6 August, and one presumes that by the end of the meeting Le Corbusier had established his control over all parts of the project. On 10 June 1956 Rietveld wrote to Kalff, obviously bitter yet understanding the character of the architect with whom he was supposed to have collaborated. Like Frank Lloyd Wright and other giants in the profession, Rietveld wrote, Le Corbusier thought he knew everything and didn't want to collaborate with any other architect, but in actuality, his ideas about painting and music were rather ordinary. Rietveld also suggested how to develop the concept for the pavilion. "If you use the visual aspects as a starting point, color should come first. The architect should stand in the background."[16] By early summer Rietveld's participation in the pavilion's design had evaporated. Kalff wrote to Bakema, one of the architects of the Dutch Pavilion, informing him that Rietveld would not be working on the Philips design, because Le Corbusier was against giving any real shape to the exterior: the "building" would constitute only an "interior" in which to present the electronic spectacle.[17]

The letter contract of 13 October 1956 stipulated that Le Corbusier would design the pavilion and conceive the program within. Engineering and construction, however, were not included in the agreement and were thus assumed to be the responsibility of the Philips organization or its assigned agent. Perhaps because the company wanted to use engineers close to home, Philips agreed to find a contractor willing to tackle the project, with the understanding that the required engineering and construction documents would be prepared by the construction firm or its consultants. At this time Le Corbusier was deeply involved in the design of the capital city of Chandigarh in India, the major work being its High Court building. He would spend more than a month of both 1956 and 1957 in India reviewing the progress of construction, developing details, and laying out the work that was to be accomplished in his absence.[18]

Given the quantity of work in his Paris studio in 1956, Le Corbusier required the assistance of a knowledgeable and talented staff. A young Greek refugee, Iannis Xenakis, had joined the office as an engineering assistant in 1951, working with the engineer Bernard Lafaille on the structural calculations for Le Corbusier's second major housing block, the *unité d'habitation*, at Nantes-Rezé.[19] Le Corbusier was impressed with Xenakis's intelligence and skill, although the young engineer had no formal architectural training. Logic, a knowledge of mathematics and engineering, dependability, and dedication seem to have been more desirable qualities than formal flair to Le Corbusier [figure 2-3]. He described Xenakis as "three favorable vocations [engineer, composer, architect] united in one man."[20] In 1956, Le Corbusier had invited Xenakis to work on the monastery of La Tourette, applying the Modulor proportional system to plan the "musical" spacing of the glass and mullions of the expanses of windows that faced the fields. The transition to a project rooted in geometry and music was natural.[21]

Impressed with the numerical proportions of the Modulor, Xenakis applied its divisions to a musical composition almost two years before the work on the Philips project began in earnest. His composition *Metastasis* of 1953–54, a seven-minute piece for sixty-one instruments, explored the dissolution of the "uniform and continuous" duration of notes. In accord with twentieth-century ideas of relativity "[t]he succession of tempered intervals is a geometric progression. The duration will be a geometrical progression too." Using Modulor relationships, the durations become additive: "A duration can be added to another and their sum is perceived as such." As to which numerical relationships would best define these durations: "Among all the geometrical progressions, there is only one whose terms possess this additive property. It is the progression of the Golden Section."[22] More germane to the forming of the Philips Pavilion, Xenakis's graphing of the string glissandi of *Metastasis* resembled the ruled surfaces of hyperbolic paraboloid shells [figure 2-4].

That Xenakis was also developing as a composer must have attracted the senior architect to his young assistant. "I am a musician at heart," Le Corbusier wrote, "but not at all by profession."[23] He would, in fact, assign Xenakis responsibility for composing the music accompanying the two-minute interlude between the performances at the pavilion in Brussels. During the two years of work until the building was inaugurated, Xenakis kept the project on a steady course, proposing a formal rationale for the building, dealing with the engineers, developing details, and, most important, coordinating the production of the artistic team, the members of which were at certain moments dispersed on three continents. That Xenakis would demand credit for his work late in the course of the pavilion's design is completely understandable—although it was not seen that way by Le Corbusier at the time. In the end, the master did credit Xenakis as the co-author of the building's architecture, but "l'incident Xenakis," as Le Corbusier termed it, ultimately

2-3
Iannis Xenakis. Probably in Eindhoven, February 1958. [Philips]

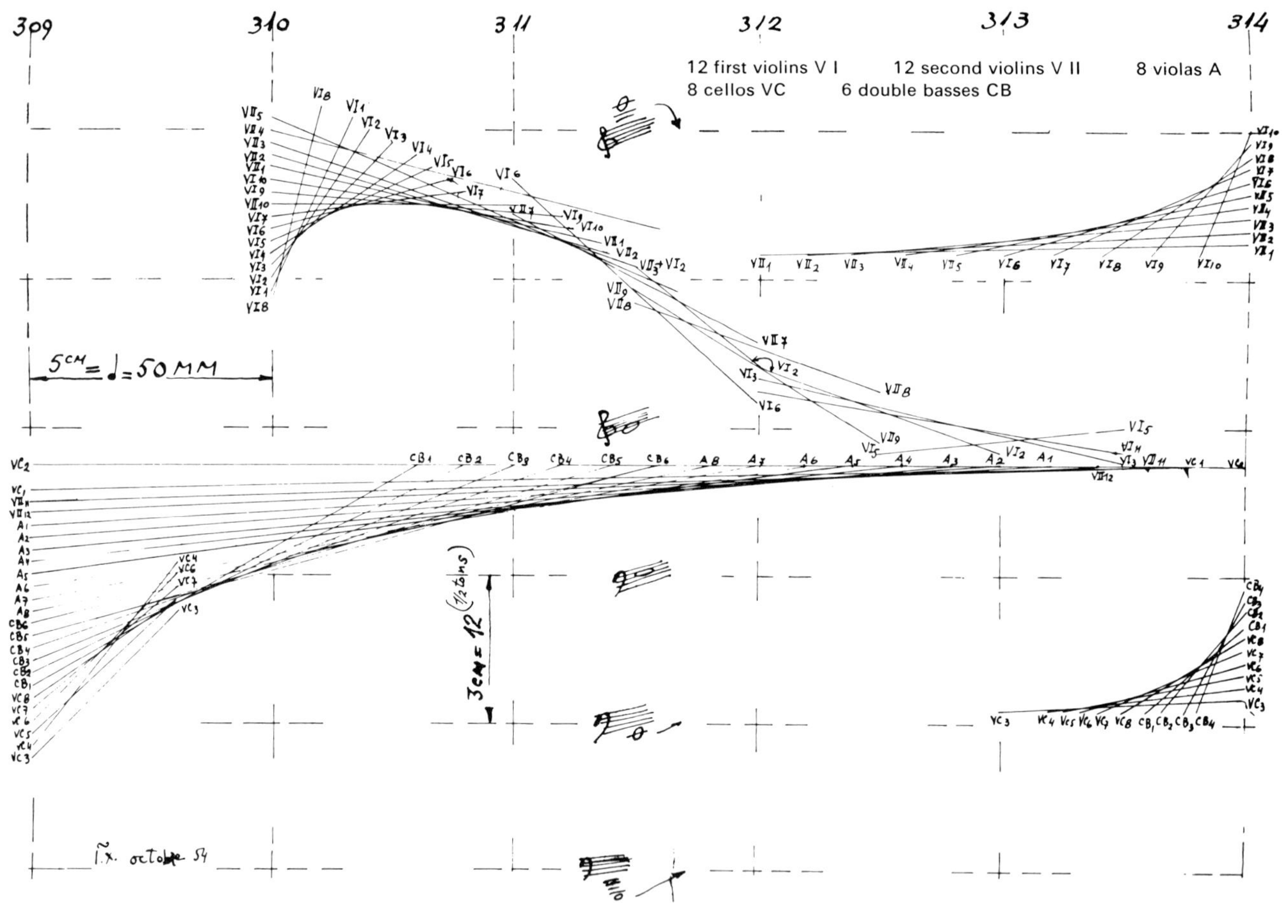

2-4

Diagram of glissandi from *Metastasis.* 1954.
Iannis Xenakis.
[Courtesy Iannis Xenakis]

ruptured their relations and encouraged the younger artist to work almost exclusively in music.[24] But that storm was in the future, and for the moment, Xenakis was intently involved with preparing the necessary documents within a very short period.

From the very beginning of the project, Kalff felt the weight of time on his shoulders. Wedged between an autocratic architect and a mundanely oriented board of directors, his role was never enviable. His correspondence to Le Corbusier and Xenakis is filled with plaintive requests for material that had been promised weeks, or even months, before. But from the beginning of negotiations, he never held a strong bargaining position. Once Philips was committed to the architect, Kalff had to bluff, chide, and even use psychological blackmail to get results. But bullying Le Corbusier was extremely difficult, and in this, Kalff never completely succeeded.

Born in 1897, trained as an architect and engineer, Louis Kalff graduated from the Quellinusschool in Amsterdam in 1923. Two years later, he entered the Philips organization in the corporate communications department and gradually rose to the position of artistic director for the corporation.[25] Lighting was both an interest and an expertise, and Kalff designed some of the illumination for buildings at the 1937 World's Fair in Paris, including the sensational effects worked on for the Trocadero Palace facing the Seine.[26]

In letters, Kalff always displayed enviable tact and patience, perhaps understanding that threats were not a satisfactory means for extracting the desired materials from the renowned French architect or his collaborators. In his introduction to the book prepared by Jean Petit at the end of the exposition, Le Corbusier was retrospectively gracious in his description of Louis Kalff, a man who had been his antagonist during a good part of the pavilion's design:

> *From the beginning to the end, during two and a half years, Mr. Kalff of Eindhoven (1.95m [6 feet] tall) has devoted himself to this work, completely, calmly, kindly; hiding his anxieties, confronting the unknown, juggling finances; the captain on board, smiling at his "shipowners," the gentlemen directors of Philips. Mr. Kalff is one of those particular products of the Netherlands—patient, persevering, believing, courageous—who have given their country such a fine history. Each of the collaborators had his nose to the grindstone, perhaps, forgetting to appreciate that someone was continually on guard. And he, during all this time responded to his management, that extremely vast Philips organization: "It's moving along, gentlemen, it's moving along quite well."*[27]

The original conception for the Philips building as an artistic rather than a purely commercial venture seems to have been very much Kalff's invention, verified by early internal memos and later correspondence. Like Le Corbusier, Kalff had a vision for the Philips participation in Brussels, but his vision was formulated in terms of an overall accomplishment rather than one invested in any particular form: "[T]he object . . . is to demonstrate the capabilities of modern technology in some of Philips' major fields of endeavor—illuminating engineering, electro-acoustics, electronics, and automatic control techniques—and also to give an impression of the way in which these technical facilities may in the future be turned to artistic ends."[28] He seems to have also understood the propaganda value of fairs such as these, and that notoriety of one sort or another can generate mountains of free advertising worldwide. It was with this strategy alone, in fact, that he could argue with the Philips management for a pavilion as a synthetic work of art. Without a client such as Louis Kalff, an architect like Le Corbusier and a composer like Edgard Varèse would never have been allowed to create works such as those at the exposition. And although one senses in letters that Kalff himself did not really appreciate the appointment of Varèse as composer of the music for the electronic poem, he earnestly tried to satisfy all the parties involved.

Compared to Benjamin Britten, Edgard Varèse was a virtually unknown quantity to the Philips management. Born in Paris in 1883, Varèse spent his childhood years in Turin before returning to the capital for musical studies.[29] After two years in Berlin and subsequent military service, in 1915 Varèse moved to New York to find a living and artistic situation more conducive to progressive musical ideas than were the capitals of the Old World. While his work became reasonably well known among musicians and received sporadic performances primarily by small ensembles specializing in contemporary music, Varèse was not part of the musical mainstream. In 1924, he returned to Paris for six months, where he was affiliated with the studio of Fernand Léger. It may have been at this time that he met Le Corbusier.

Kalff was only partially successful in persuading the Philips management that Varèse was the appropriate person to compose the music for the pavilion. They never warmed to Varèse's score of gongs, synthesized sounds, and distorted human voices making strange noises; they did, however, come to appreciate the public clamor that the pavilion and its sound score received. To Kalff, perhaps, management's acceptance was not the real issue, however. The building had to be a totality, and that totality needed a vision. Kalff appears to have believed in Le Corbusier as an architect with the proper vision for Philips. And despite his troublesome absences, the assignment of daily control of the project to Xenakis, his use of other collaborators, and his failure to master completely the building's design and its electronic spectacle, Le Corbusier realized his vision for a synthesis of the arts in Brussels.

At the initial meeting with the Philips representatives, Le Corbusier had proposed not only the idea of an electronic poem, but also the shape of the building in which it would be contained. Six hundred people would attend each show, fixed at ten-minute intervals. The actual electronic spectacle would last only eight minutes—precisely 480 seconds—with a two-minute interval to allow one group to exit the building while the succeeding party entered. Then the doors would close, the lights would dim, and the cataclysm would begin anew. Le Corbusier proposed that there would be lighting effects, images, rhythm, and sound, if not "music."[30]

Although Le Corbusier did not establish the precise form of the walls and roof in the early meetings, he soon fixed the shape of the building's footprint as that of a "stomach." The plan derived from a more or less circular enclosure for the visual presentation, extended with corridors for entrance and exiting that modulated the light level. To avoid a direct passage into the auditorium, and to dissipate the full daylight encountered when exiting, the circulation spaces were curved. Subsidiary spaces for the equipment, equipment control, and for the technical staff were accommodated within the overall sweep of the plan form. The schematic layout was thus determined early in the project; the walls and roof, however, were only vaguely conceived.

Le Corbusier first saw the pavilion as a neutral "bottle" supported by metallic scaffolding. The internal theater was the *raison d'être* for the project, but this auditorium was only a container, necessarily darkened for viewing the projections and acoustically isolated for listening to the Varèse score. Little, if any, development of that basic idea occurred through the summer months of 1956; Le Corbusier's first sketchbook entries date only from 25 September despite verbal assurance from Philips that the project would proceed [figures 2-5, 2-6]. Most architects, like other creative types, speak of a gestation period necessary to develop the embryo of an idea, and surely Le Corbusier must have dealt with his clients by giving this excuse. The metal scaffolding necessary to support the "bottle" remained in the next version of the scheme, which translated the container into an integrated structure of plan and space.

From the beginning of the project, Le Corbusier was determined that mathematics, being a central part of human creation, would play a fundamental role in the building's design:

> *Mathematics is the majestic structure conceived by man to grant him comprehension of the universe. It holds both the absolute and the infinite, the understandable and the forever elusive. It has walls before which one may pace up and down without result; sometimes there is a door: one opens it—enters—one is in another realm, the realm of the gods, the room that holds the key to the great systems. These are the doors of the miracles.*[31]

2-5
From Le Corbusier's sketchbooks. Early September? 1956.
[FLC; ©1996 ARS, New York / SPADEM, Paris]

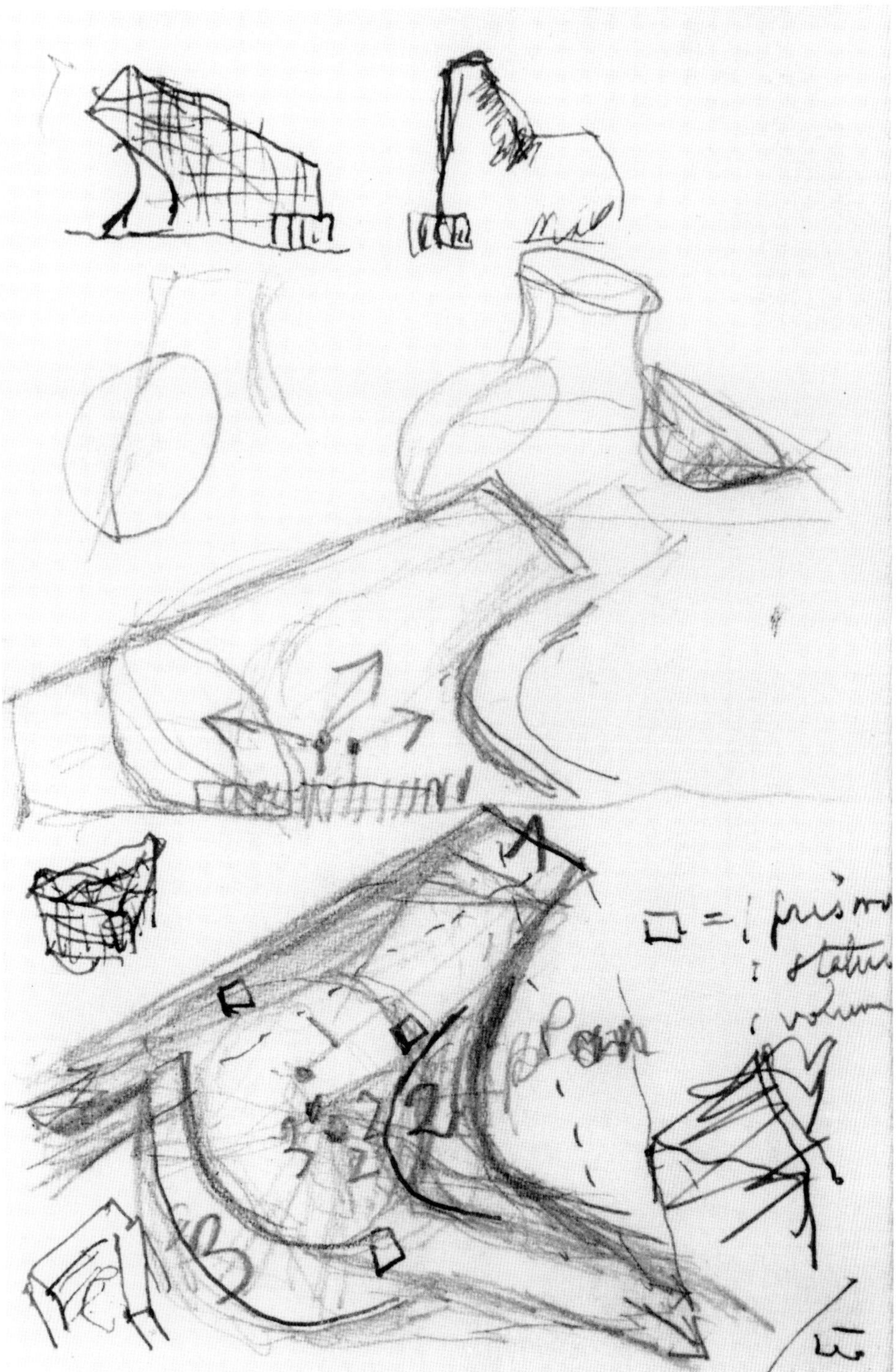

2-6
From Le Corbusier's sketchbooks. September 1956. [FLC; ©1996 ARS, New York / SPADEM, Paris]

While he might deprecate his own knowledge of mathematics, his interest in geometry—"the creation of the mind"—was longstanding: "Geometry and nature, the perpetual gauge of our equilibrium," he exclaimed in 1948.[32] Even in his early writings, Le Corbusier had extolled the influence of number upon our senses. In *Towards a New Architecture* (1923), for example, he insisted: "The Engineer, inspired by the law of Economy and governed by mathematical calculation, puts us in accord with universal law."[33] The mind expressed its accord with the spirit of the universe through geometry. As such, geometry represented the correct vehicle for creating an architecture to contain the Philips spectacle, Le Corbusier's electronic hymn to humankind:

> *Everything is geometrical to our eyes (biology exists only as organization, and this is something that the mind understands only after study).* Architectural composition is geometric, *an event primarily of a visual nature, an event implying judgements of quantities, of relationships; the appreciation of* proportions. *Proportions provoke sensations; a series of sensations is like a melody in music. Erik Satie used to say: the melody is the idea, harmony (in music) is the means, the tool, the presentation of the idea.*[34]

The interest in geometry may have been the primary reason that Le Corbusier chose Xenakis—who had worked principally in the world of mathematics—as his project architect. It would be Xenakis, in fact, who would contribute the strategy for planning the building and determine its geometry and structure. Seen cynically, Le Corbusier's interest in mathematics lacked depth; but one needs to distinguish a belief in the affect of geometric form from a profound knowledge of mathematics. Near the start of the project he wrote letters to the rectors of the university and the technical institute in Zurich, soliciting the names of books containing illustrations of solid geometric figures which he might adapt to architectural purposes. He wrote:

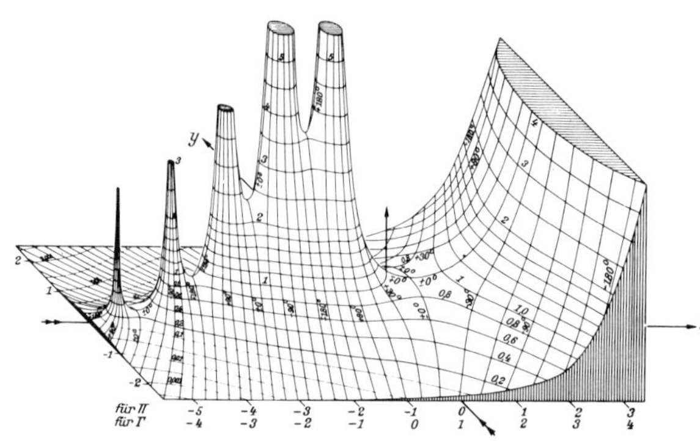

2-7a
"Relief of the factorial function."
[Eugen Jahnke and Fritz Emde, *Tables of Functions with Formulae and Curves*]

> *I would like to request your kind assistance to tell me if your institution could either lend or suggest some books—if such things exist—that would illustrate three-dimensional representations of mathematical functions, from which I might profitably choose. I must admit that concerning this subject I am a complete ass and that it is, therefore, a necessity to light a spark in my lantern. It is not necessary to make a savant out of me; I am simply an architect who wants to consult a "catalog of forms" and to make my choice without any concern for mathematical sophistication.*[35]

His letter was cordially answered by the rector and by Professor B. L. van der Waerden, Director of the Institute of Mathematics at the University of Zurich. The latter recommended *Tables of Functions with Formulae and Curves* by E. Jahnke and F. Emde, citing specific pages that present mathematical functions important to electronics. He noted that the book gives not only perspective views of the mathematical models, but also a chart of the various heights that would help when constructing them in space [figure 2-7 a, b].[36]

In all probability, the seminal ideas for the pavilion's geometric surfaces existed long before they emerged on paper. Le Corbusier's library contained a catalogue of models of mathematical solids, illustrated with line drawings of the ruled surfaces that determined the forms [figure 2-8 a, b, c]. These images—which the architect may or may not have held consciously in mind at the time of design—represented just the sort of geometric figures about which he inquired of van der Waerden in 1956.[37] Seen in relation to the warped surfaces of the pavilion as executed, the correspondences are striking.

If Le Corbusier's grasp of mathematics was not deep, it was, on the other hand, rather broad and long-standing, particularly regarding order and geometry. Christopher Green stated that Le Corbusier "saw himself as a geometrician acting *on* nature."[38] The durable association with the painter and co-editor of *L'Esprit Nouveau*, Amédée Ozenfant, had nurtured an interest in platonic figures and solids as the basis of plastic investigation. In a jointly au-

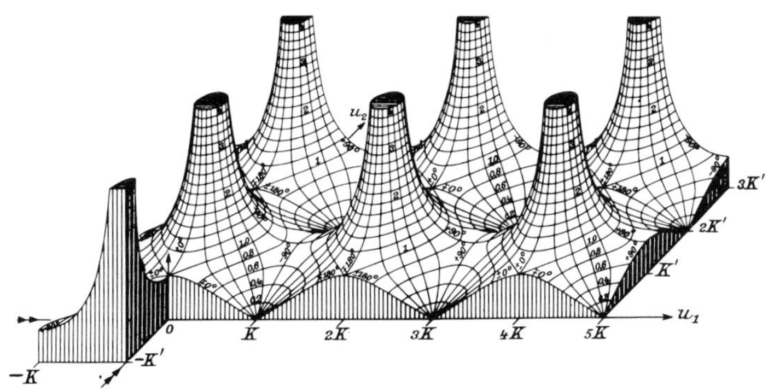

2-7b
"Relief of the doubly periodic function cn *u* for *k*=0.8." [Eugen Jahnke and Fritz Emde, *Tables of Functions with Formulae and Curves*]

thored article in the inaugural issue of *L'Esprit Nouveau*, for example, they stated:

> The primary elements are the straight [line] and the curve.
>
> When these constitute perfectly defined figures, ordered, closed figures that themselves contain each one of the individual emotions, they conjure a direct feeling of something defined, an impact clear, immediate and constant, a cause of well-being, a feeling truly sculptural.[39]

"The spirit of man is geometric," they later wrote. "Man's senses, his eyes, are led more than ever by geometric instincts."[40] Geometry thus underlay their vision of purism, and they agreed with Cézanne that: "Everything is spheres and cylinders" [figure 2-9].

Although Le Corbusier balanced the straight and the curved line in his purist paintings, in the world of his early architecture the orthogonal prevailed. The curving wall contained a functional incident such as the bathroom or solarium, but never the primary volume of the house itself. Geometry governed Le Corbusier's panels; commonplace objects served as the primary subjects of his paintings from 1918 to 1924, their curved shapes like the meandering walls ultimately enclosed by a rectangular frame. Elevation and plan views prevailed in these representations of machine-made objects "joined by a common contour."[41] After 1925, the human figure came to occupy greater prominence in Le Corbusier's painting, reflecting an increasing engagement with natural forms free from the restrictions of geometric composition. Straight lines and rigorous proportioning endured in all his panels, however, but their relationships were more complex. The natural and the made, the intuitive and the relational, joined on the painted surface and met in architecture.

First published in 1955, Le Corbusier's *Poem of the Right Angle* celebrated the work of the human spirit. "Erect on the terrestrial plain of things knowable you sign a pact of solidarity with nature," he wrote. "This is the right angle. Vertical, facing the sea there you are on your feet." From this basic stance comes a more sophisticated method for creation:

> *To place at the tips of one's fingers*
> *above all in one's head an*
> *agile instrument capable of swelling*
> *the harvest of invention*
> *clearing the path of thorns*
> *clearing the way will confer*
> *freedom on your freedom.*
> *Spark stolen from the flame*
> *the Gods nourished to*
> *make the world play . . .*
> *Mathematics!*

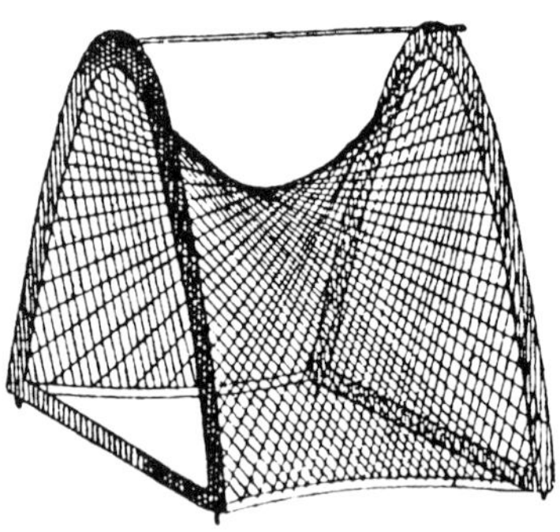

2-8a,b,c
Three illustrations of mathematical models from a German catalog in Le Corbusier's personal library. [*Mathematischer Modelle für den höheren Mathematischen Unterricht*]

Structure becomes the vehicle for achieving freedom, the freedom of the spirit. Proportion "orders the relations with our surroundings." Thus, geometry manifests human accomplishment, and as such, it provided an important underlying principle in the creation of the *Poème électronique* and its pavilion. "To make architecture is to make a creature";[42] in Brussels, the creature was a product of geometry.

Notes on Le Corbusier's sketches assign the task of translating these abstractions into architectural form to Xenakis, whose familiarity with the work on shell structures by Lafaille and whose interest in geometry were manifest in his proposals to use warped surfaces for the construction of the pavilion.[43]

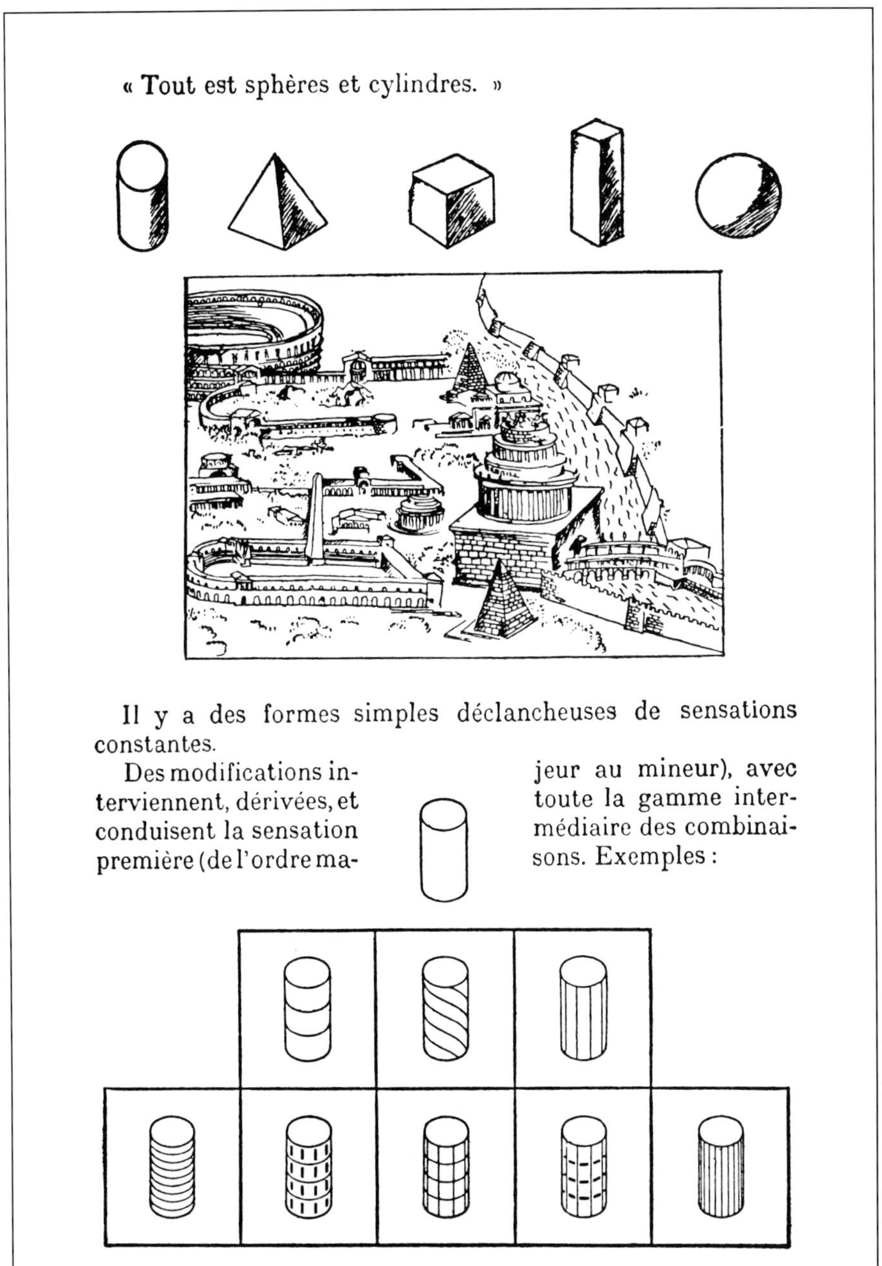

« Tout est sphères et cylindres. »

Il y a des formes simples déclancheuses de sensations constantes.

Des modifications interviennent, dérivées, et conduisent la sensation première (de l'ordre majeur au mineur), avec toute la gamme intermédiaire des combinaisons. Exemples :

2-9
Platonic solids. Amédée Ozenfant and Charles-Edouard Jeanneret (Le Corbusier). 1920. [*L'Esprit Nouveau*]

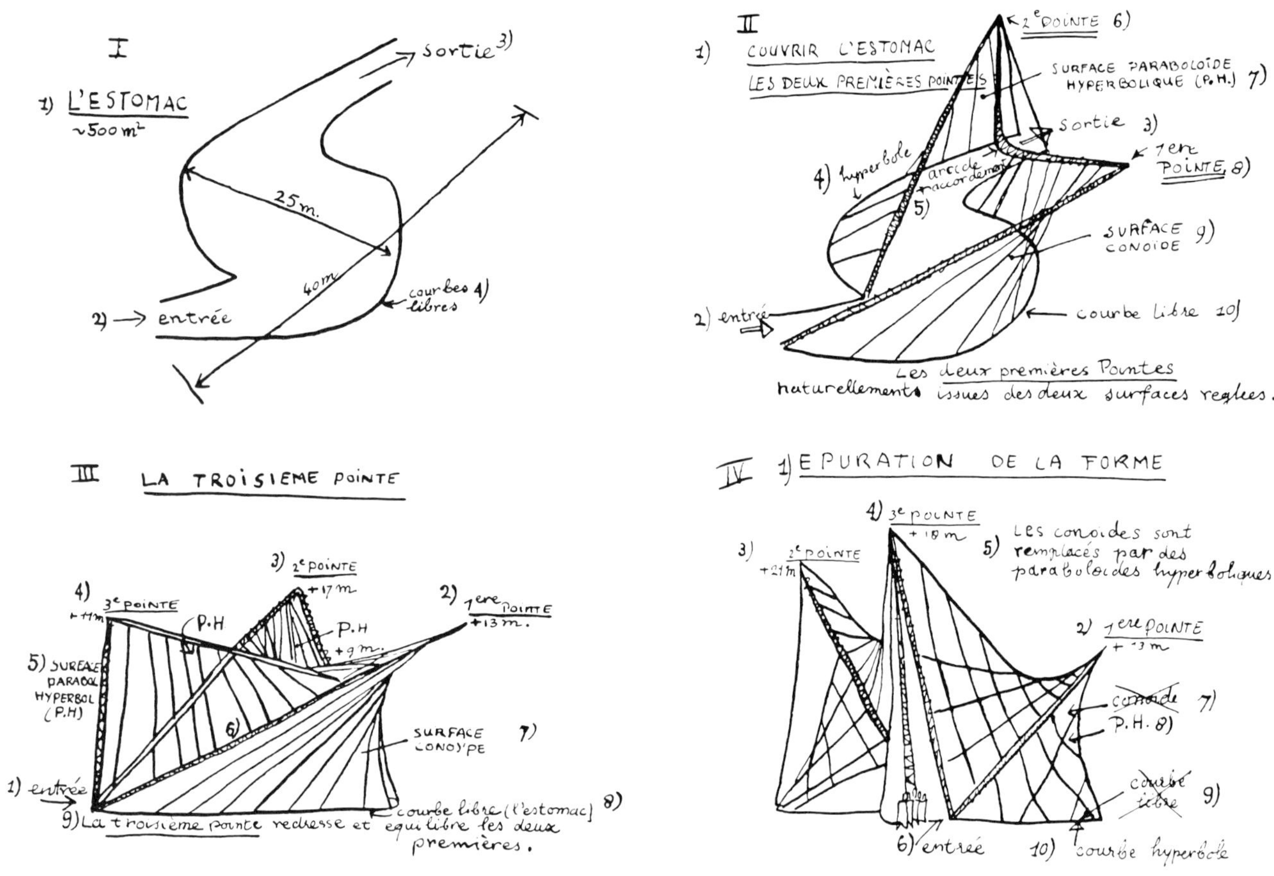

2-10
Conceptual sketches by Xenakis showing the development of the pavilion's geometry. [*Philips Technical Review*]

In beginning to investigate the superstructure which would complement the "stomach" shape of the plan, Xenakis used both reason and intuition [figure 2-10]. Various functional criteria, such as sight lines and projection surfaces, suggested a continuous form in which the distinction between ceiling and walls would disappear. In his first studies, Xenakis "went through several unfruitful attempts based for the most part on Le Corbusier's design, which [he] was trying to respect as much as possible."[44] With his continued attention, a form began to emerge: "From this point on logic, even stochastic, stops working. Intuition's arbitrariness takes over."[45]

Possibly recalling the graph of the glissandi from *Metastasis*, Xenakis first proposed a combination of hyperbolic paraboloid and conoid surfaces to describe the superstructure of the pavilion. By using these forms, the roof and walls would be seamless, and the building could appear as a totality, an integrated play of shells. The curving forms also solved the problem of sight lines: given the arching shapes, the planned projections could be seen from any point on the floor—above the head of one's neighbor, as Le Corbusier phrased it.[46]

The tent would be an expedient solution to the problem of limited time; construction could be less expensive and removal simplified. Le Corbusier had resorted to a tent structure for his Pavillon des Temps Nouveaux at the 1937 Paris World's Fair. Denied participation in the fair until almost the final hour, Le Corbusier was left with few possible alternatives: certainly standard construction was not feasible. The 11,000-square-foot building used trussed pylons to support a roof of canvas made waterproof and translucent by tallow. The network of guy wires that stayed the tent produced an effect resembling the exterior of the "primitive temple" illustrated in *Towards a New Architecture*. Like Joseph Paxton's 1851 Crystal Palace, the Pavillon des Temps Nouveaux resolutely demonstrated that talented designers can overcome the exigencies of time and money to produce innovative and handsome solutions—in fact, they might benefit by stringent design constraints.

The first scheme for the Philips Pavilion resembled a tent with stiffened fabric in which a steel armature supported a network of cables that in turn carried curving planes of gridded metal lath. Drawing on the technology of sprayed cement developed during the late 1930s—referred to as "*ciment au canon*" or "*gunitage*"—the pavilion could be easily formed, quickly enclosed, and demolished without great effort.[47] The drawings dated 16 to 23 October 1956 developed these ideas in sufficient detail to present on the following day, when the definitive meeting was planned with the Philips representatives. There was not that much to show. The drawings appear to have come from Xenakis's hand, and they establish the elements and positioning of the site

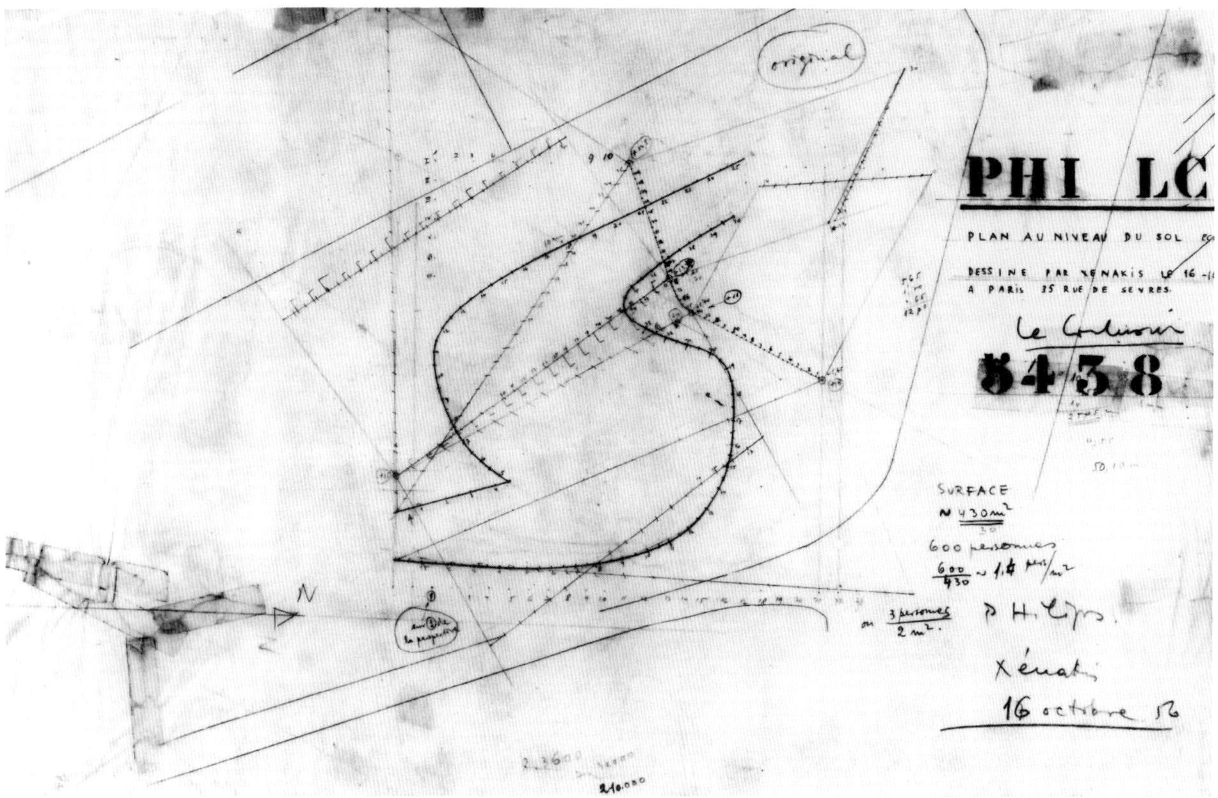

2-11
Floor plan of the first scheme. 16 October 1956. Iannis Xenakis. [FLC 28581; ©1996 ARS, New York / SPADEM, Paris]

2-12
Perspective of the first scheme. 23 October 1956. Le Corbusier. [FLC 28583; ©1996 ARS, New York / SPADEM, Paris]

2-13
Perspective of the first scheme. 19 October 1956. Iannis Xenakis. [FLC 28585; ©1996 ARS, New York / SPADEM, Paris]

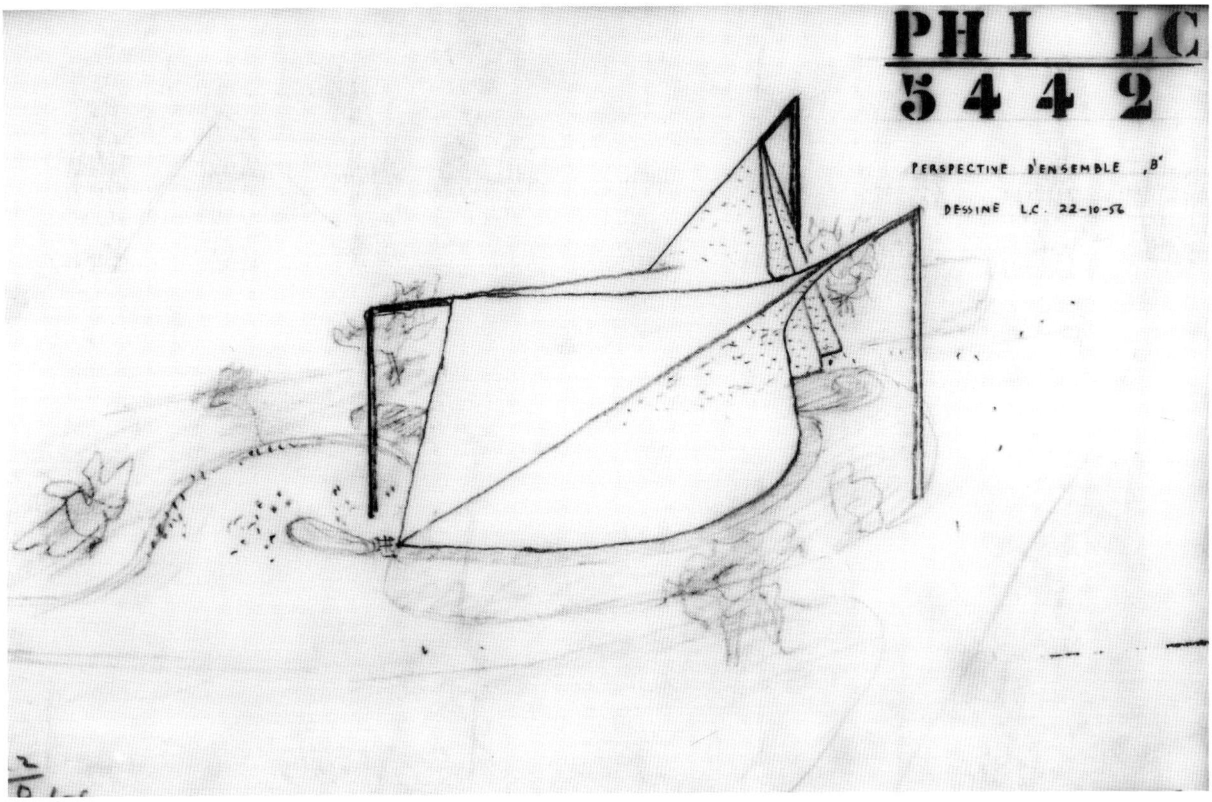
PHI LC
5442
PERSPECTIVE D'ENSEMBLE „B'
DESSINÉ L.C. 22-10-56

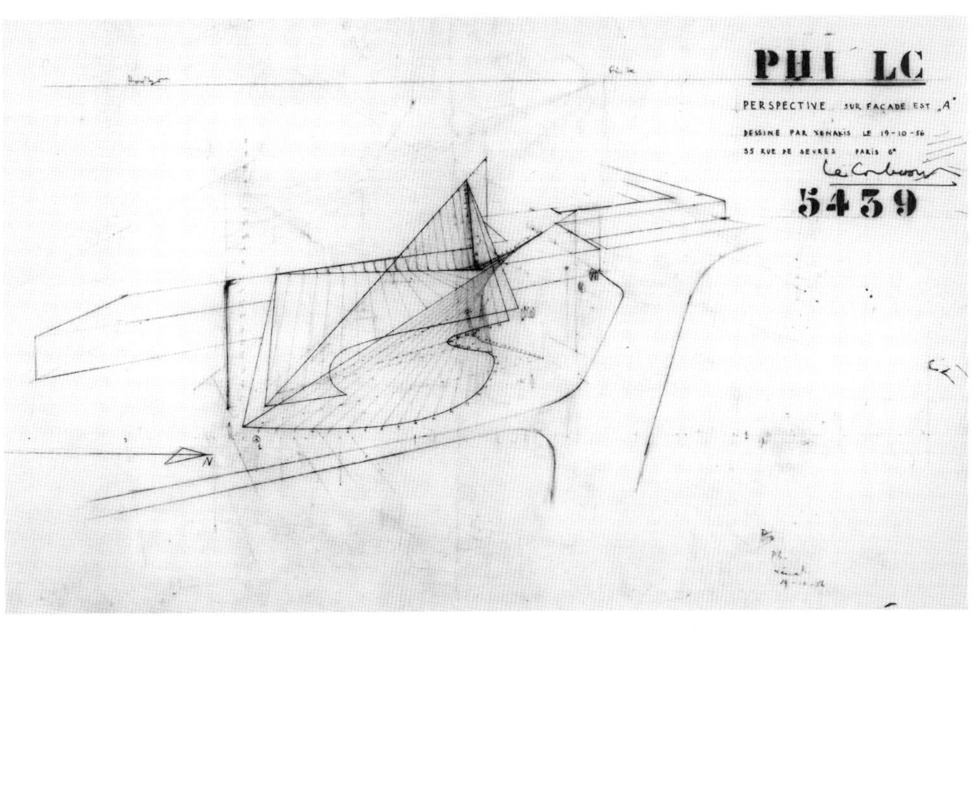
PHI LC
PERSPECTIVE SUR FAÇADE EST „A"
DESSINE PAR XENAKIS LE 19-10-56
35 RUE DE SEVRES PARIS 6e
Le Corbusier
5439
N

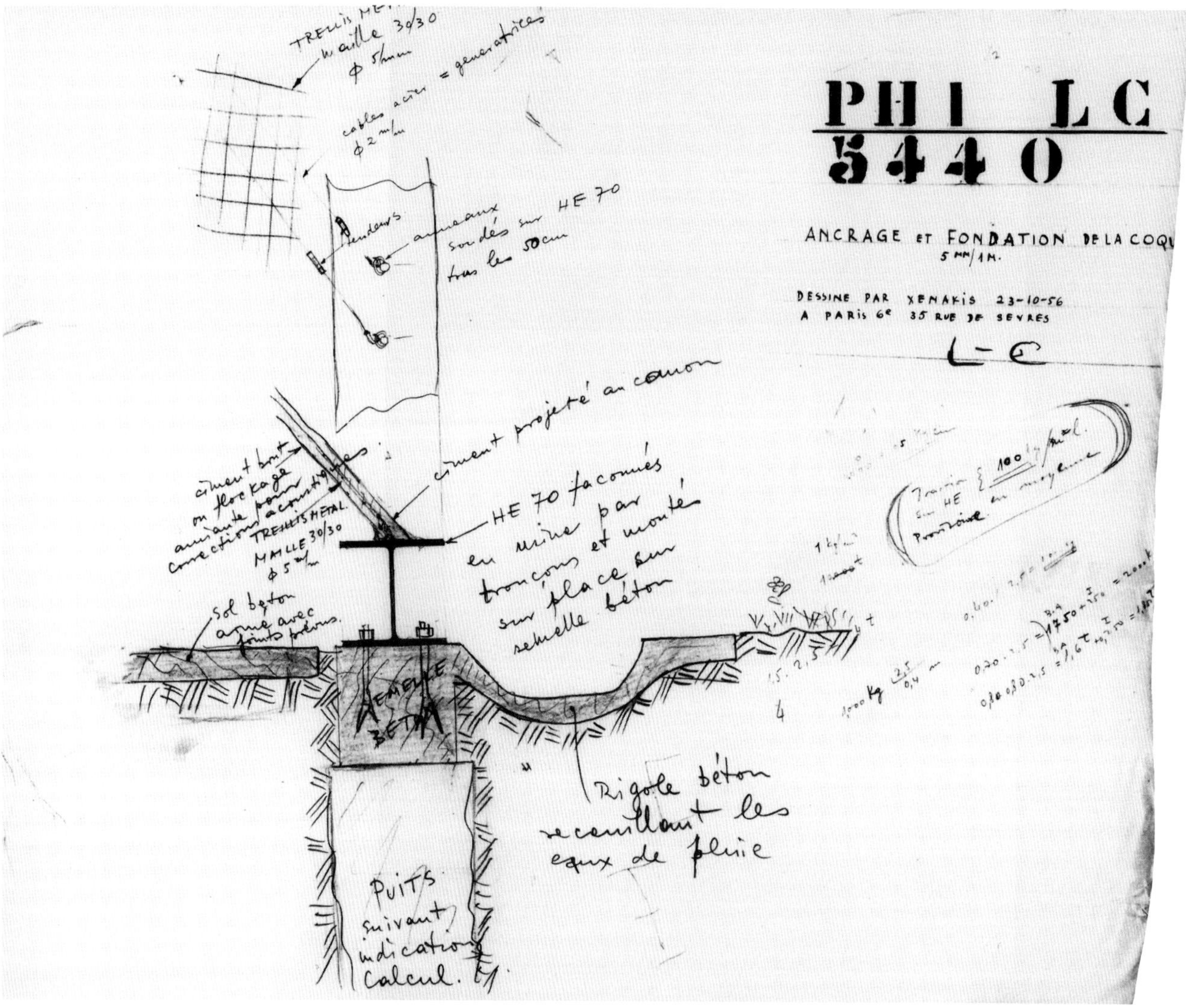

2-14

Preliminary study of the anchorage of walls to the foundation. 23 October 1956. Iannis Xenakis. [FLC 28582; ©1996 ARS, New York / SPADEM, Paris]

2-15

Wire and thread model of the first scheme. Iannis Xenakis. Late 1956. [FLC; ©1996 ARS, New York / SPADEM, Paris]

plan, the building's geometry, and the structural section of the building [figure 2-11]. Le Corbusier's contribution (drawing PHI LC 5442, figure 2-12)[48] is a single perspective which appears to have been simply a freehand tracing of the view meticulously laid out by Xenakis on drawing PHI LC 5439 [figure 2-13].[49] Other details, such as joining the walls to the ground and water drainage, received their initial study [figure 2-14].

According to Xenakis's recollection, it was his idea to utilize the curving surfaces of the hyperbolic paraboloid / conoid combination, and no evidence contradicts this assertion. In any event, the master was working in India when the revision was proposed. Le Corbusier was so pleased with the proposed direction that he tried to validate the concept by claiming it as his own.[50] He instructed Xenakis to prepare a model using wire and thread to explain the determination of the shapes and their relation to the stomach shape of the plan. This first model established the approach to be followed through the realization of the pavilion in 1958. "To make the model I had to stretch threads from along the ridges where the surfaces met, and materialized these ridges by piano wires bent and fitted into the wooden base" [figure 2-15]. While essentially a means to build the model, the use of the structural wire—also to embody the original idea of a bottle within scaffolding—was to have serious repercussions: "Up to the final project, they diverted the attention of the consulting engineers, almost all of whom missed the possibility of a self-supporting structure."[51] This crude model provided the first glimpse into the pavilion's complex geometries, and the potential of using doubly warped surfaces as the building's principal structural and architectonic idea.

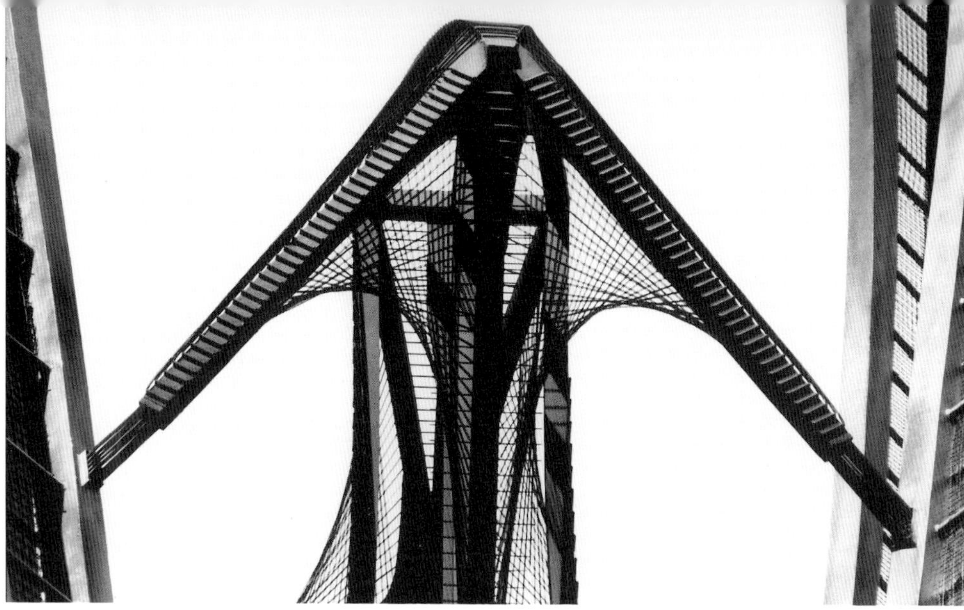

While the Philips project was probably the most unusual application of the hyperbolic paraboloid at the fair, it was far from the only one, nor was it a novel idea within the spatial arts. As a form, the hyperbolic paraboloid had been a central element in the spatial sculptures of the brothers Naum Gabo (1890–1972) and Antoine Pevsner (1886–1961), among others. One beauty of the mathematical form was its fluid development into a three-dimensional volume from a two-dimensional shape, thereby implying a movement or unfolding over time [figure 2-16]. Space-time had been a critical ingredient in the formulation of modern architecture; the hyperbolic paraboloid thus suggested a means to overcome the limitations imposed on flowing spaces by the orthogonality of the box. As the viewer moves around the hyperbolic-paraboloid surfaces of a Gabo work, the sculpture appears to regenerate its form in continual movement before the eyes [figure 2-17].

The calculations necessary for determining the dimensions and performance of shell structures in architectural construction developed at a great pace after the war, especially in Mexico and South America. In the late 1930s, the Spanish engineer Eduardo Torroja exploited the properties of folded-plate concrete construction in his celebrated Madrid Hippodrome, thereby suggesting a direction that crossed architecture with engineering. Pier Luigi Nervi in Italy surpassed Torroja's designs in both span and construction complexity in a variety of projects that ranged from sports gymnasia to hangars. Many of these were extensions of post and beam technology applied to concrete, utilizing the possibilities of the continuous joining of compression and bending members through a single pour. Closer to home, Eugène Freyssinet's airship hangars at Orly outside Paris (1916) illustrated early on the potential of using folded reinforced concrete shells—in this case, parabolic in section—to create large-scale spaces of great structural elegance.

Folded-plate and, later, surface structures, on the other hand, exploited structural properties restricted to concrete alone. Vaults and domes, each curving in only one direction, tend to operate in pure compression; they are prone, however, to extreme bending forces should their slopes be too shallow. The wooden formwork used to construct these concrete structures also needs to be curved, but designs based on a single radius reduce to some degree the complications of forming. In contrast, warped structures such as hyperbolic paraboloids gain stiffness by curvature along two axes; their surfaces are generated by straight lines alone, greatly simplifying the process of construction.

In Mexico during the 1950s, Félix Candela was building structures of incredible thinness and elegance. Working with Jorge Gonzalez Reyna, he reduced to less than an inch the thickness of the shell covering the Cosmic Ray Pavilion at the University of Mexico in Mexico City from 1952. Structural in-

2-16
Construction for the Bijenkorf, Rotterdam. 1954/55.
Naum Gabo.
[Marc Treib, 1993]

2-17
Linear Construction in Space No. 1 (Variation). 1943–48. Naum Gabo.
[Phillips Collection]

tegrity was achieved by convoluting the surface plane and reinforcing the edge rather than by thickening the concrete. It was only the first in a series of concrete shell structures that were published extensively in architectural periodicals around the world.

Shell structures are able to span considerable distances with minimal thickness, but certain inherent structural limitations constrain their performance. One of these concerns the edge of the shell which is usually strengthened against failure. The thickness required for a given span does not always determine the final thickness of the shell; the minimal concrete covering of the reinforcing steel may demand a dimension far greater than that required for structural performance. The quality of the workmanship becomes more critical as the thickness is reduced, as surface forces transferred to an unreinforced edge can lead to tearing and failure. During the 1950s, the problem of the edge hampered the widespread application of shells as sheets: the edges of the shells were instead treated as beams used to collect, distribute, and ultimately ground the forces generated through the shell structure. It was a common solution to an ubiquitous problem, and rib beams would also be used at the Philips Pavilion to transfer loads from one shell segment to another and from the shells to the ground. This problem came to light only as the design of the pavilion progressed, however, and the structure changed from the suspended "tent" to bearing walls of preformed units.

Hyperbolic paraboloids were very much in evidence in Brussels 1958. The French pavilion, designed by Guillaume Gillet with consulting engineers René Sarger and Jean Prouvé—who had collaborated with Le Corbusier on several projects—employed two shallow intersecting shells with edges reinforced by truss beams [figure 2-18]. Its structural arms extended some 240 feet and were webbed with a roof also based on the hyperbolic paraboloid. Typical of these applications of structural technology, the pavilions were often more interesting as engineering than as architecture. Professional criticism commented on the discord between the expressive structures of the pavilions and their internal contents.[52] The spans achieved by the Philips pavilion structure, however, were minuscule, with the shell forms effectively eliminating most planes that could be termed "roof." One assumes that the spatial, rather than the structural, properties of warped structures were what really interested Le Corbusier and Xenakis.

Like many of Nervi's works, the buildings at the fair appeared at their best just as the scaffolding was removed and before they were enclosed. Architectural details such as glazing, window mullions, and stair railings frequently detracted from the sweeping power of a pure span or a clean space. Shell technology was capable of spanning great distances, but it created formal conditions that were difficult to resolve, from the basic problem of

2-18
French Pavilion, Brussels, 1958. Guillaume Gillet, architect, with René Sarger and Jean Prouvé, consulting engineers. [Lennart Olson]

enclosing a space under a vault, to the more abstract challenge of fitting a flat planar surface to a curving one. Like other pavilions at the fair, the Le Corbusier-Xenakis project was plagued by instances where the conflicting geometry of the shells could not be resolved, and which remained formally awkward, especially when seen from behind or from the Atomium. But the building was never intended to be seen from the air, and from the ground it was virtually impossible to perceive more than two of the shell segments from any given point. Recall also that the building was an "exterior" structure constructed to enclose an "interior" presentation; to Le Corbusier, perhaps the form was not of great consequence as long as the interior matched his desired specifications.[53]

The dominant challenge facing Xenakis was to envelop the stomach form of the plan with a series of shell segments that would become structural and aesthetically unified. This was not an easy task. The hyperbolic paraboloid was structurally sophisticated because the strength of its warped surface derived—almost mystically—entirely from straight lines. Rather than requiring a surface made up of units with differing degrees of curvature, the casting of concrete hyperbolic paraboloid shells could rely entirely on standard sections of wood. The majority of shell structures constructed at the exposition used a complete form—for example, a warped square or rectangle—as the predominant element, often the building's roof. The plans of these buildings retained orthogonal or regular geometry. For the Philips Pavilion, however, segments of varying warped surfaces needed to be established, cut, and repositioned in a way that would replicate the curving forms intended by the building's designers.

The first approach to developing the building constituted a literal exegesis of Le Corbusier's original concept of a bottle held within scaffolding. In Xenakis's recollection of the process, the design proceeded thus:

> In October, 1956, Le Corbusier asked me to draw up these ideas and try to "translate them using mathematics." He handed me a sketch.
>
> Le Corbusier [proposed that] the building would be a "bottle" containing the "nectar of the visual presentation and the music."
>
> For the film spectacle, he wanted flat vertical surfaces.
>
> For the spatial effect, he wanted a tapering "bottleneck" high up in the ceiling of the pavilion where the projected images would disappear.
>
> For the desired luminous colors, he wanted concave and convex surfaces. . . . [H]e agreed with the requests of the architects of the adjacent Dutch pavilion: "We would recommend a simple and convex surface as the back of the Philips pavilion, so that it does not overpower the garden and the greenery which surround the Dutch buildings."

As it is night inside the "bottle," beauty is not really a concern.[54]

While the sketches for this first project are few and imprecise, Le Corbusier wrote that they first explored the possibility of using the lath and plaster common in stage design and exposition architecture of limited life-span. The scheme was discarded rather early, by the end of 1956. A number of problems limited the potential of this approach, not the least of which were the fabrication and point loads created by the metal armature. One senses, in addition, that Xenakis, taken with the sweep of shell structures and fascinated with the eloquence of the hyperbolic paraboloid and other geometric surfaces, wanted to make them work in Brussels [figure 2-19].

Xenakis later wrote that the essence of construction in concrete was continuity, and as such, it eschewed the inherent elementality of post and beam construction. At work on the pavilion design, he asked himself: "What is the geometrical form which the enclosure must take in order to use the minimal amount of material?" Nature had provided the architect with suitable models: the shells of marine and terrestrial animals, egg shells and crab shells, among others—all of them forms that had interested Le Corbusier for several decades.[55] In 1948, Le Corbusier had instructed the readers of *Architecture d'Aujourd'hui* to go to the Museum of Natural History to learn its lessons. He described the displays of sea shells, with the typical shell—monochrome, of a neutral material, but "the expression of perfect equilibrium"—and he added, "Friend, to us, nature is saying 'pssst'."[56] For Xenakis, the resolution and integration of structure, geometry and space would remain the central pursuit; for Le Corbusier at that point architecture was primarily the vehicle for containing the poem.

An unsigned memo to Xenakis (probably from Kalff) outlined the conditions for the pavilion and its presentation, and the conditions under which the project would be accomplished. First, it was imperative that the Dutch delegation be informed immediately about electrical and other service requirements. One must coordinate construction of both buildings so that neither would impede the progress of the other. By 1 November, the main lines of the pavilion must be known; by the end of December, the scenario must be complete so that the colored lighting, sound, and projections can be developed in time for the pavilion's inauguration. Philips, again, reserved the right to veto the design and jettison the entire project if everything seemed to be unworkable. The definitive scenario, ideally accompanied by the storyboard for the film, was due no later than April 1957.[57] The presentation of the project would take place, as agreed, on 24 October.

Following the meeting, Kalff wrote Le Corbusier to offer Philips's reaction to the design presentation. They were in accord with the basic ideas of the design, but they had concerns and questions regarding the details. Some

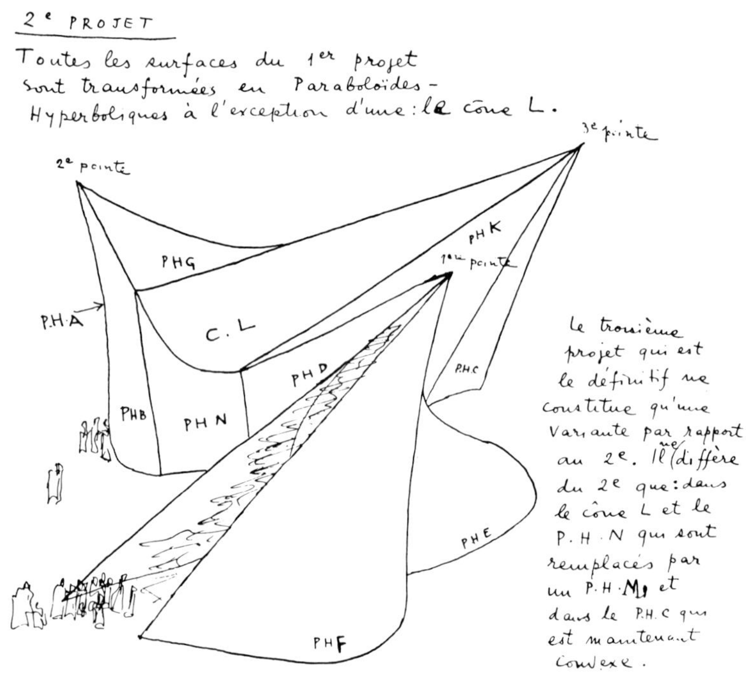

2-19
The geometry of the second scheme. Late 1956. Iannis Xenakis. In the third scheme, the building would be comprised entirely of hyperbolic paraboloid shells.
[*Philips Technical Review*]

2-20a,b
Model in wire of the second scheme. Late 1956–early 1957. Iannis Xenakis.
[Lucien Hervé; ©1996 ARS, New York / SPADEM, Paris]

items concerned the building's siting in relation to the Dutch pavilion and the neighboring Agricultural Pavilion. A connection with the Dutch pavilion would be desirable, but if it would create undue complications, it should be dropped (this is what, in fact, happened). After deliberation, the plan of the building was "flopped," that is, the entrance and exit points were reversed. The Philips site was quite limited, and the distances between the two buildings remained tight. But the small scale of the curious pavilion mitigated most chances of any major clash. Since the authorities of the fair had to approve all the architectural schemes, realizing the convoluted forms of the proposed pavilion, Kalff requested that the architect prepare a model within two weeks.[58] A second wire model by Xenakis, using tracing paper to enclose the surfaces, provided a more tangible illustration of the pavilion's complex surfaces [figures 2-21, 2-22].

According to plans, the Dutch pavilion would be under construction by December 1956. Due to the onset of winter and the slowing of chemical reactions by cold temperatures, the time needed to cure the concrete for the Dutch pavilion would be increased. The excavations for the Philips building must not disturb the Dutch project during this critical curing period; in other words, the Philips excavations had to be completed by mid-December. Kalff intended that all Philips construction should be completed by September 1957 to permit planting trees, grass, and tulips around the pavilion. Kalff remained eternally optimistic—or extremely naive. But he cautioned Le Corbusier and Xenakis that the delivery time for steel members could control the pace of construction, as it would for the Dutch pavilion. Thus, he advised that steel should be kept to its most simple forms, a prohibition which would more or less eliminate the structural steel skeleton initally proposed, and condemn the original scheme.[59]

Because the established budget of about 50 million French francs would be difficult to meet given the nature of the current design, Kalff requested that the pavilion be "deflated" about 10 percent to reduce its projected cost. He reminded the architect that service spaces for the automation and projection equipment, as well as for the personnel, needed to be included within the building, but these should not "sacrifice the form of the pavilion."

After the definitive meeting in October, Xenakis returned to work on the scheme, and examined the functional aspects in greater detail. Would Kalff provide him with the floor area necessary for the equipment and the toilets?[60] From what direction were the prevailing winds on the site (one assumes for determining the wind load or the optimum ventilation)? Kalff replied with words and a sketch [figure 2-23], describing at some length the volumes and equipment required for the work. Xenakis then met with "a great French firm" who would undertake "the calculations and eventually the realization of the

2-21
Model of the second scheme in wire and tracing paper.
Iannis Xenakis.
[Philips]

2-22
Model of the second scheme covered with tracing paper.
Iannis Xenakis.
[FLC; ©1996 ARS, New York / SPADEM, Paris]

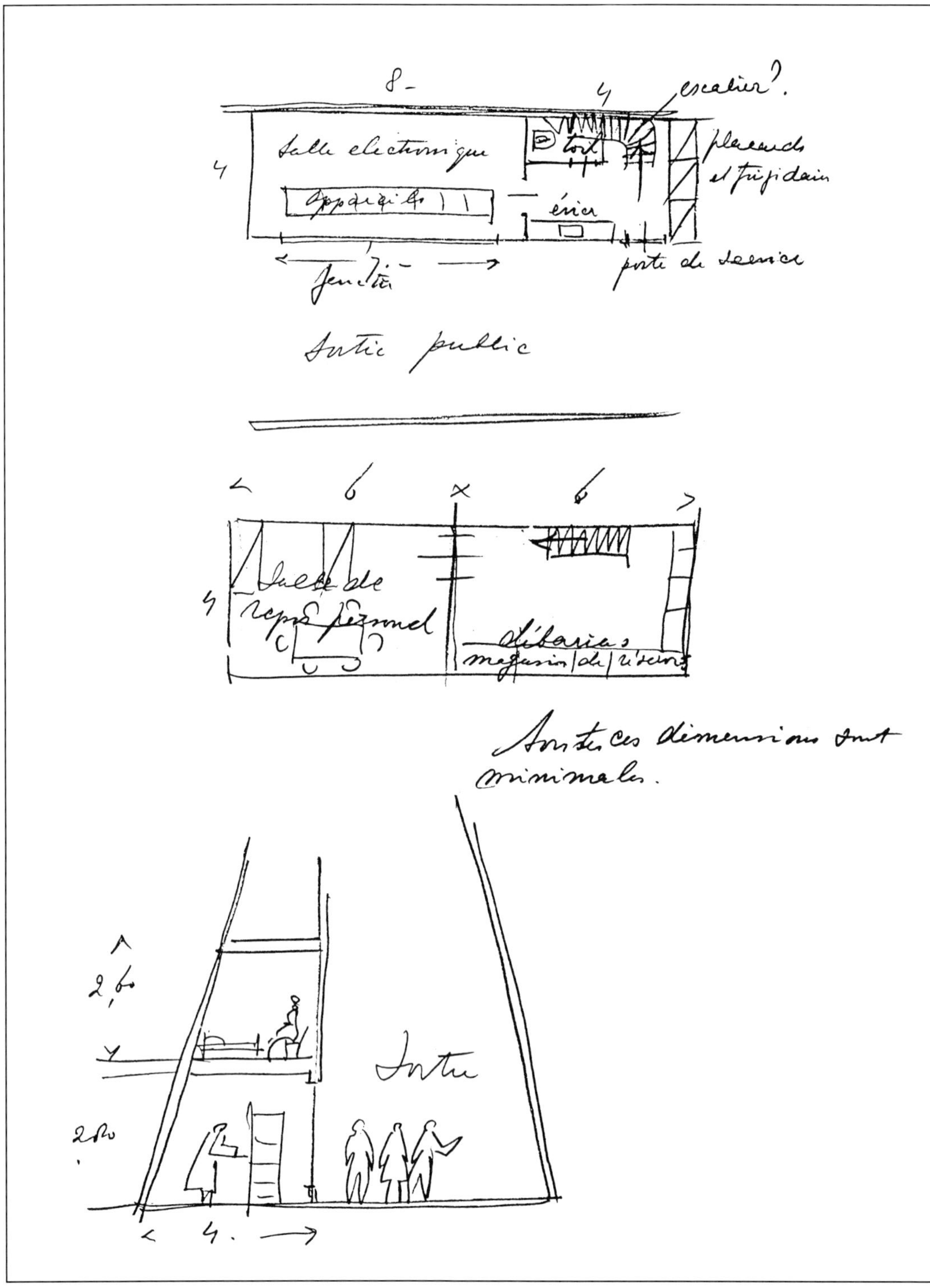
8-
4
escalier?
Salle électronique
4
placards
et frigidaire
évier
porte de service
Sortie public
6
6
4
Aussi ces dimensions sont minimales.
2,60
2,80
Sortie
4.

pavilion" (the Anciens Etablissements Eiffel).[61] They agreed with Kalff's prediction that the size of the pavilion would have to be reduced to come in under budget. Xenakis proposed two options for meeting the December deadline: either Kalff could show the Brussels authorities a model based on the old scheme, or he would have to postpone the meeting to allow some time for the engineers to make their contribution to a revised scheme.[62] It was decided to wait; a meeting was scheduled for 23 November at which the revisions would be discussed, so that foundation work could commence in short order.[63]

Following the meeting with Xenakis, Messrs. Lorin and Vallée of the Eiffel office, the engineers and potential builders, responded with a detailed letter. They reviewed the notes from the discussions and concluded that a metal frame was preferable to one of reinforced concrete: less weight, simpler erection, connections between the elements more classic and more certain, simpler attachment of the walls to the frame.[64]

Two networks of cables of opposing curvature would compose each screen wall segment; in turn, these would be sandwiched between two skins of bituminous *ruberoid*. The engineers were confident of their ability to execute the calculations without delay, as they knew that time was a consideration. The use of *gunitage* was dismissed because it was inelastic, and given the flexibility of the frame, the sprayed cement was sure to crack. In spite of their favoring a system of cables suspended from a structural frame, the engineers also allowed that it was possible to realize the walls in reinforced concrete. While it was technically feasible, the inclination of some wall segments was far from vertical, undermining the structural efficacy of the hyperbolic paraboloid as a structural form. "We believe," the letter concluded, "that the ideas expressed in your model correspond to the construction rationale upon which we have decided: rigid structural frames of metal upon which are supported flexible surfaces in the form of hyperbolic paraboloids and whose transverse surfaces comprise metallic cables set in opposing curves."[65] If these constraints were acceptable, they assured the architect that the work could progress on schedule, with the delivery of the main pieces on site in September 1957.

At this stage the hyperbolic paraboloid was the principle warped surface used in the segments of the building envelope, but one conoid section joined the three principal peaks. At first, the Eiffel firm engineers believed that the pavilion could be engineered with existing methods of calculation, and could be built for the 50 million French francs allotted for construction. A critical meeting took place in the atelier on rue de Sèvres in early November, at the end of which all parties seem to be in accord: the project could be realized as projected. Several days, later, however, the Eiffel engineers changed their

2-23
Sketches for services and control spaces. 17 November 1956. Louis Kalff. [Getty Center]

minds. There were problems with the structural frame, they wrote, and there would be problems in calculating—and constructing—the conoid. Several of these problems could be eliminated by discarding the conoid section, developing instead a conglomerate of shells that employed hyperbolic paraboloids exclusively. Xenakis agreed to this solution, and worked on developing the geometry of the shells.[66]

Xenakis wrestled with the implications of these new engineering developments. Le Corbusier was already in India, and all decisions to him were filtered through the Paris office. On 26 November, Xenakis wrote to the master, noting the Eiffel office's change in attitude during the previous two weeks. He noted that at the first meeting, he and the engineers agreed on many points: "1) the form of the model, which can be decomposed into conoids and hyperbolic paraboloids; 2) a structure of metal; 3) cables and *gunitage*; 4) a proportional reduction to safeguard the form and permit a reduction in cost to 50 million; 5) that the engineers would provide the precise techniques for construction as soon as possible, above all the foundations." A week later, however, Vallée and a Mr. Dausy came to the office and declared that: "1) the form cannot be realized; 2) arcs must replace the straight lines; and 3) one must replace the *gunitage* with flexible sheet materials, two layers of *ruberoid* for example."[67]

Many questions were asked, and little by little, Xenakis was able to convince the engineers that if the conoids were converted into hyperbolic paraboloids, the structure could become determinate. He found the use of arcs untenable, but accepted the substitution of new geometries for the conoids [figure 2-24]. In principle, the other conditions outlined in the Eiffel's assessment were admissible, although Xenakis was concerned that the *ruberoid* would not provide the acoustical isolation necessary for the sound presentation. He suggested that a layer of lath and plaster inside the flexible roof could be applied in such a way that it would not crack, thus giving the roof/walls the required acoustical properties. He made the various adjustments to the design, however, and constructed a new model. Philips was not entirely convinced of the efficacy of the structural approach offered by the Eiffel office. From meeting with Kalff came the suggestion that Xenakis request the services of the Dutch construction firm erecting the adjacent Dutch pavilion, but this was beyond Xenakis's authority, and Le Corbusier was still in India.

Other external factors forced a modification of the initial design ideas. The team designing the adjacent Dutch pavilion—Van der Broek & Bakema, Boks & Rietveld—were far ahead in their preparations for the fair; foundations would begin by December 1956. During the course of preconstruction studies, soil tests revealed that the building site afforded poor bearing capacity and that structurally stable material was found only twenty

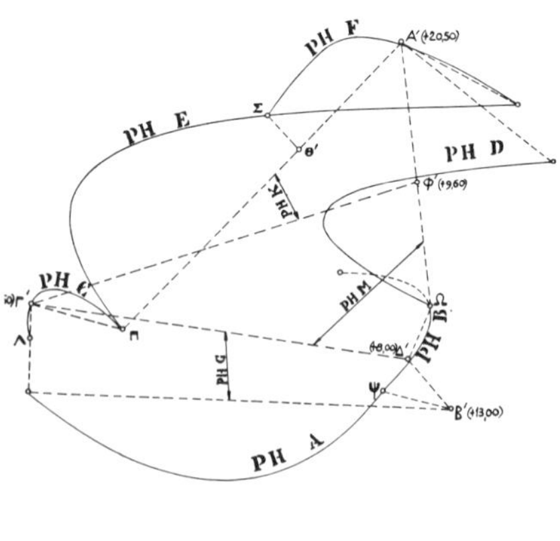

2-24
Definitive geometry for the pavilion. Late 1956–early 1957. Iannis Xenakis. [*Philips Technical Review*]

meters below the surface. Because a steel frame building exerts concentrated loads on its foundations, piles would be required to transfer these forces to a level that would provide sufficient support. This would be a costly process. Complicating matters, as Kalff explained in a letter to the architect, was manufacture of the steel. Fabricating major steel members such as those proposed for the Philips Pavilion would require as much as eight months' lead time—the Dutch pavilion had ordered their steel three months prior to Kalff's writing. At this time, the Philips design was barely underway, and Le Corbusier was again abroad [figures 2-25, 2-26, 2-27]. There was no way that steel could be ordered with sufficient leeway to allow for any hitches in delivery. And given that literally everything depended on the structural framework, problems with the steel could seriously hamper every aspect of the pavilion's construction.[68]

Preliminary analysis by the acoustical engineers at Philips had also revealed that the fabric solution would not provide sufficient acoustical isolation for the pavilion's sonic spectacle. Exterior noise would be audible within the pavilion, and the din projected for the Philips presentation would permeate the surroundings. Calculations established that at least five centimeters of concrete would be necessary for acoustical separation, and that other materials would not perform to the proposed level of satisfaction. The poor bearing capacity of the soil would require deep pilings to support the point loads exerted by the steel frame; a structure that would distribute its load along its perimeter would be preferable, and might even be treated as a raft floating on the spongy earth.[69] Although Xenakis believed that the foundation problems could be minimized, he was open to the possibility of a building of concrete.[70] In fact, from this point on, thinking about the pavilion as a simple form, perhaps nonstructural in itself, set within a structural framework, evolved into thinking of a building that would be self-supporting. Le Corbusier returned to Paris from India on 15 December; Varèse arrived in Europe five days later, and a meeting of the collaborators took place that very day.

At his house at Cap Martin for the Christmas holidays, Le Corbusier wrote to Kalff, explaining that his ideas were fixed in his head, but recorded only as "notes tossed onto paper." Obviously at a loss for documentation to verify his thoughts at the December meeting, Le Corbusier told Kalff that words can't really explain these ideas; which seemed to be only too true. "All of these things are mature, clear and sharp in my mind; there is no hesitation. On the contrary, there is a[n internal] schedule which orders it in one way and not in another."[71] He, again, was stalling for more time to make his selection of images, and to coordinate the work with the cinematographer Philippe Agostini.

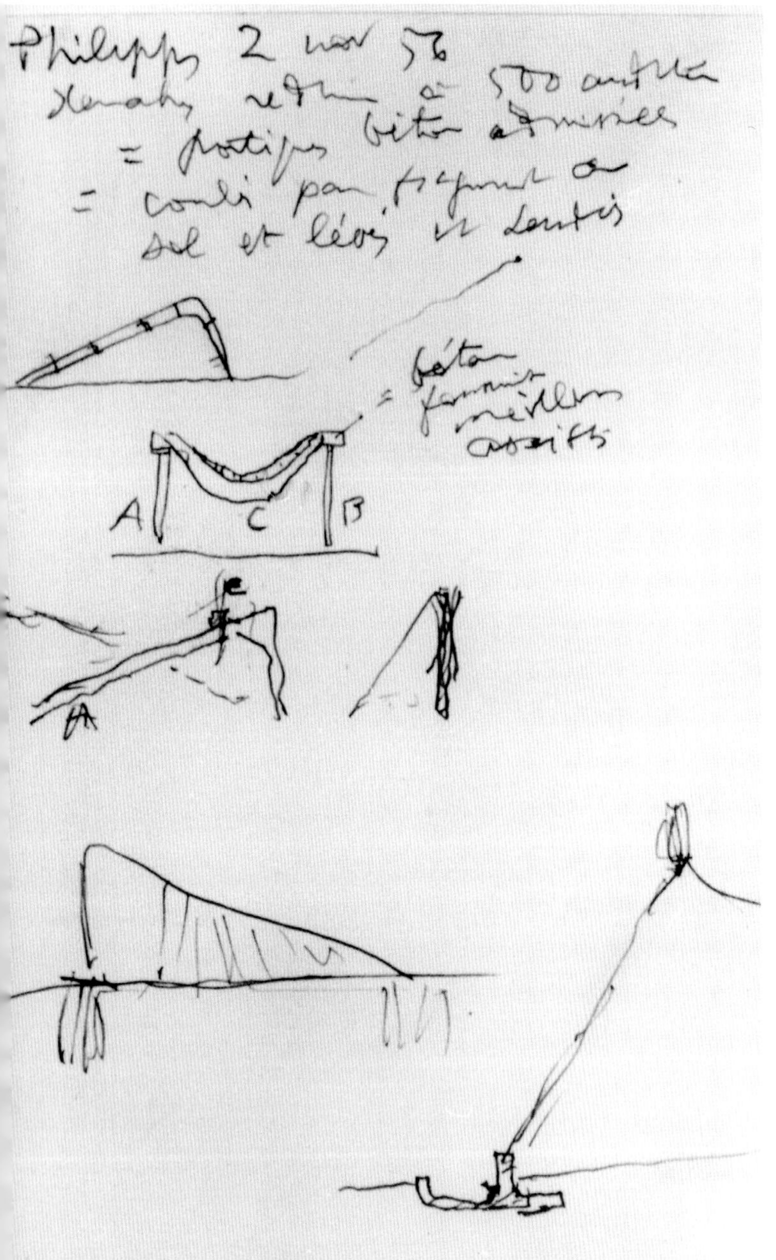
Philipps 2 nov 56
Xenakis
béton
A
C
B
A

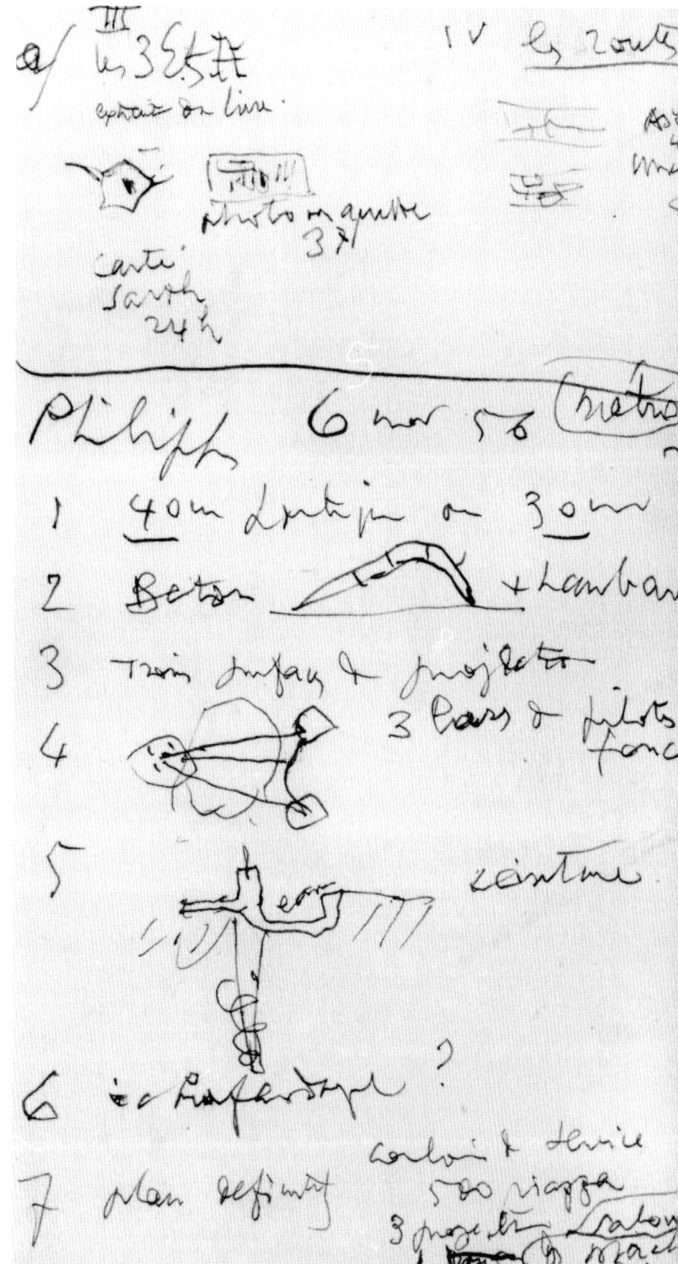
III
IV
photo maquette
Philips 6 nov 56
métro
1 40m
2 Beton
3
4
5
6
7 plan definitif
500
3 projecteurs

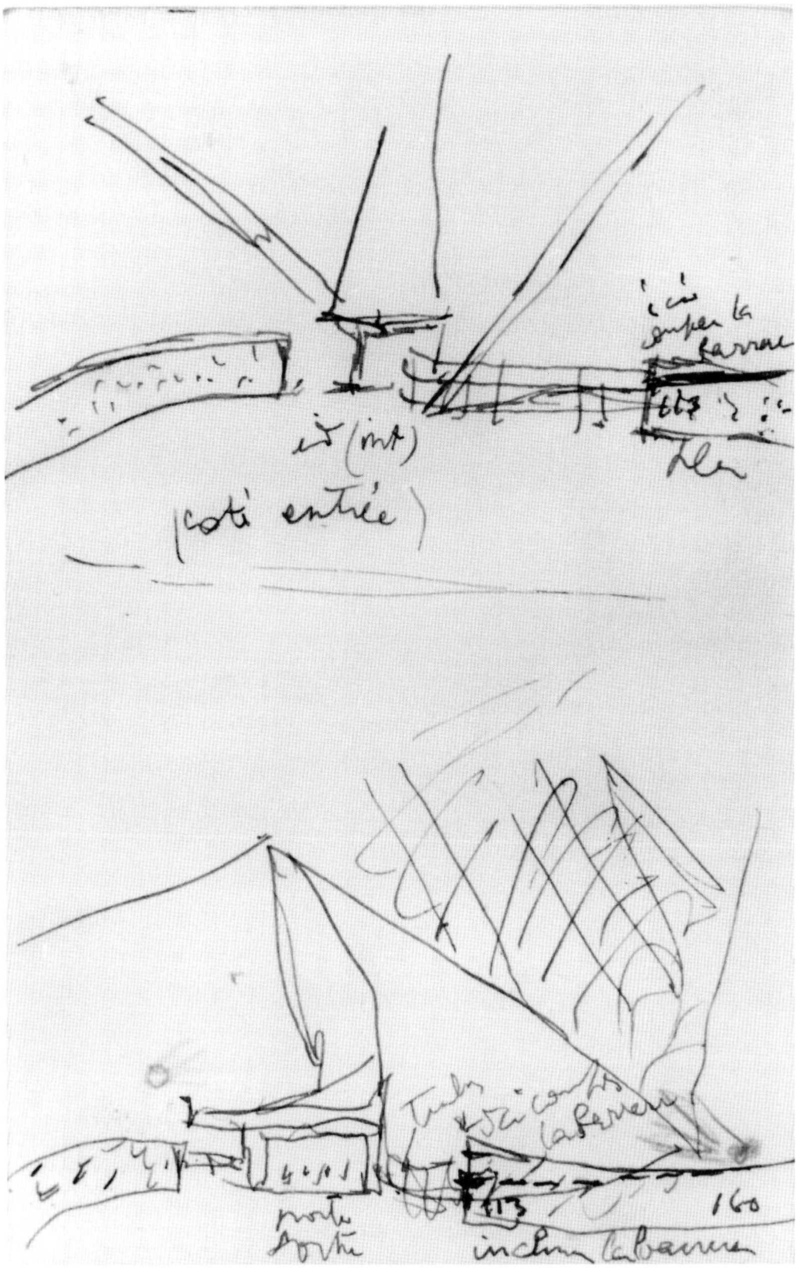

2-25
Sketchbook study of the pavilion's structure. 2 November 1956. Le Corbusier. [FLC; ©1996 ARS, New York / SPADEM, Paris]

2-26
Sketchbook study for the pavilion, with subjects ranging from structure to projection. 6 November 1956. Le Corbusier. [FLC; ©1996 ARS, New York / SPADEM, Paris]

2-27
Sketchbook studies for the pavilion's interior, with notes on the wall screening the projection equipment. April 1957. Le Corbusier. [FLC; ©1996 ARS, New York / SPADEM, Paris]

At the end of the year, the Eiffel office again affirmed their interest in engineering and constructing the pavilion, and reaffirmed that, under the conditions of the first scheme, they could still operate within the established budget. But if the design moved in a direction in which the proven means of calculation could no longer apply—requiring the use of structural models, for example—there would be no way that the contractors could advance a firm bid.[72] Le Corbusier's reply assured the engineers / builders that a definitive meeting of the major parties concerned would take place in the near future. He would propose that the budget be fixed at 50 million and he would get them together with Kalff to discuss the matter and get things rolling. "We have no more time to lose," Le Corbusier closed, "and I am counting on you. Your pavilion will be honored in the best places; it will be a masterwork."[73]

On 19 January 1957 a meeting was scheduled on the worksite of the fairgrounds in Brussels, with Le Corbusier, Xenakis, Kalff, and other Philips representatives in attendance. They were to meet with the architects of the Dutch pavilion—each with model in hand—to reveal the details of their respective schemes. The noted modernist architect Ben Merkelbach, consultant to the Dutch section, had cordially written to Le Corbusier in order to avoid any possible conflicts, and to request the delicate cooperation of the architect in reviewing their scheme. We are a small country, Merkelbach explained, and we want only a quiet building to display the exhibition on water as our friend and enemy.[74] Presumably, their discussions also raised issues of coordination on the construction as well as the respective idioms of the two pavilions.

As time passed, Le Corbusier was gradually turning the sketches in his notebooks into ideas that required increased precision and technical support. He wrote to Kalff to establish the limits of the electronic technology that Philips could provide, suggesting that at their next meeting, the artistic director be accompanied by technicians specializing in light, color, static image projection, film, and sound. He also confirmed his contact with Agostini.[75] The meeting would take place at the rue de Sèvres office; Willem Tak would serve as expert on acoustics, and S. L. de Bruin the problems of automation. Kalff himself was a specialist in lighting and color.[76]

A Xenakis memo made after the 11 February 1957 discussions established the definitive framework for the light and image presentations, and the technical means by which the intended effects would be produced. Using vertical cylinders of colored filters, it would be possible to wash the walls with varying chroma. The cylinders could be covered by continuous bands to maintain a consistent hue, or each horizontal band could be divided into panels producing a continually changing projection of color as the cylinder rotated. The projectors for the effects would be mounted within the shell; with the "setting sun" lighting effects created by lamps of diminished intensity hidden behind a parapet. They would also experiment with other forms of illumination including ultraviolet "black" light.[77]

Two booths would control the projected images, each of which would contain two reels of three hours' duration, allowing eighteen repetitions, and rewinding in spells. Two of the warped surfaces roughly 20 meters long and 8 meters high received the projected images. Within a central radius of 4 meters there would be no optical distortion. As a rule, the images in the film would visually move from left to right or from bottom to top. Since the projectors would be fixed—and this was before the era of zoom lenses—all movement and variation in size had to be accomplished in the filming. This was the primary reason for using film rather than slides, despite the fact that all objects in the film were essentially immobile. The film was planned to be in black-and-white and color.[78]

Because all sound was to be recorded, the interior of the shells would be sprayed with an acoustically absorbent asbestos, painted soft white. Speakers would be positioned on a conceptual grid of 2 meters square, with additional units at the summits of the shells. Given that the area of the shell surface was calculated at 2,100 square meters, they would theoretically need 500 speakers to carry the sound. In reality, they adjusted the number to around 300—there were no exact counts made, and the estimates vary wildly, from 150 to 450, depending on the source. The most accurate estimate seems to be 325. Prefabricated baffles would also be necessary in some places; insets in the shell were projected to be of about 30 centimeters in diameter.[79]

A showdown between the Eiffel office and Philips made the position of the former untenable. The Eiffel engineers appeared inflexible and remained doggedly committed to a structural steel framework, which had been, after all, the historic specialization of the firm. They addressed the problem of acoustic deficiency by proposing two layers of flexible skins, held 80 centimeters apart by a network of truss beams. "This construction is not any more reasonable [than the first proposal]," Kalff retorted, "and requires much more space than that available."[80] If this system were used, the diameter of the pavilion would need to be increased by 2.5 meters to maintain the interior space required for the pavilion's spectacle.

Returning to Eindhoven disappointed after his Paris meeting with the Eiffel firm, Kalff decided to interview other contractors for building the pavilion [figures 2-28, 2-29]. The large construction firm Hollandsche Beton Maatschappij was under contract to erect the Netherlands pavilion at the exposition, granted with the legal constraint that they would undertake no other projects at the fair. Kalff found a meeting with representatives of the Societé de Travaux en Béton et Dragages (contracted to the acronym "Strabed"), a filial of the Hollandsche Beton Maatschappij, particularly promising. Because

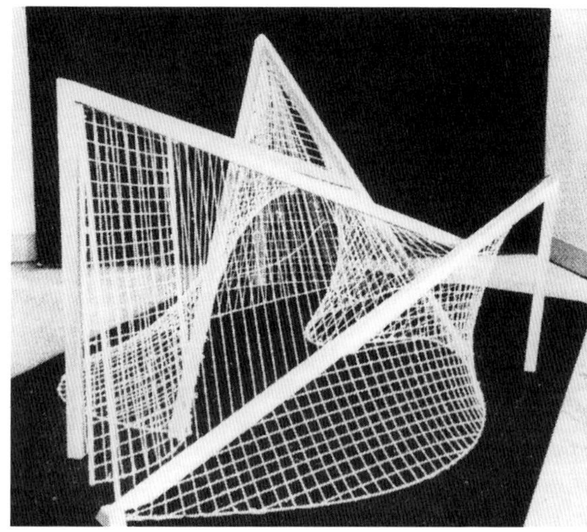

they were an independent Belgian subsidiary the restrictions at the exposition could be circumvented.[81] Working with Strabed's director, H. C. Duyster, and the engineers Vreedenburgh, Bouma, and Lightenberg of the Toegepast Natuurwetenschappelijk Ondezoek in Delft, new possibilities for realizing the construction of the pavilion arose. The consultants felt that it would be possible to retain the basic lines of the design by erecting bearing walls of concrete stiffened by concrete ribs, the geometry determined by connecting the three highest points of the pavilion: these were fixed at 42, 60, and 65 feet. Kalff was excited by the new direction: "[The use of a prestressed structure] could result in a spectacular building, unique in the world, which would certainly cost less than the Eiffel proposal and which would demonstrate an interesting advance in the science and technique of building, realized by Dutch engineers."[82]

Le Corbusier replied immediately, giving his consent. But he was not going to let anyone join the project without his first acquiring at least minimal background knowledge. He wrote his engineer friend Le Lionnais outlining the project upon which he was at work and asking him if he knew Vreedenburgh and, if so, what he thought of him. Le Corbusier wrote, "This guy appears to be fully capable. He has made a proposal supported by tests on models, proving that he knows the subject well."[83]

Most architects prefer to take a project from conception to construction, with control over the details of the design and its realization. In the case of Philips, however, the Le Corbusier office operated, in effect, as the design architects for the building; Philips, that is to say Kalff, was responsible for its realization. Kalff sought a contractor who would agree to tackle the problems of this small but complicated building. Luckily, they selected Strabed. More important, they discovered its director, chief engineer and construction expert H. C. Duyster, who would devise an incredibly ingenious solution to the difficult problem of constructing the pavilion: its surfaces would not be poured in place, but would instead be constructed of precast concrete panels set within a frame of round concrete beams, the entirety laced together by steel wires. That simple description, however, says little of the eloquence of the construction idea or the means of its execution.

2-28
Alternate structural proposal for the Philips Pavilion. Geerdink Aannemersbedrijf. 1957. [Courtesy Bart Lootsma]

2-29
Alternative construction proposal for the Philips Pavilion. 1957. [Courtesy Bart Lootsma]

3. REFINEMENT AND CONSTRUCTION

To construct a building of poured concrete, one must first construct two buildings of wood to serve as its formwork. While the use of repetitive forms has simplified the building of many other concrete structures, the shapes and curvatures of the Philips Pavilion's shells precluded such rationalized method. The difficulty of constructing in situ two precise wooden hyperbolic paraboloid forms in which to cast the concrete shells exceeded the technical ability of average workmen; the costs of the project would have become excessive. Instead of simplifying either the system of forming or the building's configuration, Strabed proposed an alternative method that reduced both the amount of formwork and building cost: constructing the building of precast panels. This was, in fact, a brilliant solution to a complex and intractable problem; but no one knew for sure whether the shells of the pavilion would function structurally, and whether they could be fabricated as units and thereafter fixed together as cohesive shells using prestressing wires. The task fell to Strabed and its engineer-director Hoyte Duyster.

By all accounts, Duyster was an exceptional engineer with an exceptional ability to see through construction problems and to logically resolve them. Born in 1907 and trained as an engineer at the technical institute at Delft, Duyster had worked through the 1930s for the Hollandsche Beton Maatschappij (which became the Hollandsche Beton Groep in 1968) in the Dutch Indies.[1] In 1937, he returned to Europe, first to Holland but then to Belgium where he assumed the directorship of "Strabed"—a subsidiary of HBM—in Brussels. He served in that capacity for twenty-one years. During the war, when building projects were few, more theoretical studies engaged his attention, particularly those concerning prestressed concrete and anchoring systems. He engineered the precast elements for the Eekloo bridge in 1946, and the following year designed and built a hangar at the Melsbroek Airport near Brussels, which for ten years boasted the world's longest prestressed primary structural elements. Appreciated for his sense of construction design, he later taught building in the architecture department at the technical university in Eindhoven.

For the upcoming World's Exposition in the Belgian capital, Duyster was commissioned to design and engineer a service station for Volkswagen. The resulting structure featured two "umbrellas" of prestressed concrete, each consisting of four hyperbolic paraboloids [figure 3-1]. The enclosing skin of glass with metal mullions was elegant and minimal, allowing the concrete structure to dominate the building's architectural expression. Although commissioned in 1957, this relatively small and straightforward project provided a testing ground for some of the prestressing methods that would be applied to the Philips building.[2] That Duyster discovered unexpected deflections along the edges of the shells, and could view firsthand the performance of pre-

3-1
Service Station, Brussels. 1958. H. C. Duyster, engineer/builder. [HBG]

3-2
Service Station, Brussels. 1958. H. C. Duyster, engineer/builder. [HBG]

stressed concrete hyperbolic paraboloids, may have caused him to be more cautious in resolving the construction and the engineering for the Philips structure; for example, he ascertained that the diameters of the ribs joining the shells needed to be sufficient to withstand the requisite forces of bending and torque [figure 3-2].

In his recollection of the Philips project, Duyster explained that there were essentially two possible approaches to executing the building in concrete: reinforced and prestressed. In reinforced concrete, steel increases the concrete's resistance to tensile forces; surrounded by the concrete, the steel bonds with it mechanically as the material sets. Prestressing, on the other hand, grants the concrete an increased resistance to compressive stress, allowing it to resist bending forces and to maintain the element in pure (or nearly pure) compression. Given the thinness of the Philips shells, it would have been nearly impossible to cast them in place; the complications increased exponentially when reinforcing was required. "There are thus objections to the *principle* of reinforced concrete," Duyster stated [figure 3-3].

> With prestressed concrete, on the other hand, there are no such inherent difficulties: the prestressing wires can be applied not inside the concrete but on its surface, if desired. In fact the *shape* of the walls of the Philips pavilion lent itself exceptionally well to this treatment owing to the fact that hyperbolic paraboloids may be generated by straight lines, a property possessed in common with ruled surfaces in general; this made it possible to apply all or most of the prestressing wires such that they would run *straight*. This would make it easier to introduce the appropriate prestresses in the material.[3]

In a uniformly loaded hyperbolic paraboloid shell, these forces are transferred as membrane stresses to the edges of the surface, which then direct them toward the foundations. The ribs stiffen the shells, and also absorb the planar forces as well as those created by the interaction of shell segments. In the case of the Philips project, the diameter of these ribs was increased due to the complexity of the shell configuration and the uncertainty of the direction and magnitude of the forces.

From the beginning of his work on the project, Duyster had wanted to construct the building of precast elements held together by wires and performing as a cohesive, if not homogenous, entity; he wanted it to operate structurally under compressive forces alone. A prestressed construction system would be the perfect means for holding the constituent fragments together as if by a number of giant claws.

Although the calculation of stresses on engineered shell structures had become more sophisticated during the late 1950s, it was simply not possible

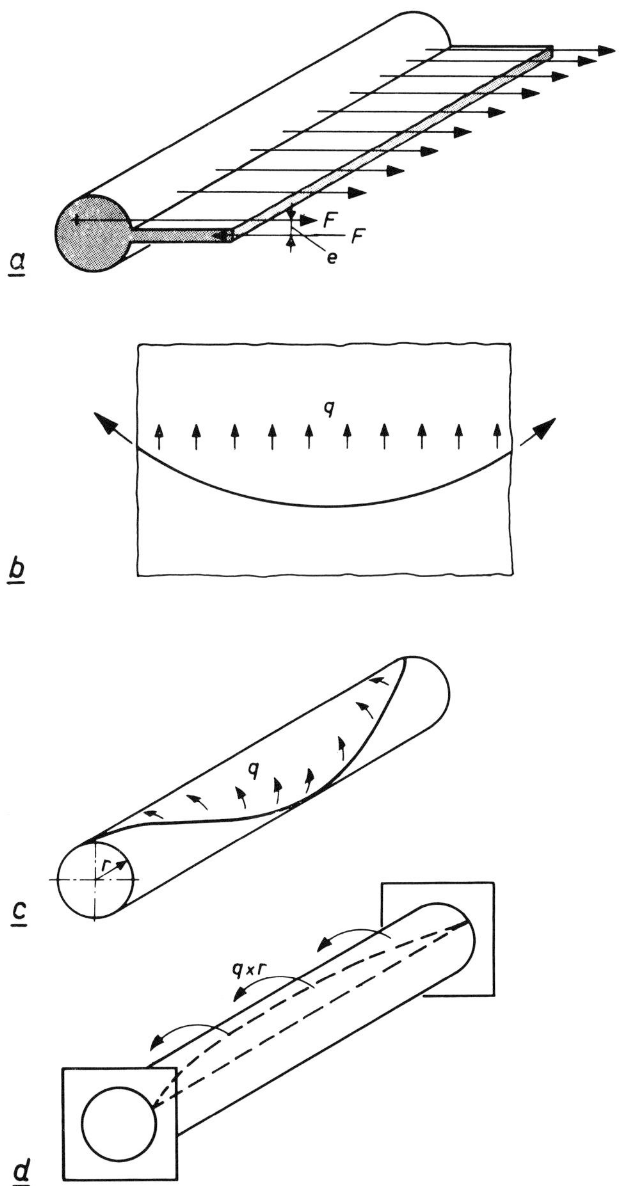

3-3
"At places where the prestressing forces do not act centrally through the shell (*a*), the rib is loaded by a torsional moment, which gives rise to bending moments in the shell. These can be eliminated by introducing torsional prestressing in the rib. For this purpose it is necessary to tension wires along a path that would be parabolic on the surface of the rib. This may be explained as follows. Consider first a plane on which a parabolically curved wire is fixed as shown in (*b*). When the wire is tensioned, the plane is loaded by a uniformly distributed force *q* kg/m. If this imaginary plane is now wrapped around a cylinder (c), the latter is loaded by a uniformly distributed torsional moment *qxr* kg.m/m when the wire is tensioned. When *q* is given the proper value, this moment will exactly compensate for the distortion caused by a uniformly distributed torsional moment *Fxe* (cf. *a* and *d*)."
[*Philips Technical Review*]

to determine through mathematics alone the stresses of the agglomeration of planar fragments projected for the Philips building. The hyperbolic paraboloid is an inherently strong form, which gains strength from its double curvature and translates forces perpendicular to its curved shell into membrane forces.[4] The more pronounced the curvature, the stronger the shell; conversely, the flattest parts of the shell are inherently the weakest. The pavilion's surfaces were, for the most part, quite steep and thus structurally strong. The majority of the forces were compressive and could be distributed within the shells. Vertical stanchions originally proposed as structural supports—the wire frame of the early models—were found to be unnecessary. But forces received on or near the edge had to be mitigated along the ribs. Stress differences between the rib and the shell could give rise to "edge disturbances," their determination being one of the most trying tasks of shell construction.[5] While the composition of the ribs could address these forces, the Philips design was a complicated melange of interacting shells, each of which transmitted forces to the ribs. In the days before the computer, there was no way that the structural design could be established by mathematics alone. Instead, the engineers relied on a direct and venerable technique: they built and tested structural models.

An early and innovative employ of structural models in this century was central to the designs of the Catalan architect Antoni Gaudí. Believing that even the flying buttresses of the Gothic period were unacceptable "crutches" applied to resolve the thrust of the pointed arches upon the side walls, Gaudí instead proposed arches planned along catenary curves—a line most easily envisioned by thinking of the curve a chain assumes when supported by two hands held apart. Because the forces distributed by a catenary enter its support (the ground, for example) perpendicular to the plane of the surface, no outward thrust is exerted on the support. Gaudí projected a structure of stone for the Güell Chapel outside Barcelona at Colonia Güell [1898; 1908–15] that utilized tiers of catenary arches diminishing in height and width. The entity would comprise a complex arrangement of interacting arches and vaults, all built of stone, ideally all acting in compression alone.

While Gaudí understood—and could predict with some certainty—the structural properties of a single catenary arch of stone taken in isolation, the stacking and interaction of multiple arches created conditions too complex to be solved with rules of thumb or existing engineering formulae. Instead, Gaudí relied on models [figures 3-4, 3-5]. The plan of the church traced upon a board became the base of the model. Wires of proportional lengths represented the arches of the church and joined the points of support established by the plan. The wire construction was then inverted, mounted on the ceiling of the atelier, and loaded with weighted bags of sand based propor-

3-4
Wire structural model, loaded with proportional weights, for the chapel at Colonia Güell. ca. 1900. Antoni Gaudí.
[Arxiu de la Sagrada Família]

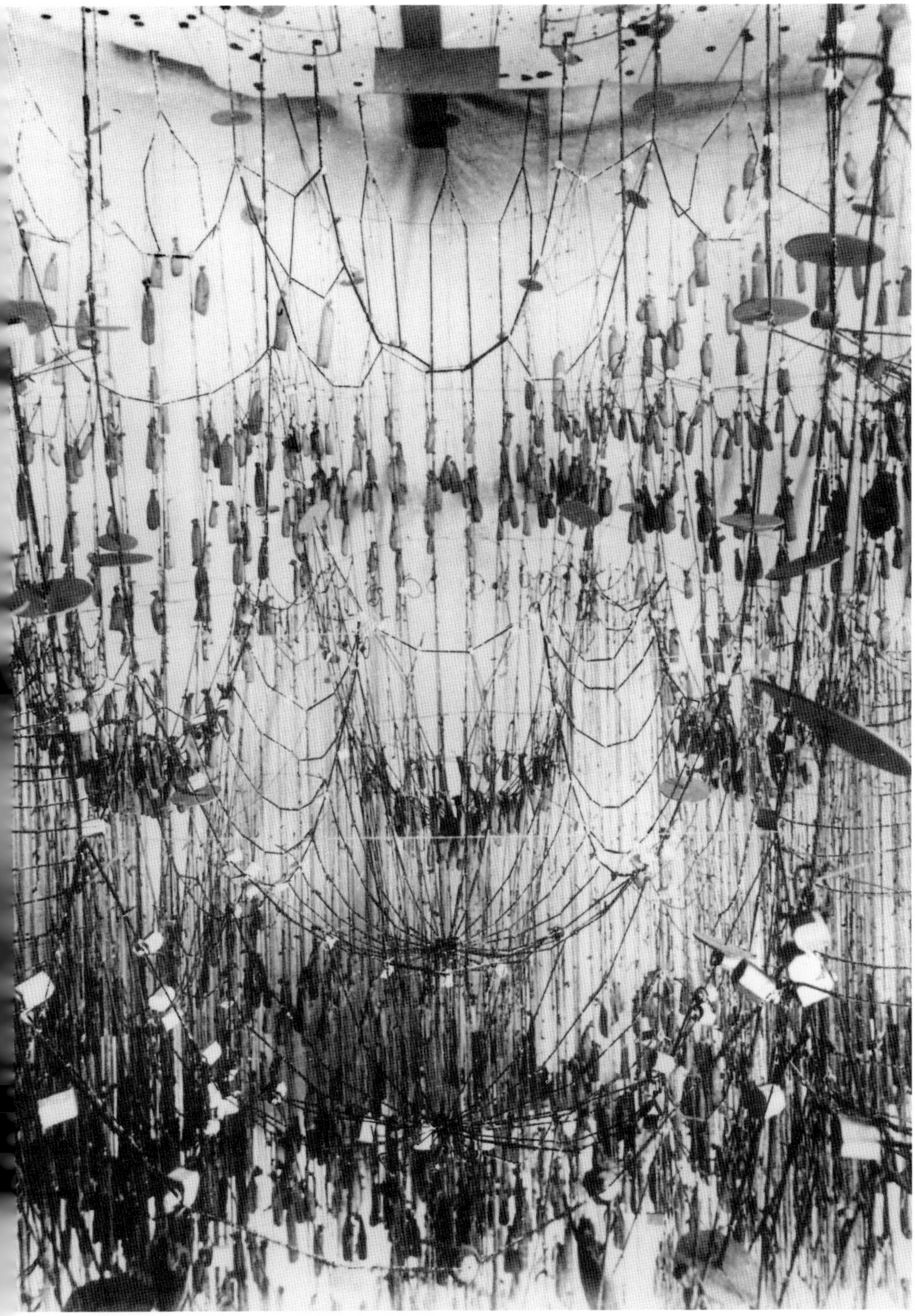

tionally on the projected weight of the stone. Under the downward pressure of the sandbags, the wires deformed into segmented catenary curves. Even the performance of arches upon arches could be demonstrated using this method, and by wrapping the arches with pieces of cloth, the wire arches were transformed into vaults. When the loading and deformations were complete, photographs of the model were taken, and inverted once again, making the images right side up. In this way, the model provided an accurate rendering of the structural skeleton for a stone building whose building blocks were under compressive forces alone.[6]

The Philips model studies echoed these methods pioneered by Gaudí, although the equipment for loading and measuring the deformations of the structure in Holland were far more sophisticated, including the use of polarized light to ascertain the stresses within the panels.

With about only a year remaining before opening day, the engineers and contractors were under considerable time pressures. For the preliminary study, Kalff and Strabed gave their consulting engineers only a week to determine whether the entire approach to the pavilion's structure was feasible. In February 1957 a model 1⁄25 actual size was built at the TNO, or Toegepast Natuurwetenschappelijk Onderzoek (Applied Scientific Research), an independent research institute in Rijswijk [7] [figure 3-6]. Fabricated of metal tubes and wire mesh covered with plaster, the model was test loaded to derive a general idea of the performance of the proposed structure. Three stanchions supported the peaks of the pavilion. Although composed of mixed materials, the structure was regarded as monolithic for testing purposes.[8] It should be noted that at this time, Duyster and Vreedenburgh were hoping that the prestressing wires would be necessary only on the interior surfaces, leaving the exterior to appear as a homogenous skin. The model also served other purposes, among them evaluating the design and using it for publicity, and no doubt, as propaganda for Kalff in his continual quest to convince the Philips management of the wisdom and excitement of the scheme.

Tests by Vreedenburgh's associates Adolf L. Bouma and Frans Lightenberg on a second model built at a scale of 1:10 in the Stevin Laboratory of the Department of Civil Engineering at the Delft Technische Hogeschool further investigated the performance of the structure under three conditions: loading under its own weight; under a vertical load on a single "fairly horizontal surface;" and in response to forces used to simulate wind pressure on the shell structure.[9] Although the stresses in some places were considerable, the tests determined that the structure was feasible, and that the stanchions were essentially redundant [figure 3-7]. They were subsequently removed for tests of a revised design where it was determined that the shells alone could support the necessary loads. From these tests, Strabed learned at what points

3-5
Structural model (inverted), using cloth to simulate vaults. Chapel at Colonia Güell. ca. 1900. Antoni Gaudí.
[Arxiu de la Sagrada Família]

3-6
Wire and gauze model, 1:25. Delft. January 1957. [*Philips Technical Review*]

3-7
Plaster model, with level system for simulating wind loads, 1:20. Delft. January 1957. [*Philips Technical Review*]

3-8
Plaster model, with measurement systems, 1:10. The extensive wiring connects the strain gauges to the measuring devices partially visible in lower right. TNO, Delft. January 1957. [*Philips Technical Review*]

the prestressing required intensification, but the results were still quite general. "The Belgian building authorities had laid down that the building must be able to withstand loading by its own weight plus a double wind load (150 kg/m2) with complete safety."[10]

To determine the effectiveness of the structure with greater precision, researchers loaded the 1:10 plywood model to simulate the actual performance of the building on site [figures 3-8 through 3-12]. Since the general nature of the structural behavior had been determined in the plaster model, the engineers felt that it would be sufficient to test only a portion of the building at enlarged scale. For this purpose they selected two problematic sections, one of which was almost flat, emphasizing the "visual observation of how the model behaved under diverse conditions."[11] The model, which measured 2 by 3 meters, was constructed analogously to the actual pavilion, with primary ribs, prestretched nylon thread, and hundreds of three-ply plywood diamonds.[12]

The plywood slabs were first held together using adhesive tape, and then uniformly prestressed. Because the prestressing was executed on only one side of the structure—as was intended in the actual pavilion—the ribs that were free on one side were eccentrically loaded, and thus buckled. The torsional forces upon the ribs were corrected "by fixing a large number of spokes at right angles to the ribs, and by arranging for forces to act on the ends of these spokes in directions perpendicular to both the ribs and the spokes"[13] [figures 3-13, 3-14]. In addition to the effects of surface forces, the performance of the substructure was also studied at this time: during the prestressing process, the shells had to remain structurally independent so that all of the force would be taken by the shells rather than being absorbed by the foundations and floor slab. The model performed as predicted and served to confirm the original design and engineering assumptions. As a final gesture, five members of the laboratory staff climbed on the model, adding one and a half times the dead weight to the load, without collapse[14] [figure 3-15].

Once the selections of the engineer and contractor were confirmed and the structural tests initiated, the pace of the work and the process of refining the design quickened. By 2 April the architects fixed the forms of the entrances/exits and directed more thought toward the design of the site as a whole. At this point, for example, there appeared free-form pools to catch the rainwater shed by the building, which would have no gutters or drainpipes [figures 3-16, 3-17, 3-18]. The engineers continued their detailed studies of various structural issues, such as the connection of the pavilion to its foundations.[15] The frequency of the communication among Kalff, Duyster, and the architects increased.

At this stage, Xenakis was centrally involved in the finalization of key details. The February meeting had established the directives for the visual and

3-9
Plywood model, 1:10.
TNO, Delft. January 1957.
[*Philips Technical Review*]

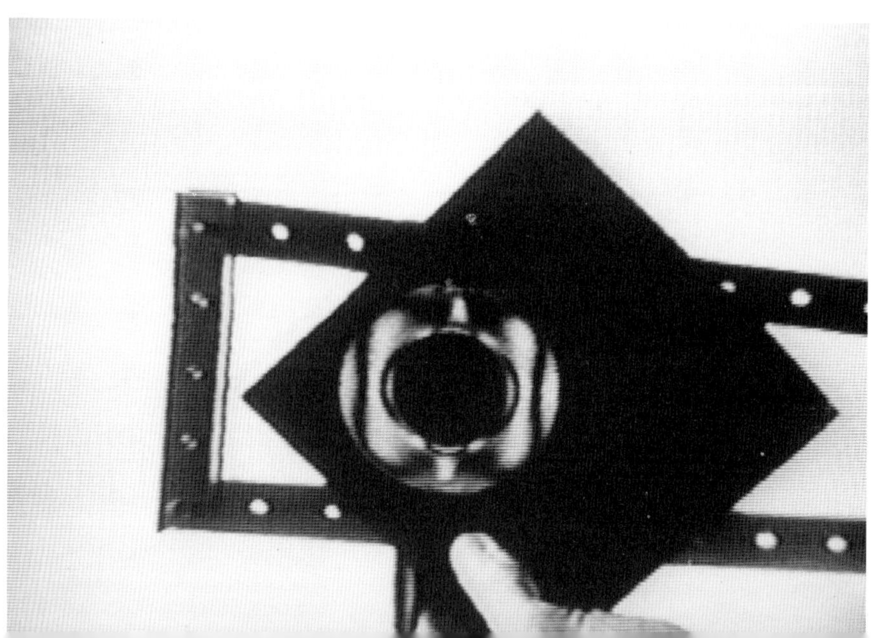

3-10
The plywood model, 1:10.
TNO, Delft.
January 1957.
[*Philips Technical Review*]

3-11
Detail of the plywood model, 1:10, "illustrating the bending of each plywood slab to fit a certain part of a saddle surface. In the center . . . is a photo-elastic measuring element."
TNO, Delft.
January 1957.
[*Philips Technical Review*]

3-12
Explaining the use of polarized light to trace stress in the structural models. From the film *De Bouw van het Philips Paviljoen*, Strabed, 1958.
[HGB]

3-13
"In the free ribs (to which no second hyperbolic-paraboloidal surface abuts in the model), spokes are fitted by means of which moments can be applied to the ribs to oppose the torsional moment exerted on the spokes by the wires. This is done by exerting transverse forces on the spokes by wires passing over pulleys and from which sandbags are suspended."
[*Philips Technical Review*]

3-14
The plywood model loaded with weights proportional to the projected loads.
[*Philips Technical Review*]

3-15
"The plywood model severely overloaded."
[*Philips Technical Review*]

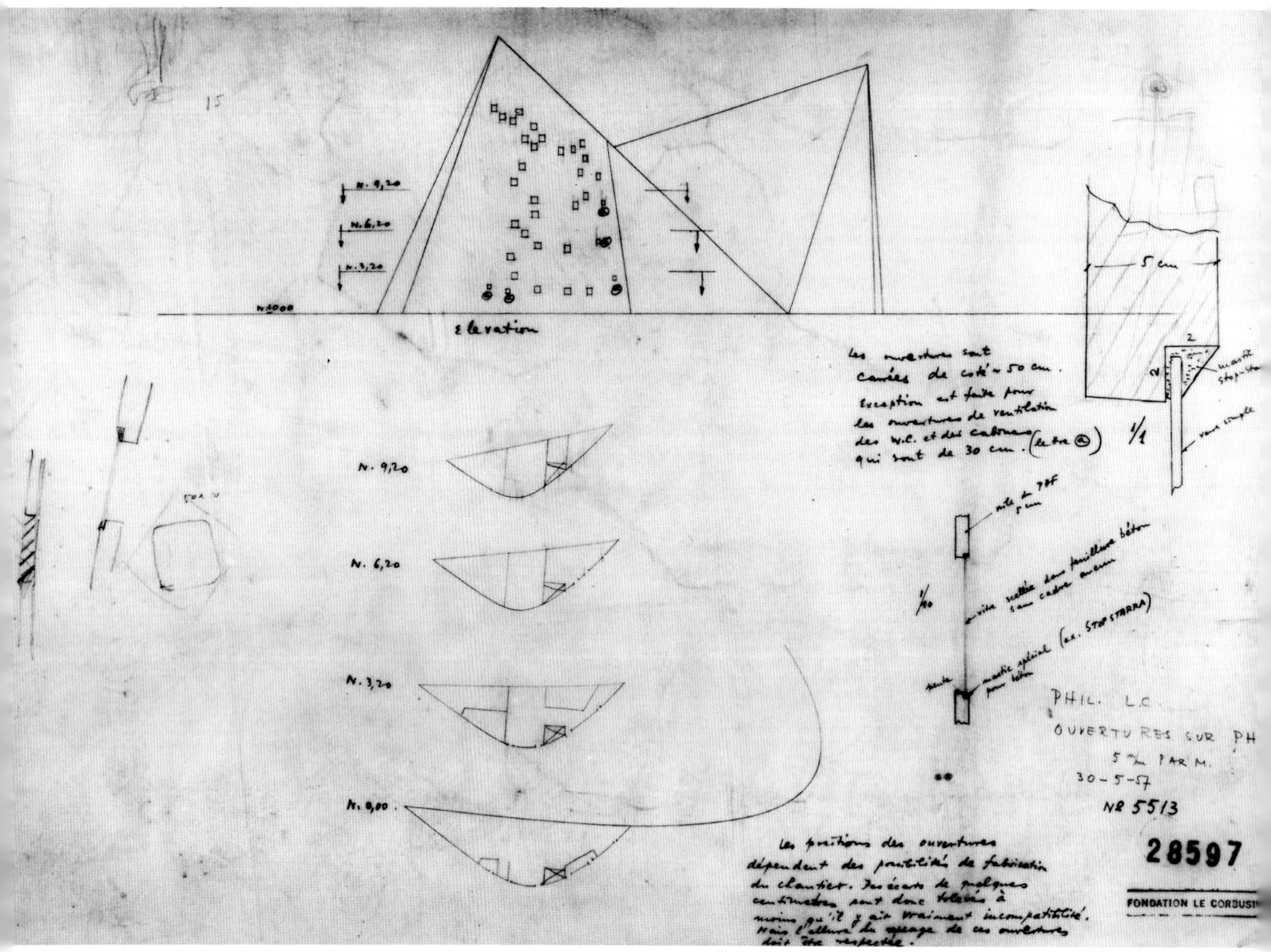
15
N. 9,20
N. 6,20
N. 3,20
N. 0,00
Elevation
Les ouvertures sont
carrées de coté ~ 50 cm.
Exception est faite pour
les ouvertures de ventilation
des W.C. et des cabines
qui sont de 30 cm. (lettre ⓐ)
5 cm
2
1/1
1/10
PHIL. LC
OUVERTURES SUR PH
5 m/m PAR M.
30-5-57
Nº 5513
Les positions des ouvertures
dépendent des possibilités de fabrication
du chantier. Des écarts de quelques
centimètres sont donc tolérés à
moins qu'il y ait vraiment incompatibilité.
Mais l'allure du passage de ces ouvertures
doit être respectée.
28597
FONDATION LE CORBUSI

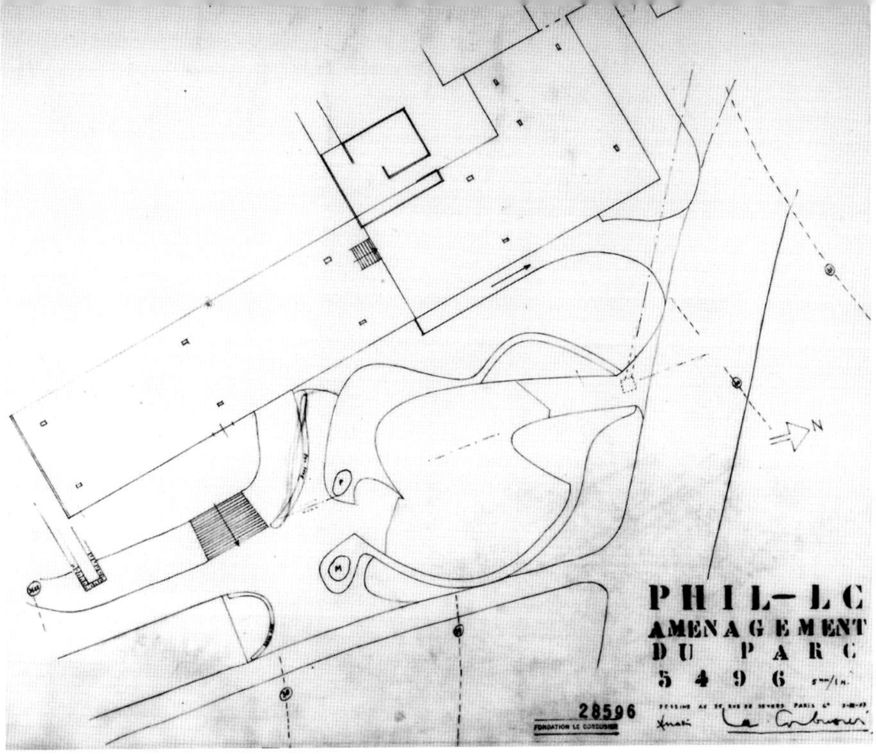

3-16
Studies for openings in the pavilion surface.
30 May 1957.
Iannis Xenakis.
[FLC 28597; ©1996 ARS, New York / SPADEM, Paris]

3-17
Site plan.
19 June 1957.
Iannis Xenakis.
[FLC 28596; ©1996 ARS, New York / SPADEM, Paris]

3-18
Definitive site plan.
19 June 1957.
Iannis Xenakis.
[FLC 28598; ©1996 ARS, New York / SPADEM, Paris]

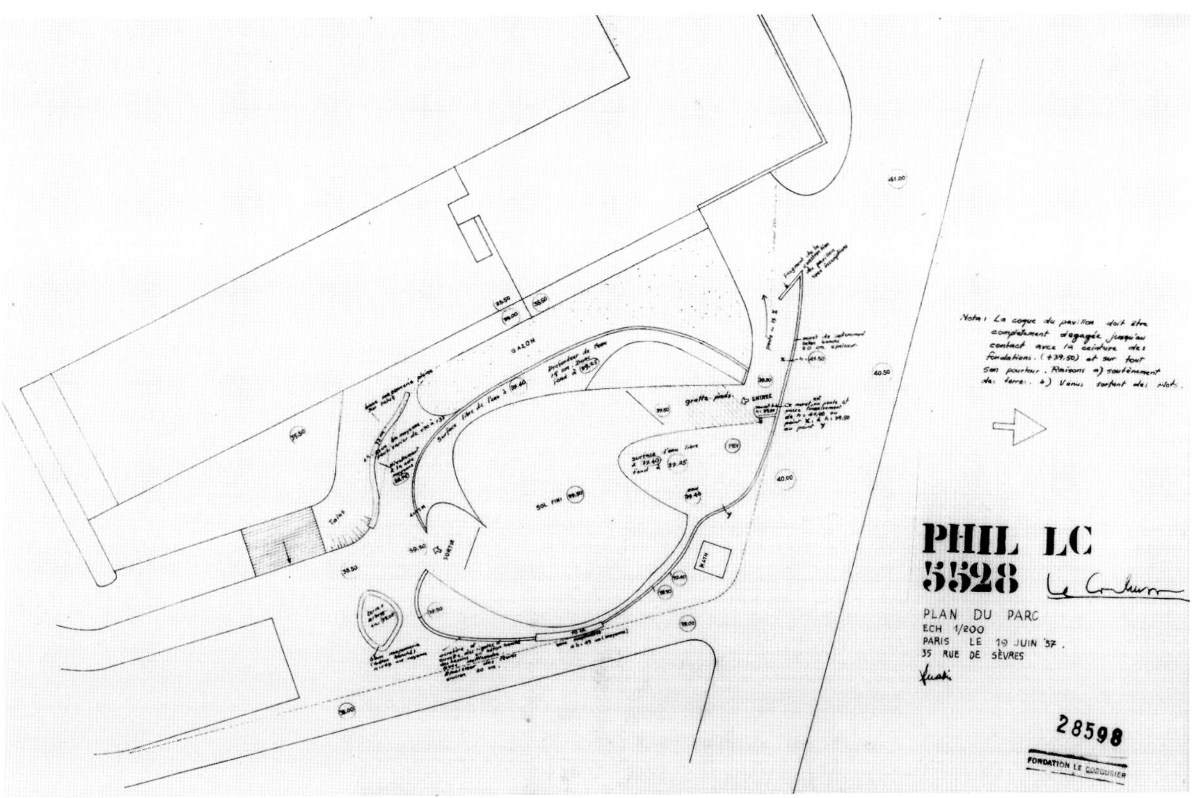

aural aspects of the pavilion's spectacle, but these were general goals rather than specific means. Given Xenakis's studies in music and work as a composer, it is not surprising that the resolution of the emission of sound throughout the pavilion was one of his most intimate concerns. A meeting with Tak in Eindhoven had raised the issues of the baffles and the loudspeaker design. Given that the pavilion was to have some three hundred speakers, they would constitute a prominent visual feature of the interior space or the pavilion's exterior appearance. At one point, there had been a proposal to mount the loudspeakers flush with the interior surface, allowing the mechanisms to extend through openings in the concrete shells to the exterior. The idea was discarded quite quickly on grounds that the necessary openings would require extensive weatherproofing while undermining the structural integrity of the shells.[16] Xenakis cautioned Tak that "[t]he Pavilion must not give the impression of a bolted or riveted construction or the impression of a badly decked-out rag doll."[17] In the end, structural and aesthetic considerations prevailed, and the speakers were mounted on the inside surfaces of the pavilion.

Since the high- and medium-frequency sources were multidirectional, they did not really require speaker enclosures. To reduce their bulk, Xenakis proposed mounting the speakers directly within absorbent cylinders throughout the interior, except where they would conflict with the function of the wall as projection screen. To maximize the effect of the nondirectional speakers for the bass frequencies, they were to be dispersed and fixed directly to the pavilion floor. The same parapet wall screening the lighting and projection equipment would hide them from view. For the most part, these guidelines governed the final designs of the speaker units and their distribution.

In May, Le Corbusier was again in India, at work on the Assembly Building in Chandigarh.[18] Xenakis informed him that work on the foundations would begin on 6 May, that Strabed had already begun casting the concrete panels—and that they were greatly relieved that no holes for loudspeakers would mar the integrity of the shell. In collaboration with the architects and engineers, Duyster had been able to limit the number of shell segments to nine[19] [figure 3-19]. A test of the sonic effects would probably take place in the "garage" in Eindhoven sometime in the near future.[20]

The method for casting and assembling the concrete panels was perhaps the most interesting aspect of the entire project, as curious and radical—and yet primitive—in its own way, as the form of the building was as architectural and structural design. As Duyster noted in retrospect, to have cast a building of this complexity in situ would have taken skill, time, and expense far beyond that allowed by the schedule and budget. The form of the hyperbolic paraboloid, on the other hand, was ideal for construction in units, because it was generated by straight lines. But how could the contractor realize

3-19
Schedule of shell segments. Redrawn from memo. Strabed. March? 1957. [Philips].

in concrete the desired curves, some of which were quite acute?

Strabed had at its disposition in the Brussels area a large shed, termed a hangar in the written documents. Working on the ground, under cover, the construction team could more conveniently and more accurately produce the required units. Mounds of sand were deposited on the hangar floor, creating a gigantic sandbox in which to sculpt the shapes of the shells [figure 3-20]. By erecting boards following the boundary lines of each shell segment the builders were able to generate a sand surface corresponding exactly to that of the desired plane [figure 3-21]. It was not necessary that the sand surface lie at the same angle to the ground as it would when a part of the completed building, as long as its surface curvature was identical to the final wall section. Within the shell boundaries, workers positioned wooden strips that followed the regulating lines of the hyperbolic paraboloids, thus dividing the surface into hundreds of rhomboidal units. The strips also kept the panels discrete after the pour, and could be used as guides for screeding the sand within [figure 3-22]. The hundreds of precast panels would be rejoined on the construction site, laced together into a coherent structure.

The conditions within the shed to some degree determined the method of execution: "The extent to which each shell had to be subdivided," Duyster explained, "depended in the first place on the available height in the shed and on the permissible steepness of the slopes of the sand bed (shuttering would have been necessary if the slopes were too steep)."[21] Because of the limited height of the hangar, some shells had to be divided into two forming units, somewhat increasing the complexity of the pours—which were executed using wheelbarrows of cement, hand-poured from buckets into the forms. To reduce the deformation of the battens and to allow worker access, the panels were cast as alternating units, as a checkerboard [figures 3-23, 3-24]. Since the structural design guaranteed that the final shell of panels would be almost completely in compression, little reinforcing steel was needed for structural performance. Transit and handling, however, required at least a minimal amount of reinforcement, and a light steel mesh was used to strengthen the concrete and to provide for expansion and contraction. This simple idea of Duyster's—to precast and prestress the shells—brought the pavilion's construction into the realm of possibility [figures 3-25, 3-26, 3-27].

In contrast to the prefabrication of the panels, the ribs between the shell segments were cast in place [figure 3-28]. "It was rational to make the ribs cylindrical, because the shells 'turn' about the ribs; with any other shape it would have been troublesome to join the shells to the ribs." To handle compressive, bending, and torsional forces directed upon the edges by abutting the shells the ribs were also prestressed. "The torsional prestressing wires were disposed helically around the axis of the ribs," in contrast to the radial or lon-

3-20
Sand form with bounded edge, determining the configuration and slope of the shell segment. Strabed hangar, Brussels. 1957. [HBG]

3-21
Placing the dividing boards within the shell. Strabed work site, Brussels. 1957. [Philips]

3-22
Although the work crew was quite small, casting continued on several shells simultaneously. Strabed work site, Brussels. 1957. [Philips]

3-23
Presumably for worker mobility, to allow for some expansion while containing deformation in the separating boards, the shells were cast in alternating panels. [*Philips Technical Review*]

3-24
The casting method was remarkably primitive, essentially techniques used to pour concrete sidewalks.
[HBG]

3-25
The completed segments awaiting disassembly and re-erection on the exposition site.
[Philips]

3-26
Disassembling the completed pours, scraping excess sand from the surface of the panels. From the film *De Bouw van het Philips Paviljoen*, Strabed, 1958.
[HBG]

gitudinal positioning for compression and bending. To compensate for differences in forces, the ribs were partially prestressed before the panels were connected to them, "in such a way that when the final prestressing was applied, the deformation then occurring in rib and shell were to all practical purposes equal"[22] [figure 3-29].

The differential in stress between the shell as an entity and its foundation structure paralleled the differential between shell and rib. To mitigate the problems of the foundation absorbing a portion of the prestressing forces or of the shell deforming during the process, the shell structure was separated from the foundation with two layers of bituminous felt [figures 3-30, 3-31]. A continuous series of anchor blocks reinforced the concrete perimeter beam, giving the impression of a curving dentiled line. These blocks, anchored to the foundation and running along the interior of the building, were not visible from the exterior [figure 3-32]. After all the panels were in place, the prestressing executed, and the deformation completed, the interstices between the blocks were filled with concrete, yielding in effect a continuous flat horizontal beam.[23]

An elaborate wooden scaffolding supported the concrete panels as they were assembled; wooden wedges used as shims brought them into alignment, "[t]he operation thus being a rather odd kind of bricklaying process"[24] [figures 3-33 through 3-37]. After all the concrete panels were in place, the joints were grouted and the process of prestressing began. Seven-millimeter wires, each with a maximum tensile strength of 3300 kilograms, traced the generating lines of the hyperbolic paraboloid and created a deformed grid with wires roughly 50 centimeters apart [figure 3-38].

Strabed had developed a prestressing system sufficiently simple to manage the extreme variations of curvature of the Philips shells. In several other prestressing systems, reinforcing wires passed through apertures in the members or panels, making connections to superior beams or posts more difficult, and final grouting to cover the joints more troublesome. In the Strabed system, an anchorage plate with a threaded section of the prestressing wire—crimped to hold it in position—was cast into the rib or into the substructure, but free of any of the constituent panels (it was 2 to 3 centimeters from the shells in the case of the pavilion). Much like a turnbuckle, the wires were threaded into special sleeves, "so designed that no slip can occur during this fixing process."[25] A jack developed by the company gripped the wires from the side, "[t]he amount of tensioning applied can be read accurately from a dynamometer fitted to the jack,"[26] as one might tighten the spokes of a bicycle wheel [figure 3-39]. The genius of the system was its relatively easy application to a great variety of conditions in curvature, position, and length of prestress wires. Thus, the stressing of the assembly of concrete panels was

3-27
The panels, stacked and coded, ready for shipment to the building site.
[HBG]

3-28
Casting the ribs that stiffen the shells, their steel reinforcing used to counter bending and torsion.
[Philips]

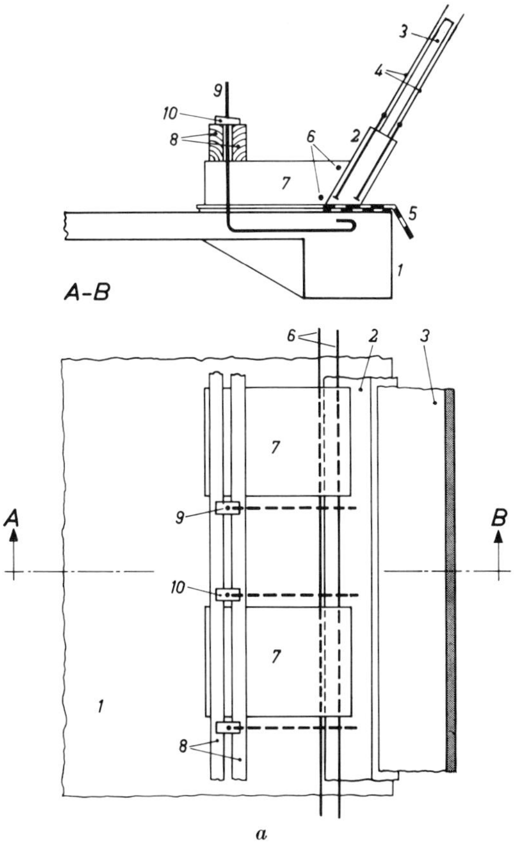

3-29
The jungle of wooden scaffolding necessary to cast the ribs and mount the panels.
[Philips]

3-30
The floor slab of the pavilion; reinforcing bars for the stiffening blocks can be seen on the perimeter.
[Philips]

3-31
Section and plan segment showing the joint between wall and slab (left flexible until the building was prestressed), and the location of reinforcing and prestress rods.
1 foundation;
2 substructure; 3 shell;
4 prestress wire;
5 bituminous felt;
6 perimeter prestress wires; 7 stiffening blocks;
8 temporary wood beams; 9 flexible rods cast into the foundations between tailblocks;
10 clamps.
[*Philips Technical Review*]

3-32
The blocks used to strengthen the shell perimeter. Although seen as a convex line in the photograph, the blocks are actually within the shells of the pavilion.
[*Philips Technical Review*]

3-33
Upon the scaffolding, within the ribs, the walls of precast panels begin to rise.
[Philips]

3-34
The assembly was "an odd bricklaying process." Comprised of panels, the shells lay one upon another and begin to suggest a continuous surface.
[Philips]

3-35
As the panels are installed, the sense of enclosure becomes more apparent. [Philips]

3-36
The prestress wires follow the installation of panels. [Philips]

3-37
The joints are grouted; the pavilion is painted with a sealer, and then with aluminum paint. The prestressing wires are in place, but still relaxed. [Philips]

3-38
Using a special device developed by Strabed, the wires, on a grid roughly 50cm × 50cm, are tightened one by one.
[*Philips Technical Review*]

3-39
Detail of the prestressing wires and their tension devices after painting.
[Philips]

3-40
The shells completed; new scaffolding permits access to the pavilion's exterior surfaces.
[Philips]

WOLF-ZONDERVAN

much like the gradual tightening of a multitude of belts applied to both the inside and outside of the pavilion. Although attached to the foundation and substructure, the pavilion was essentially bonded as a monolith by its prestressing wires, and acted almost as the hull of a ship inverted upon a concrete slab.[27] The completed concrete work, like a barrel, was a compression structure lashed together by steel in tension [figures 3-40, 3-41]. A waterproof paint was applied to seal the concrete surfaces and protect the wires, although the assembly of the structure was preestablished at seven months. The entirety was finally coated with aluminum paint, unintentionally fooling some into thinking the pavilion was built of metal, or that it might be a tent.

As the months passed, those modifications and developments that characterize, or dog, every building also affected the Philips project. Shifts in the siting of the Dutch pavilion forced changes in the position and level of the Philips pools, and caused a complete reversal of the visitor's path: what had been the entrance became the exit and vice versa.[28] This change allowed for the pools to be widened, forming an ideal location for the pavilion's sculpture/sign. And could a grill be added at the entry upon which the visitors could wipe their feet?[29] Issues large and small required attention as the deadline for construction approached.

The most important development to influence the appearance of the pavilion was the decision to use prestressing on the exterior as well as the interior surfaces. At the beginning of the project Duyster had hoped that the straight generating lines characteristic of hyperbolic paraboloids would allow the surfaces to be joined by prestressing wires on only one side, the interior. While this was possible in theory and had been confirmed in the model tests, the extreme torsional forces exerted on certain ribs would have required extensive reinforcement and an increase in their thickness. Kalff telegraphed Le Corbusier on 11 June: "We can avoid the necessary, but offensive, reinforcements by placing half of the 7 mm steel cables, the same color as the shell, on the exterior, forming a mesh roughly 50cm x 50cm. Do you agree?"[30] Le Corbusier replied immediately, in agreement, and on the same day followed his telegram with a letter to avoid any ambiguity. In Brussels, the construction work continued under the supervision of a foreman named Matthijs and the team of twenty-five workers who would realize the project from casting to erection.[31]

3-41
With most of the prestressing completed, the final adjustments to the exterior are made. Early 1958. [Philips]

As construction continued, Kalff continued his efforts to extort the definitive scenario for the electronic spectacle from Le Corbusier. His frustration was apparent. He wrote to Varèse informing him about the progress of construction, the current formulation of the sound and light presentations, and the composer's work schedule in Eindhoven. He admitted that there were still few fixed details regarding the scenario, noting that "You can show your

annoyance, but losing your composure won't get [Le Corbusier] to move any faster."[32]

From the beginning, it was proposed that sculpture would grace the exterior of the pavilion near the entry, probably a carry-over from the original conception of the pavilion as a collaboration of artists. In Kalff's handwritten undated record of a meeting (possibly as early as February 1956), he noted that there would be two sculptures at the entrance to the pavilion: one, to be executed by "Mr. Pevsner"; the second by the "mathematician Le Lionnais."[33] Kalff's note referred to Antoine Pevsner, brother of Naum Gabo discussed earlier in the text. Both of the Russian brothers worked with constructed forms that employed surfaces that used ruled figures such as hyperbolic paraboloid. Why Pevsner rather than Gabo was suggested was probably a matter of proximity and expediency: Pevsner was living in Paris.[34]

An internal memo suggested that it was a question of time rather than artistic preference alone that generated the choice. Pevsner's sculpture *Phoenix* stood in the Musée d'art moderne in Paris and was both familiar to the Philips team and appropriate in character to the architecture of the pavilion. A bronze casting of the sculpture was permanently installed outside Detroit at the General Motors Technical Center (designed by Eero Saarinen) and had acquired a certain degree of celebrity. Kalff hoped that it would be possible to order a new casting directly from the original, as time was too short for creating a completely new work.[35] Earlier in the year, Kalff had again raised the issue of the sculpture, which he and Le Corbusier had left unresolved. The architect had replied: "Concerning the sculpture on the surface of the water, I would like to point out that in my mind, there are two sculptures, sisters, to be positioned: one is purely geometric; the second is purely artistic."[36] (These became the two *volumes* suspended near the apexes of the interior). Only one sculpture—geometric, and in Xenakis's hand following Le Corbusier's idea—greeted the visitors as they waited for the next show. The efforts to acquire the Pevsner were stillborn.

Le Corbusier also proposed that the "epitaph" identifying the building and providing credits should take the form of a concrete stela cast with inset letters.[37] Like the credits of a film he said, the sign would list the team members and their assignments. Other questions that normally require the architect's attention, study, and design, also accompanied the Philips project: Where and how would the doors to the control booths best be positioned? What would be the nature of the reinforcement required by the openings in the shells for ventilation? A visit to Eindhoven 24 July 1957 familiarized Le Corbusier with certain technical aspects of sound, light, color, and images necessary for the *Poème électronique*, and with these in mind, he retired for his annual vacation at Cap Martin. He wrote Agostini at the end of the month, realizing that time was rapidly becoming precious, perhaps also acknowledging the limits of his knowledge of film, projection, and cinematic montage.

The calm frenzy of construction and finalizing of design was disturbed by the question of attribution regarding the pavilion's design. An article on the pavilion had appeared in the 30 September 1957 issue of *Combat*, accompanied by a photograph of the model. The article attributed the design to Le Corbusier alone, typical of the casual manner in which publications assign credit. Rather few architects execute buildings on their own, and perhaps not even a member of the general public believes that building design is the work of one person. But the contract establishes a firm name, and it is usually the firm that is legally regarded as the author. Whether this is equitable is hardly the question; it was (and to some degree remains) normal practice. But to Xenakis, who had been intimately involved with the design and production of the building from its inception, and to whom the majority of the credit for its design could be justifiably assigned, the omission of his name came as a slight.

On 3 October, his dignity and feelings wounded, Xenakis wrote to Kalff, bemoaning the attribution of authorship in the article. Having received the assignment from Le Corbusier:

> *Within several days*, I designed a project which he accepted straightaway, and I pursued this scheme until the agreement with the contractors "Strabed" was finalized.
>
> It is I who entirely conceived the form and the mathematical expression of the Philips Pavilion. Le Corbusier never denied this; and that he accepted in its entirety the creation of one of his collaborators, is a grand gesture for which I am very grateful.
>
> . . . It is very painful to me to see that all my loyal efforts, to produce a Pavilion worthy of attention, have been completely ignored. As an example of the dedication with which I have worked on this project, I would remind you of the report I had submitted. This report suggested removing completely the concrete enclosures for the 300 loudspeakers that Mr. Tak had recommended. In the end, you adopted my proposal, with which you saved several millions.
>
> I now demand very strongly that your press office mention my name in the architectural creation of the Pavilion, next to the name of Mr. Le Corbusier, as he was the architect chosen by Philips. This is the minimal gesture of justice and truth that Philips owes me for the mental and moral energy that I have given.[38]

As one might imagine, neither Kalff nor Le Corbusier were pleased by this explosion of moral indignation.

Kalff replied to Xenakis, calmly, but firmly. As an experienced architect, corporate representative, and mediator, he realized that the project was more important than any single aspect—or person—taken in isolation. Certainly, he was in no position to alienate Le Corbusier with so much of the project still hanging in the balance.

> We were a bit surprised in receiving your letter of 3 October. You say that the article in "Combat" did not tell the truth and that as a result you demand that our press office mention your name next to that of Mr. Le Corbusier.
>
> If we were to believe you, Le Corbusier would have had nothing to do with the project for the pavilion. According to our knowledge, this is not at all true and Mr. Le Corbusier never said a word to suggest that the general conception of the pavilion was not the result of his own creative power. We should clarify several things. We commissioned Mr. Le Corbusier to design the pavilion for us, and we all agreed that this pavilion would not be an empty hall but instead would house a spectacle, also conceived by him, of light, color, images and architecture.
>
> Mr. Le Corbusier developed the idea of the "stomach" . . . in which one enters, digests, and exits. Above this totally free plan, he erected three metallic triangles and using strings linked these triangles to the contours of the plan. At that moment, you had the idea to define the free contour by forming it with sections of paraboloids in such a way that it became possible to design the surfaces of the pavilion's shell as hyperbolic paraboloids. That is where we position your intervention and the valuable contribution that you made to this project.
>
> The form of the pavilion having been established, we (that is to say, myself personally) sought from the contractors we consulted the construction which would most closely follow this specific form, and that would best express its characteristics.
>
> As you know, we found in Mr. Duyster the mind which best understood the potentials of these forms and who, in collaboration with Professor Vreedenburgh at Delft, found an ingenious and original solution for the construction which is an inherent part of the surprising and novel spirit of our pavilion.
>
> Thus, we believe that if we speak properly of the pavilion's design, there is just cause to think of it as the teamwork of four parties. If we cite them chronologically we have ourselves, Mr. Le Corbusier, Mr. Xenakis, and Strabed and its collaborators.
>
> On Friday, the eleventh of October, we are going to have a technical press

conference in Brussels for the editors of Dutch and Belgian journals concerned with architecture and construction. There we will certainly explain the work of each of the parties in the conception of the pavilion and you will be mentioned there with honor; but we believe that it would not be desirable [to do to this] in the popular press as the general public cannot distinguish the fine details that may have a certain value for professionals. There, we believe that the interests of Philips would be best served by mentioning only Philips and Mr. Le Corbusier and Mr. Edgard Varèse.

We hope that you will agree with this way of handling [the credits]. This said, we admit that the editing of the article in "Combat" is not very gracious. We do not know who wrote the article, but if it says that the construction, which supplies a most elegant and mathematically certain solution to the problem, was imagined [solely] by Mr. Le Corbusier, it certainly does not present the process as it truly was. We do not think that it will be possible to repair the wrong by a counterclaim, which would be too complicated and which would cause considerable harm to Mr. Le Corbusier and to us.

In the future we will make every possible effort in other publications so that your part in this great work will not be ignored. Today, we are sending a copy of this letter to Mr. Le Corbusier and we hope that both of you will be completely in accord with our explanation.[39]

A single letter could not calm the waters, however, particularly when the Philips office had heard only one side of the story. Seemingly under control, Le Corbusier answered Kalff's letter with a reasoned and calculated reply, informing him of the manner in which he worked, and reminding him of the breadth and depth of his accomplishments. Once again, the architect resorted to a barrage of names, citing his projects from various places and times. This did not diminish the validity of his argument, however. From this, the Dutchman was to understand that the Philips design could truly be considered to be by Le Corbusier and no one else:

Yesterday I received a copy of the letter which you had sent on the eighth of October to Jean [sic] Xenakis concerning the complaint he made on the subject of the Combat *article about the Philips Pavilion in Brussels.*

I regret that Mr. Xenakis has taken the liberty to contact one of my clients without my knowledge. I cannot accept the substance of his protest, given that I am the author of your Pavilion and the Poème électronique which will be presented there.

Dear Mr. Kalff, take a panoramic glance at the six volumes of the "Oeuvres Complètes Le Corbusier." Since 1922, I have not personally drawn a line on a

> *drafting table. Consequently, I authorize you to agree that it was not I who designed the Palace of the League of Nations in Geneva nor the Centrosoyus nor the Mundaneum nor the World Headquarters of the United Nations in New York nor the Villa Savoye nor the villa at Garches nor the plan for Bogotà nor the plan of Meaux nor the chapel at Ronchamp nor even the apartment in which I live on rue Nungesser et Coli. However, the following fact will appear curious: among the some 250 architects who over thirty years have formed the studio at 35 rue de Sèvres, not one has appeared on the professional horizon . . .*
>
> *All of this is quite troubling, don't you think?*
>
> *I very much regret this personal intervention by Mr. Xenakis, whom incidentally, I esteem as an artist and technician.*
> *It frequently happens that the team is convinced that it is driving the carriage.*
>
> *Let's not give this incident greater importance than that of a violent outburst of a temperament itself violent.*
>
> *Regarding the decision you have reached about publicity, I would therefore request that we agree on the way in which the authors of the Pavilion will be presented to the public.*[40]

The matter was still not settled. Four days later, one assumes after careful reflection on his part, Le Corbusier again wrote to Kalff regarding "*l'incident Xenakis*": "I have made the decision, if you will agree to it, to adopt the following attribution for the Philips Pavilion at the Exposition at Brussels: *'Philips - Le Corbusier (coll. Xenakis) Varèse'*." Thereafter, the question of attribution was not discussed, and all the names ultimately appeared on the concrete stela fronting the pavilion, and in the professional publications. Xenakis continued to see the building through construction, and worked on the stadium project for Baghdad thereafter. But the presumably mutual bitterness was never completely dissipated. At the end of the August 1959 vacation period, Xenakis and two similarly maverick colleagues, returned to find that the locks to the studio had been changed. His letter of dismissal began with the following phrase:

> *Modern architecture triumphs in France; it has been adopted. Today you may find a field of application for everything which you have acquired by yourself as well as through your work with me. I believe you have the benefit of being able to seize all the opportunities which may present themselves to you.*[41]

A year later, at the inauguration of the monastery of La Tourette, for which Xenakis had designed the rhythmic play of window mullions, Le Corbusier showed signs of late appreciation for his younger colleague's contributions. Would Xenakis return to the studio, not as an assistant, but as its architect-in-chief?[42] "I told him it was too late to go back," Xenakis recalls. "I was sure of one thing; all I wanted to do was compose, to think about the problems of music and to write about them all—but above all to compose."[43] The rupture was definitive; Xenakis left the office, and architecture, in 1959 to pursue what would become a distinguished career as a composer, achieving international celebrity on his own terms.[44]

If there were heavy interpersonal tensions within the office, there were also lighter moments, at least in retrospect. What to call the pavilion? Le Corbusier was aghast at calling it "Spatiorama," as Philips had suggested. "Decidedly," he wrote Kalff, "I do not accept a name like that for the Philips pavilion; it is too awful! I prefer to call it 'The Philips Pavilion' [Pavilion Philips] which clearly expresses its elements . . . that is to say a name which would not be laughed at by everyone."[45]

The pavilion's construction continued apace without major mishap or complications. By mid-December, the building shell was completely enclosed and the prestressing "with 3000 steel cords" was well underway. By the end of the month, the greater part of the scaffolding could be removed, after which the mounting of the loudspeakers on the walls began and a layer of acoustic asbestos could be applied to the interior surfaces around them [figures 3-42, 3-43, 3-44, 3-45].[46] Construction was moving forward, but the completed scenario for the electronic spectacle was still missing. In his letter of 12 December, Kalff attempted to pressure Le Corbusier into finalizing his work (prompting the architect's written outburst quoted in Chapter 5). Adjustments to the elevations of the levels needed to be made; the heights of the small barrier walls required more precise determination. The design of the paving warranted attention; its executed pattern design reflected the module of the precast concrete panels of which the building was made.

The final design for the geometric sculpture was still wanting, and this was brought to the architect's attention by Kalff in mid-February.[47] No documentation contradicts the claim that Xenakis defined the form of the piece, although the idea was probably Le Corbusier's [figures 3-46, 3-47, 3-48]. The form of the sculpture developed as a series of outlined geometric solids, nested one within the other. By selectively sheathing and revealing the various surfaces, the sense of interior space was heightened and the viewer's eye led within. At the heart of the piece is a neon rendering of Le Corbusier's writing announcing the *Poème électronique*. The piece was fabricated in the Philips machine shop in Eindoven. (It was recently discovered and "restored," and today stands there in front of the lighting division offices.)[48] Mounted in the small pools surrounding the pavilion, the colored illumination was reflected in the water, suggesting in some small way, the electronic environment within [figures 3-49, 3-50].

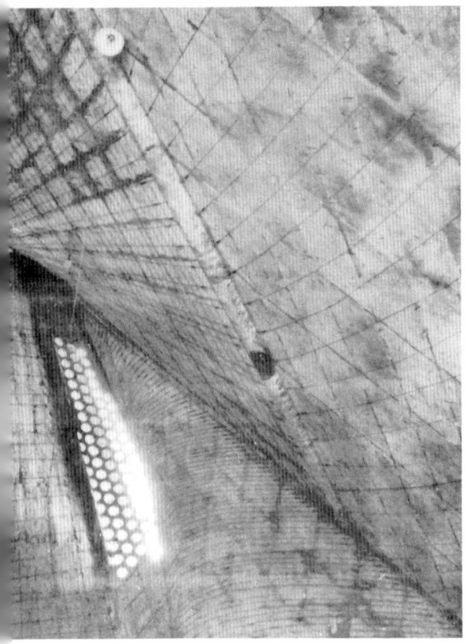

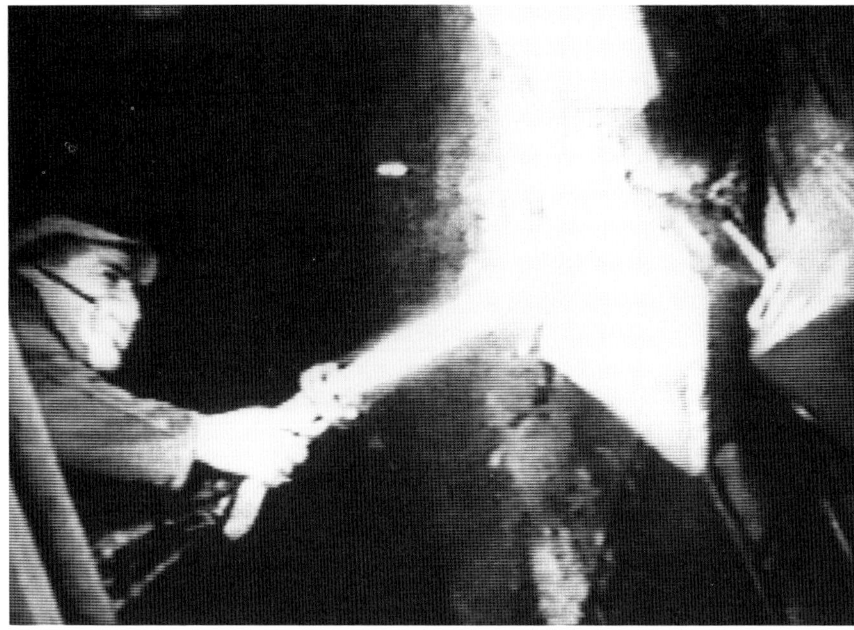

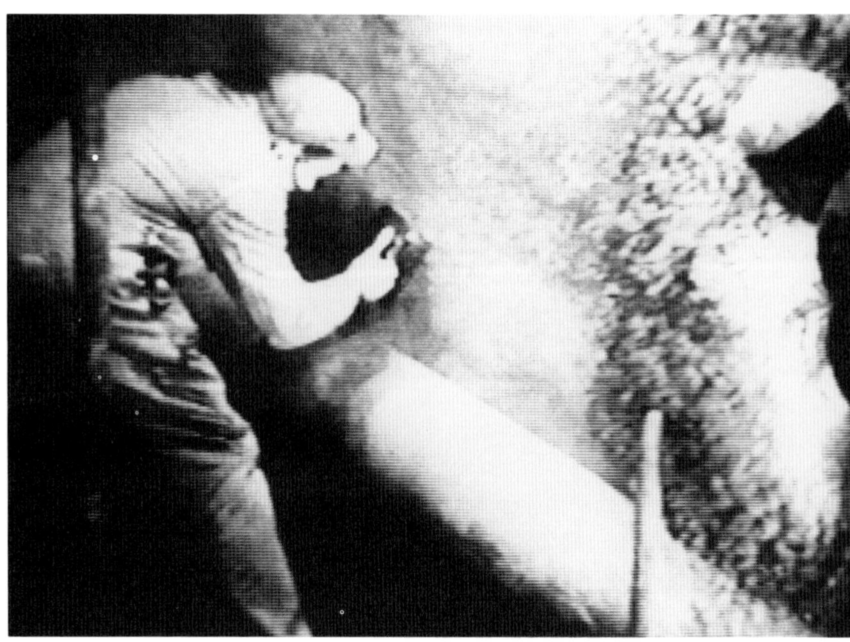

3-42
The interior of the pavilion nearing completion.
[HBG]

3-43
Spraying the interior coating of asbestos cement. From the film *De Bouw van het Philips Paviljoen*, Strabed, 1958.
[HBG]

3-44
Troweling the asbestos cement for a smooth surfaces for projection. From the film *De Bouw van het Philips Paviljoen*, Strabed, 1958.
[HBG]

3-45
With the interiors surfaces nearly finished, the speakers begin to appear on the shells.
[Philips]

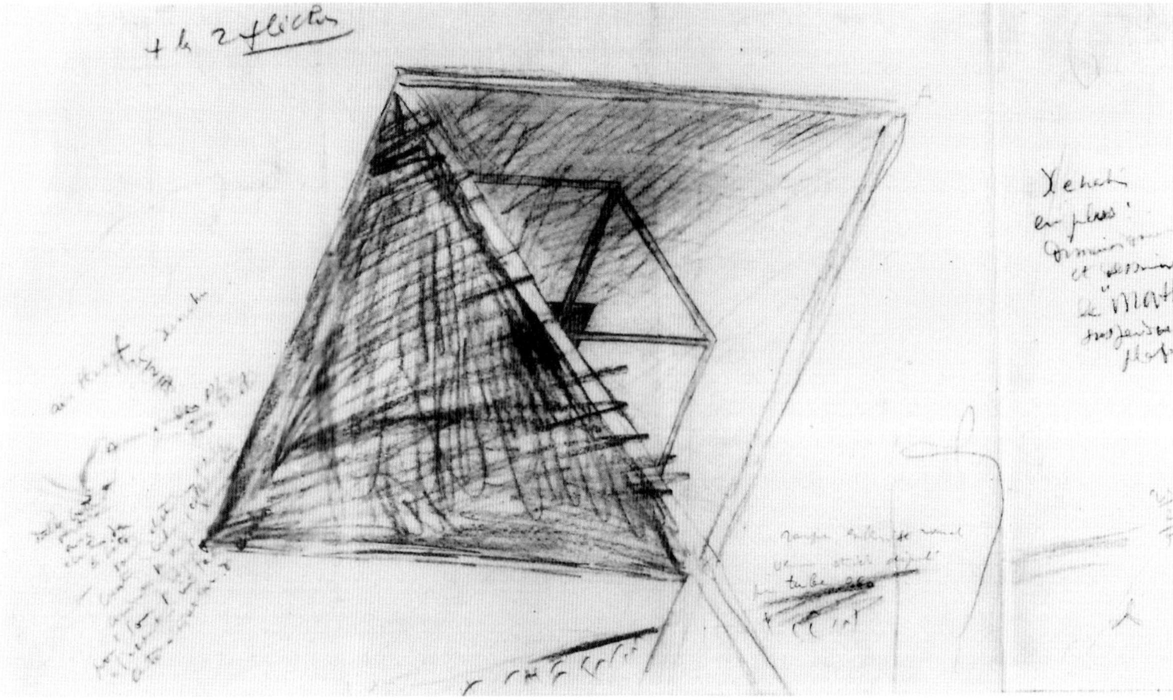

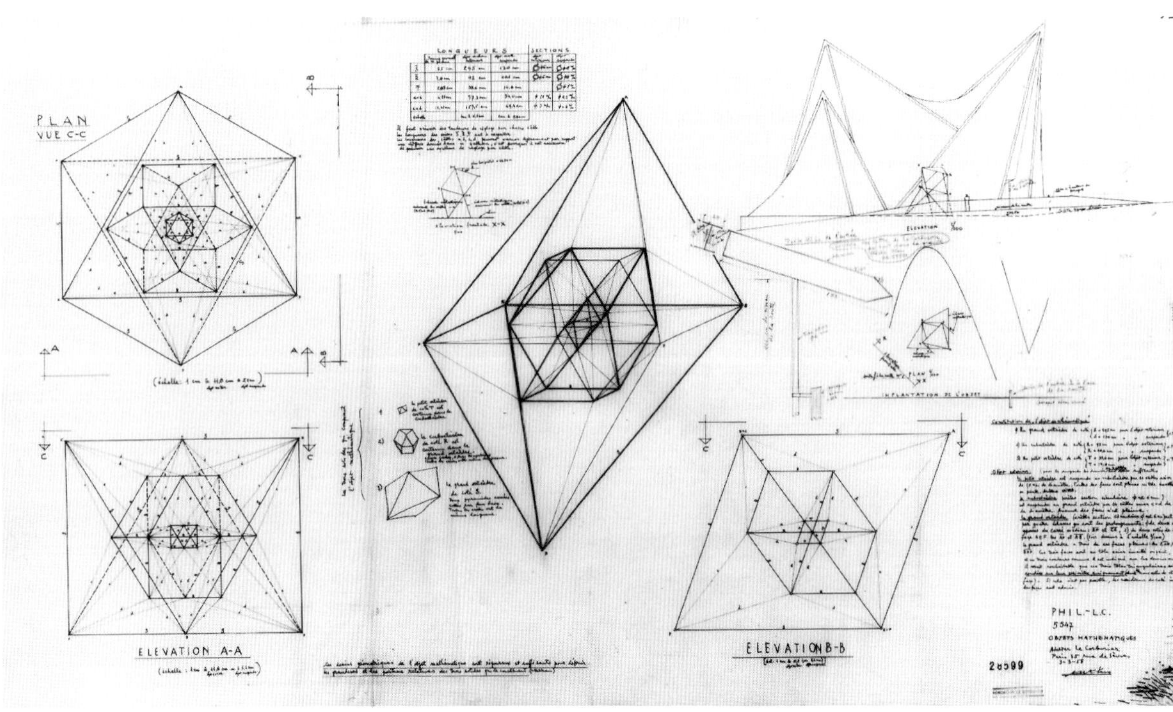
PLAN
VUE C-C
ELEVATION A-A
ELEVATION B-B
PHIL-LC.
5547
OBJETS MATHEMATIQUES
28599

3-46
Study for the Mathematical Object.
25 February 1958.
Le Corbusier?
[FLC 30622]

3-47
Definitive geometry and placement for the Mathematical Object.
3 March 1958.
Iannis Xenakis.
[FLC 28599]

3-48
The Mathematical Object.
3 March 1958.
Iannis Xenakis.
[FLC 28591]

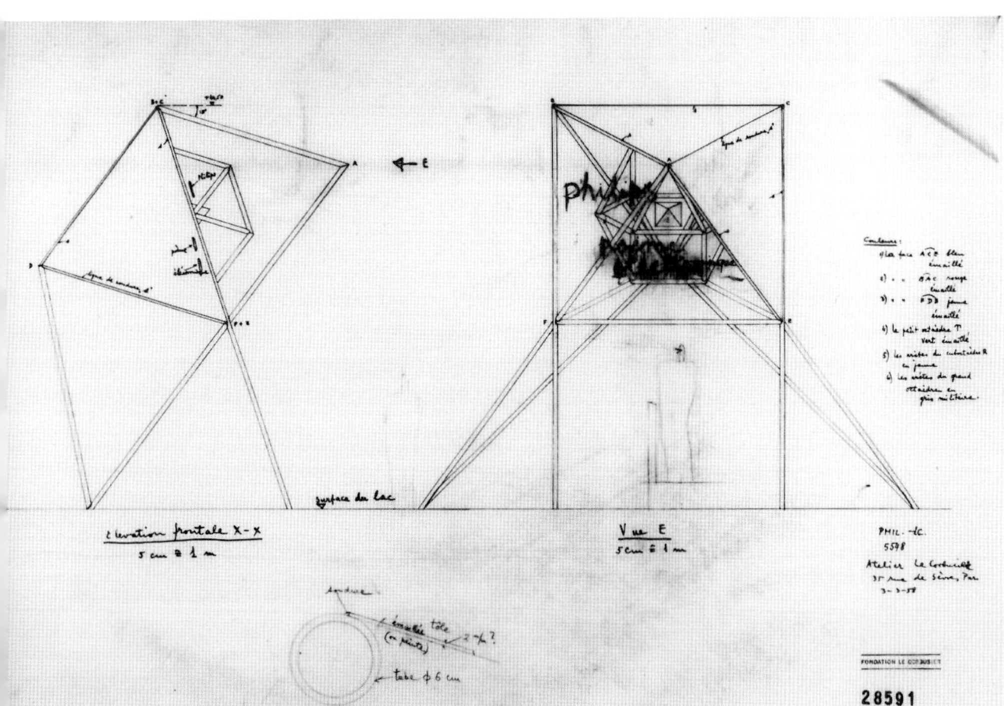

3-49
The Mathematical Object
in the shop.
[Philips]

3-50
The Mathematical Object at the fair.
[Philips]

From September—which was, admittedly, less than six months before the projected opening of the pavilion—Le Corbusier's attention, and the office documentation, had centered on the conception and production of the *Poème électronique*. Continually stalling, bluffing, providing bits of ideas as they arose, Le Corbusier leaned heavily upon Philippe Agostini and Jean Petit to help him realize what was left as only rough notes in his sketchbooks. Over the months racing by, the visual spectacle slowly took form; barely 60 of the 480 seconds required had been completed by the beginning of March. The clock was moving rapidly, and Kalff was becoming even more pessimistic about meeting the opening date. Barely a month remained.

The construction was completed and the building was opened for the official ceremonies in April 1958 [figure 3-51]. But it was closed immediately thereafter for a month of adjustments to the sound, projection, and automation systems. Yet another person, Johan Jansen, was called in during the late moments as an overall troubleshooter. To him fell the responsibility for perfecting the various electronic systems and making them operate reliably. The pavilion did not open to the general public until May. While no major faults tainted the building itself, the lateness of the visual and sonic elements of the *Poème électronique* itself precluded the perfected automation the presentation required, and adjustments were made continually.

One sees in the construction of the pavilion the result of a nearly ideal collaboration of architect and builder, at least as seen from the architect's perspective. H. C. Duyster was an engineer and builder who atypically understood both the architectural intention and the constellation of constraints necessary to realize the vision. With the engineer Vreedenburgh, Duyster devised a system simple in idea, though complex in execution. While the architecture of the pavilion was radical in its utilization of hyperbolic paraboloid shells and the method of prestressing the concrete was innovative, the day-to-day work of constructing the pavilion was a remarkably basic affair. In its very simplicity, the pavilion's construction was a tribute to the breadth of Le Corbusier's idea, the clarity of Xenakis's design development, and the eloquence of Duyster's sense of reason.

3-51
The Inauguration of the Pavilion. 22 April 1958. [Philips]

4. IMAGES, COLOR, AND LIGHT

From the very origin of the Philips project, Le Corbusier had proposed the creation of an electronic poem [figure 4-1]. This work, intended to be a new genre, would utilize, publicize—and stimulate—the technical potential of the Philips corporation, and would synthesize light, color, sound, and rhythm. Sound, for the most part, was left entirely to Le Corbusier's collaborators Edgard Varèse and Iannis Xenakis, who would compose the primary eight-minute score and the two-minute interlude, respectively. The links between the visual and aural senses would be fortuitously accidental; the moment of silence Le Corbusier proposed early in the conception of the spectacle was never to materialize. For the most part, the musical score was conceived and produced completely independently of the visual work. Their points of congruence, or at least intersection, would be determined solely by chance; the haphazard juxtapositions of image and pitch created a multimedia collage that was to test the patience and comprehension of the audience. In several ways, the composition of the artistic team and the production of their respective works suggests the oft-cited definition of surrealism coined by Lautréamont: the chance encounter of an umbrella and sewing machine on a dissection table.[1] The aspect of rhythm present in the architect's original formulation regulated the other media and would be the glue that joined them, however anemically. The single element all four pieces had in common was time, which provided the base upon which the spectacle would be constructed. It would last precisely 480 seconds.

The visual components of the *Poème électronique* were four: a film (*écran*, literally "screen"[2]) presenting images illustrating the course of human civilization and the threats to its prolongation; colored lighting (*ambiance*) within the pavilion to manipulate atmosphere and mood; simplified shapes superimposed upon the film by projectors (*tri-trous*, so named for the three holes, or *trous*, in the projection device); and three-dimensional forms (*volumes*) to be illuminated with ultraviolet light for maximum effect.

Rapid developments in artificial lighting during the early twentieth century had transformed the city as well as its interiors. Writers described the nocturnal urban environment of dismembered signs afloat in darkness; László Moholy-Nagy looked upon artificial illumination as the most potent, and as yet undiscovered element of the twentieth-century urban environment. "The night life of a big city can no longer be imagined without the varied play of electric advertisements, or night air traffic without lighted beacons along the way," he wrote.[3] The advertising logo became a ubiquitous form of modern heraldry, especially with the advent of neon illumination and its pulsating line. Gunnar Asplund celebrated light in his central mast for the Stockholm Exposition of 1930, which became a totem pole of corporate symbols in neon and tungsten. Light had achieved a prominence in modern urban life, architecture, and art.

4-1
The interior of the pavilion showing the distribution of the higher range loudspeakers; the geometrical *Figure* hangs from one apex.
[Lucien Hervé; ©1996 ARS, New York / SPADEM, Paris]

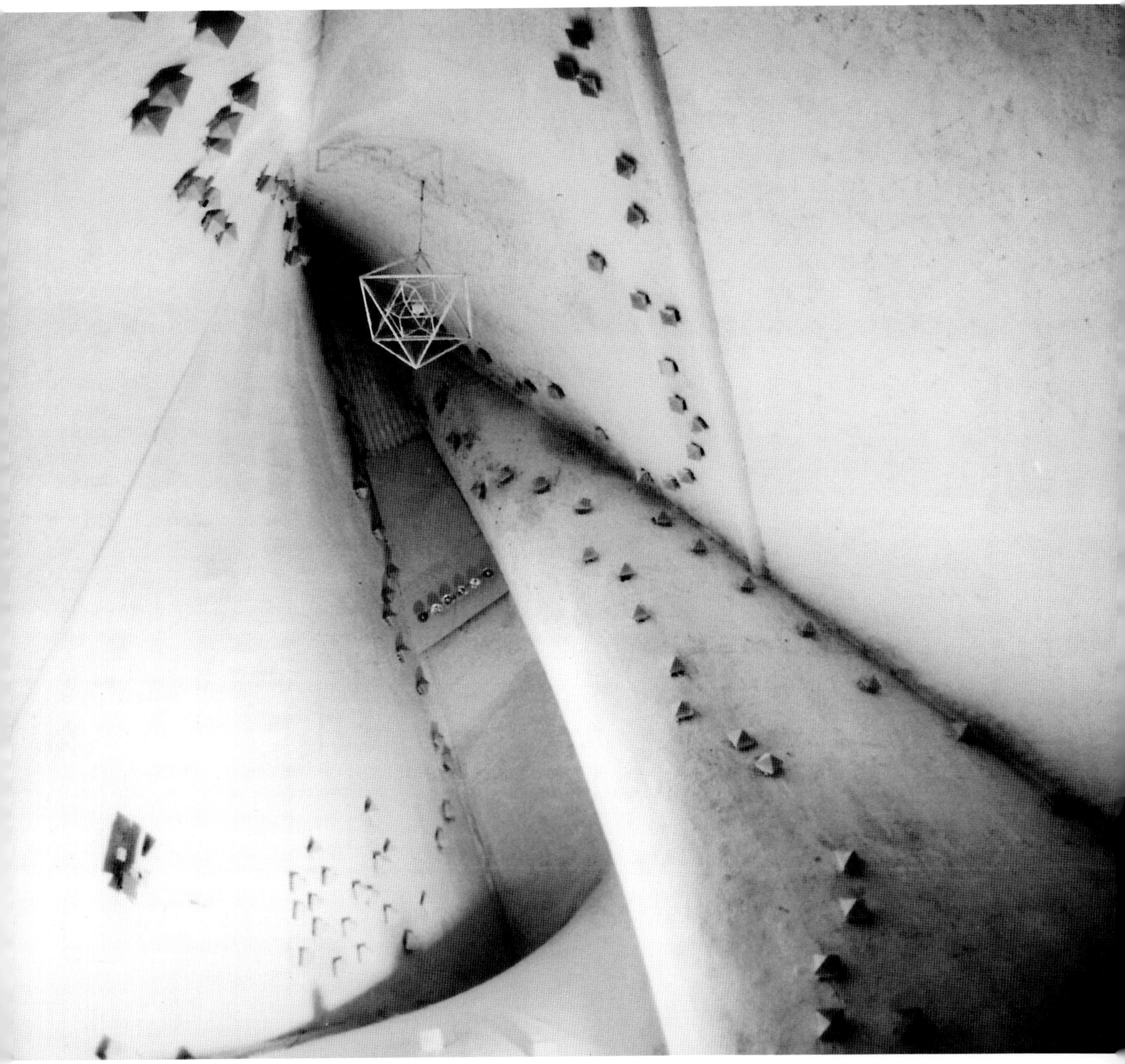

Film in Moholy-Nagy's eyes was a form of light: light treated abstractly, given and giving animation. In *The New Vision*, he had outlined a project which resembled the conceptual parameters for the *Poème électronique*: "The *film* with its unexplored possibilities of projection, with color, plasticity and simultaneous displays, either by means of an increased number of projectors concentrated on a single screen, or in the form of simultaneous sequences covering the walls of the room."[4] Abstraction, not figuration, was the desired means; harmonious form was the goal.

Of course, light had been a central concern for Le Corbusier for several decades. But during the late 1940s and into the 1950s, as his belief in architecture as a synthesis of the arts assumed greater prominence in his work, light—with space and form—became one of the principal vehicles for its realization. Perhaps the fullest realization of his idea of "ineffable space" was the pilgrimage church of Notre-Dame-du-Haut, Ronchamp (1952), a building that eschewed orthogonal geometry for a plastic modelling of space activated by light. Walls so thick they can themselves contain small chapels and tall shafts that orchestrate the flow of light dependent on the time of day establish the superstructure for this organic work. The resultant interior is dark and mysterious, with a floor that slopes toward the altar gently coaxing the celebrant toward the statue of Our Lady. Splayed windows sheathed with painted glass cut into the thick walls are less sources of illumination than colored projections, animating the "constellated" south wall. A narrow slot of light at the top of the wall transfigures the sagging roof overhead into a levitating mass. While a concrete reality of stone and cement, the mass of the chapel dissolves and its space prevails. The Philips *Poème électronique* would utilize electronic media to further the power of architecture in its quest for the unknowable: "Then a boundless depth opens up, effaces the walls, drives away contingent presences, *accomplishes the miracle of ineffable space*.[5]

Early on, Le Corbusier had determined the form of the pavilion as a warped projection screen that would allow some 500 visitors, who would remain standing for the full eight minutes, to experience the presentation from all points. Sight lines, visitor circulation, and such utilitarian concerns as light level for entry and exit, and provision for emergencies caused by electrical failure were also accommodated. To some degree, Le Corbusier considered the various elements of the presentation simultaneously; in other ways—suggested by the writings and sketches in his sketchbooks—the creation of the *Poème électronique* followed no logical sequence, and was considered and recorded in fits and starts until time was literally running out.

In executing his ideas, Le Corbusier relied heavily on the efforts of his collaborators [figure 4-2]. Jean Petit [b.1927], author, editor, graphic designer, and publisher, had been involved with the production of books on Le

4-2
Philippe Agostini, Le Corbusier, Louis Kalff. March 1958?, Eindhoven. [Philips]

Corbusier's chapel at Ronchamp and monastery of La Tourette as part of his series for Editions de Minuit called "Forces Vives." Petit knew Le Corbusier well and understood his work and way of thinking. He would also produce a book, underwritten by Philips, on the pavilion and its spectacle, at the end of the exposition. This became the primary artistic record on the building and presentation.[6]

The filmmaker Philippe Agostini, at least as revealed by the written records, was less known to the architect.[7] Beginning his career as a cinematographer, Agostini worked in the 1940s and 1950s with noted directors such as Robert Bresson and Max Ophüls, and has been credited with a cinematic stance or "look" that was widely influential in postwar cinema.[8] He shot Bresson's *Les Anges du Péché* (1943) and *Les Dames du Bois de Boulogne* (1945) with images, framing, and distance jumps that were equally striking as the turns in the script [figures 4-3, 4-4]. In the 1950s he began making films of his own, but never became a first-rate director.

In his sketchbooks, Le Corbusier noted "for useful technical talks call in Agostini film [*sic*]" as early as November 1956, and perhaps had met with the filmmaker in person or spoke to him on the telephone about that time. The earliest written record, however, indicates he solicited the aid of the *cinéaste* in June 1957, writing that "I have confirmed the plan of my Poème with a very clear visual concept. Would you be disposed to help me on this project which will not take a lot of time, but will require a sharp mind?"[9] Presumably, he received a favorable reply. Two weeks later he wrote again divulging that the Philips people were breathing down his neck and demanding from him the definitive scenario, which was far from complete. He proposed a collective meeting on 15 July, with phrasing both imploring and tempting: "*I am truly and amicably counting on you.* You do not have the right to let me hang on a matter which could open to electronics the poetic ways of rhythm, of sound, of color and of image. The Brussels affair will be something extraordinary; I am talking about the building, and the Poème électronique must be the same."[10] Agostini replied at the end of the month; but he was still a bit uncertain about the nature and scope of the proposed work.[11] Schedule conflicts would not allow a meeting before the vacation month of August; instead, Le Corbusier suggested dealing with the project early in September.[12] After his visit to Eindhoven on 24 July, Le Corbusier wrote to Agostini once again noting that his familiarization with Philips technical abilities now provided him with greater insight. Discussion with the filmmaker was necessary, however, because the film would comprise only one element of the multimedia presentation. Philips repeated their request for the final scenario, and Le Corbusier thought he had better solicit Agostini's input before making anything definite. He again requested a meeting, for early August 1957 at Cap Martin.[13]

4-3
Still from *Les Anges du Péché*, 1943. Robert Bresson, director; Philippe Agostini, cinematographer. Agostini's characteristic use of strong lighting and dense shadows. [Museum of Modern Art]

4-4
Still from *Les Dames du Bois de Boulogne*, 1945. Robert Bresson, director; Philippe Agostini, cinematographer. [Museum of Modern Art]

Le Corbusier jotted some thoughts in the sketchbooks during September, but he was still far from a definitive scenario. "I promised to give you the first minute of my scenario on 15 September," Le Corbusier wrote Kalff. "You know what sad events [the death of Yvonne Le Corbusier, the architect's wife] have intervened at that moment!"[14] Another month had passed, but he claimed that he had now made the selection of the images, or "documents" as he referred to them. He now intended to pass the whole lot on to Agostini, with whom he would meet in two days. Agostini would photograph them and prepare the montage in accordance with the architect's scenario. Le Corbusier proposed coming to Eindhoven to visit Varèse, who was at work there on his score. The purpose for the visit would be to "find the points of contact between my scenario and the music. It concerns the words which I myself will repeat four times (very short), but which must be like piecemeal holes in the middle of the music."[15]

At the end of October, Kalff finally received the letter—and the material—for which he had been so long waiting.

> *Following my letter of 21 October, I'm letting you know that I met with Mr. Agostini who has given his full approval to the outline of my Poème and to the projection techniques which concern him personally. He is awaiting your approval for the proposal he submitted to you.*
>
> *As for me, I have the pleasure to inform you that I have completed my Poème électronique. . . .* [16]

He closed by adding after his signature: "I very sincerely believe that there has been no loss of time and that everything is perfectly under control."

Abroad in Baghdad and later India, Le Corbusier continued to record fleeting thoughts in the sketchbooks. His studies of November 8–13 focused on the colors and general ambiance rather than on the selection and sequence of the images within the film. He reminds himself about the characteristics of projected and colored light acquired at the meeting with the Philips technicians:

> *if we want to have a clearly defined image, the ambiance must be of a* dark *color.*
>
> *The tri-trous (stencilled projections) are possible only on a dark ground.*[17]

The issue of authors' rights and copyright recurred several times during the project development. Le Corbusier first raised the question of the residual artistic property in his questioning of the original letter contract proposed by Kalff in October 1956. As he developed the scenario for the *Poème électronique*, Le Corbusier was well aware of the range of the institutional sources and the problems that might arise concerning the copyright of the images. At the end of November, he informed Kalff that certain images were taken from books, and listed the publishers and the works from which they were borrowed. Questions of ownership would not only be germane to the film but also to the book by Jean Petit intended to document the conception, construction, and presentation of the pavilion as a lasting record of the work after the building's demolition.[18]

During the autumn of 1957 Le Corbusier assigned a priority to the completion of the scenario, perhaps keeping in mind that the work was to be executed by his two collaborators, one of whom—Agostini—was involved in other projects that demanded his time and attention. At the end of November, he confidently informed Kalff that "[y]esterday evening, I brought to completion the entire 480 seconds of the scenario and the synopsis. All the documents are here."[19] He was about to leave Paris again, for India, and would return at the end of the year, at which time he proposed a meeting in Eindhoven with Agostini.

Before departing, he outlined the work to be completed by Jean Petit in his absence. Petit was left with the thorny problem of translating Le Corbusier's notes into a sufficiently clear, detailed, and coherent form that would allow Agostini to continue the next stage of the work. As Le Corbusier wrote after the project was completed:

> *I knew that for this task [conceiving the film] no one could replace me. But I asked Jean Petit to aid me, to help me assemble and order the products of my hunt. All of these must pass through the eye of the camera and be recorded on film. I asked Agostini to take charge of this task. But Agostini was in Paris, the films for Philips were in Amsterdam and, at the moment that the film must be made, I was in Chandigarh. How to send the instructions? It was necessary to devise a means to carry and transmit the idea. And it was only in November 1957, then in February 1958, that the poem was not written, but drawn using an album-scenario organized in vertical columns marking time with horizontal bands of one second each: total, 8 minutes, each divided into 60 increments, forming 480 horizontal bands. Once drafted and put in order by Jean Petit, the score was ready to synchronize the hundreds of events.*[20]

In an early attempt to coordinate the images and chroma of the poem, Le Corbusier arranged them radially within a circle [figure 4-5]. Reproduced in the Jean Petit record of *Le Poème électronique*, this diagram positioned the ambiances on the outer edge, the segments of the film within it, and the *tri-trous* in the central panel. One could speculate that the team abandoned this format early on because of the inherent limits of circular organization. As a fixed figure it was difficult to modify by additions and subtraction; with severely constricted space at the center, there was little room for notations.

The format of the definitive scenario, or *minutage* as it is known in

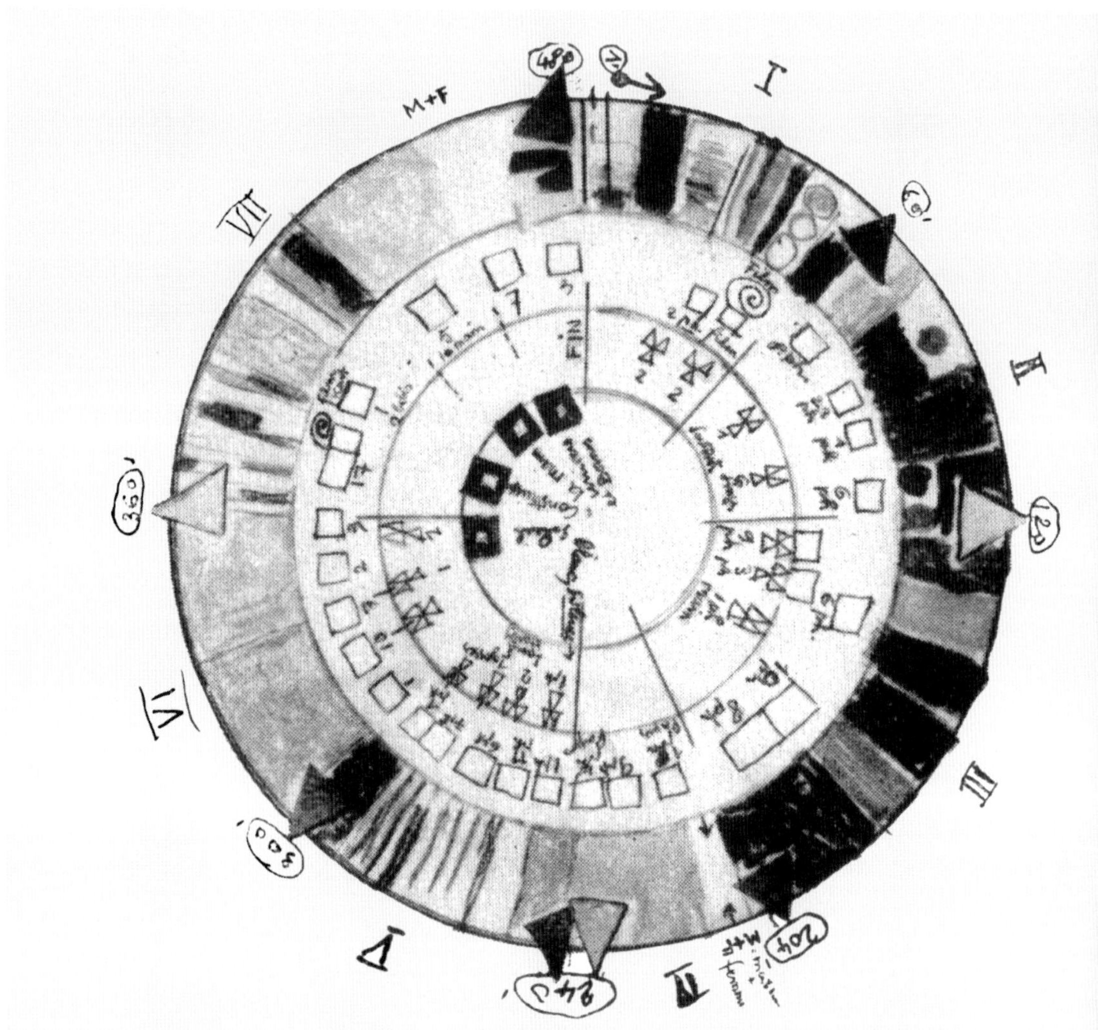

4-5
Early proposal for the *minutage*. The ambiances are cued along on the outer edge, the segments of the film within it on the next ring, and the *tri-trous* in the central panel. [*Le Poème Électronique*]

the French film industry, was as ingenious in its way as the prestressed assembly was for the pavilion's construction [figures 4-6, 4-7; Plates IIIa, b]. The work of synchronizing the sound, the images, the colors, the "stencil" projections (the *tri-trous*), and the illuminated volumes required tremendous coordination. The distances separating the collaborators, who were on different continents, added to the difficulties of jointly creating a work precisely 480 seconds long. Unlike most buildings which are measured in linear and areal dimensions, the Philips Pavilion was a space calculated in seconds.

In order to facilitate this necessary coordination, Le Corbusier devised a mimeographed graphic matrix which listed the *ambiances*, film (divided into sequence, image, and notes), and the *tri-trous*. A diagrammatic section through the pavilion was hand-colored in gouache accompanied by a verbal description of the lighting effect. Repeated throughout was a time-line divided into seconds, the common chord that tied together the constituent fragments of the poem.

Le Corbusier assigned Petit the task of filling in the boxes of the preprinted sheets and passing them on to Agostini. In addition, the architect instructed Petit to turn his rough notes into a serviceable working document:

> *Jean Petit will . . . complete sequences 1,3,4,5,6, and 7 [that is to say, the entire film!].*
>
> *How will he do this? This work must be reproduced by mimeograph. Jeanne [Le Corbusier's secretary] will send to me in India a diazo print of each of the 7 sequences for approval and from these we will then produce 20 copies of each of the 12 sheets of 12 pages, 21x27 (total 144). Three series of each of these copies will be sent to Kalff.*[21]

At the close of the year Le Corbusier was in India (telling Kalff, who had nearly reached the limit of his patience, to keep calm);[22] Kalff was in Holland, not at all tranquil; Petit was in Paris; Varèse in Eindhoven. The laboratory Alga-Cinema felt sufficiently confident with the progress of their work to propose a meeting in mid-January to screen the rushes of the first minute of the film—hopefully, under conditions simulating those of the completed pavilion.[23] Petit, for his part, was only now expressing some assurance. He had found the process of documenting and equilibrating the elements of the poem to be far more complicated and time-consuming than he and Le Corbusier had predicted. But he developed a comprehensive format and had completed his work, sending the required copies to Kalff, the Le Corbusier office, and Agostini. His mild annoyance at being left holding the bag was revealed in the closing to his letter that accompanied the materials for the film: "The only problem is that [Agostini] is not easy to reach because he is frequently traveling, and I myself have to be at his disposal." He predicted that everything

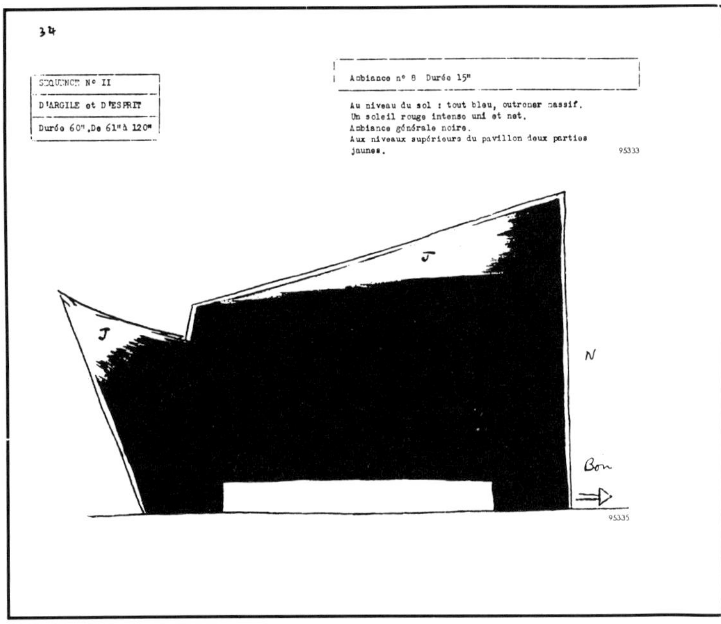

35

Sec	Volumes		Ecrans			Sec	Tritons	Paroles
	Nat	Fon	Notes	Vision	Référence			
81				Les quatre savants	T.113	81	Les quatre savants dans les trous éclairés en couleur bleu et rouge.	
82						82		
83						83		
84			Toutes ces têtes seront animées, c'est-à-dire qu'il faut à tout prix éviter la monotonie et la lenteur. Cela doit être vivant	Tête de nègre Congo	T.114	84		
85				Tête de nègre Maori	T.116	85		
86				Tête de nègre Mayogo	T.117	86		
87				Tête de fille Mangbetu	T.118	87		
88				Courbet:femme couchée FAII 227	D.121	88	Les tritrous en couleur bleu et rouge.	
89				Art Attique MAL 49	D;124	89		
90				Art Sumérien MAL 11	D.125	90		
91				Egypte MAL 68	D.128	91		
92				Dame D'Elcho MAZ 94	D.130	92		
93				Art Sumérien MAL 11	D.125	93		
94				César MAL 171	D.133	94		
95				Art Celtique MAL 212	D.135	95		

95334

4-6
Definitive form of the *minutage*. Using a grid divided into the complete 480 seconds, four columns coordinate the relative composition of the *ambiances*, *volumes*, *écrans*, and *tri-trous*. A final column allows for notes, often concerning the sound. [Philips]

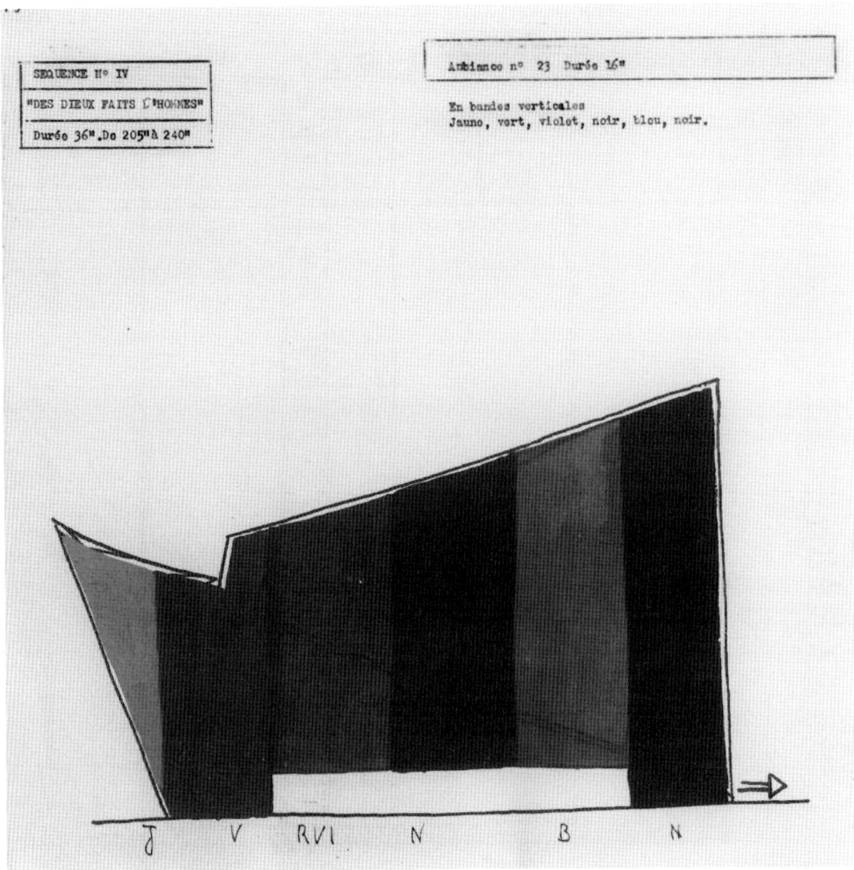

4-7
Copies of the definitive scenario were hand-colored in gouache, illustrating the desired effect for the *ambiances*. Sequence 4, "Manmade Gods," with stripes of yellow, green, violet, black, blue and black. [Philips]

would now go smoothly, however, in spite of the fact that the lateness of the project had forced him to cancel a trip.[24] Perhaps Agostini was experiencing difficulty integrating the full complement of the documentary photographs provided by Petit. Time was closing in, and many of the original images existed only in books. To keep the project workable and on schedule, he proposed eliminating a number of the selections.[25]

The crucial conference was set for 17 January 1958, with the principal players in attendance. "I will have, at that time," wrote Agostini, "the materials for the film, which I will project on your screen, in order to evaluate the technique for presenting the images. . . . After the first showing, I will determine the definitive completion of the first minute of film, which according to the results of the first trial, whether more or less satisfactory, will be completed either by the end of January or at the latest, 15 February."[26] Le Corbusier was back in Paris, and wrote to Agostini: "Jean Petit sent you the scenario for the Poème électronique Philips. It is crucial that I see you when you return before I leave for Eindhoven. I humbly think that I must be with you at the time of the first filmed transcriptions of the documents listed in the scenario. Be so kind as to call me when you return." Added, by hand, to the typescript was the appeal: "Very urgent! very important! I am counting on your enthusiasm."[27] Taking a different tack, Le Corbusier further tried Kalff's patience, assuming that a strong offense is the best defense: "I have seen Jean Petit in the meantime. You wrote to him that the scenario which I sent to you was not definitive. Wrong! It is completely definitive and it will not require any modifications due to technical [limitations], it is a simple question of adaptation." In case the Philips art director had not been able to decipher the argumentative letter sent express from Chandigarh, or in case the original had been delayed or lost because of the holiday season, Le Corbusier enclosed a typed copy so that Kalff could more easily read its message. He felt that its message—great art takes time, everything will work out, I have done this before, etc.—was still very much to the point.[28]

The two peaks of the pavilion were each to be graced with an allegorical figure. One was to contain a female mannequin; the other a mathematical figure. Presumably, this corresponded to the two sculptures Le Corbusier once proposed for the areas adjacent to the pavilion: one representing mathematics and rationality, the other art and intuition. The *volumes*, as they were termed, were to be painted in such a way that they would fluoresce under ultraviolet light, and eradicate the differences between real volumes and the virtual depth of the projected images. Le Corbusier intended to design these himself.[29] The easy solution was suggested by a sketchbook notation from early 1957: "hang from the ceiling a Galeries Lafayette bare mannequin with pretty eyes or a standard mannequin with neither eyes nor hands."[30] Small, twinkling lightbulbs would be mounted in irregular patterns within the upper reaches of the shells, suggesting stars set against the darkness.

The two figures placed high in the apexes of the shells were the only real elements in the pavilion: sound, color, and image—which may, of course, be taken as concrete—created every sensory impression experienced by the crowd. In contrast to Le Corbusier's designs for the Pavillon de l'Esprit Nouveau at the 1925 Paris fair or the 1937 Pavillon des Temps Nouveaux, there were no exhibits to proclaim the architect's thoughts on the machine, architecture, the city, and modern dwelling. All had melted into air; all became an element of a broader vision of human life in the twentieth century expressed through a synthesis of the arts. Le Corbusier's own work did appear in the closing moments of the *Poème électronique*, however, as both a moral and architectural lesson.

"We are going to ask Mr. Agostini to complete the first minute of the edited film before the 6th of February; in Eindhoven, we intend to show you, although provisional, the first minute of the film, combined with some color effects and the music of Varèse."[31] Kalff and the Philips technical staff were anxious to begin coordinating the parts of the electronic poem, realizing, as perhaps Le Corbusier did not, that they needed some room for error and time to undertake adjustments. And adjustments would be required. Many aspects of the technical apparatus, as well as the control systems, were to be developed specifically for the pavilion and stood almost completely untried. Kalff, with Mr. Numann, director of public relations for Philips, went to Paris specifically to check on the questions involved with the production of the film by Alga-Cinema. What they learned was not comforting. Even Agostini had had some hesitations about the firm's ability to produce the desired effects within the given time. The proposed work was far from typical, requiring considerable graphic preparation and photography. The studio estimated that the costs would run between 8 and 14 million francs, and that sum excluded the actual filming.[32] As a result, together with Agostini, they were examining the possibility of producing the work in Eindhoven or Amsterdam instead of Paris.

Although originally proposed for early January, the screening of the film's test segment was postponed a month.[33] While Kalff was less than pleased with the delay, he realized that he was in no bargaining position and was, in fact, at the mercy of the filmmaker and the studio. He proposed that the demonstration be undertaken with the most sophisticated means possible, and that the audience be restricted to a small group. The event took place in the "Garage" in Eindhoven on the afternoon of 6 February, with the technical arrangements for film, lighting, and sound highly improvised. While the film met with greater approval than the minute or so of Varèse's score, it did not receive critical acclaim.[34] "The reactions were not very enthusiastic," reads a company memo, although one of the committee members, quite surprisingly, found the sound composition had more apparent structure than the film. The lighting effects seem to have impressed no one, although given the makeshift conditions in the garage they did little to replicate the pavilion interior.[35]

Le Corbusier, like Varèse, contracted flu from a visit to the building site—"the weather was, in effect, miserable"[36]—and the news was not at all consoling. But there was little else to be done as it was necessary to produce the second and third sequence in order to meet the 10 April deadline. The film project was transferred to Geesink Geep in Amsterdam, who were producing the second sequence as February came to a close. While the work was proceeding "feverishly," they estimated seven or eight weeks to complete production. Several copies of the final version would be required—taking still more time—and synchronization remained to be done. Also still unresolved were the details of the mathematical figure suspended inside the pavilion, and the precise forms of the stencil projections.[37]

Ill or not, Le Corbusier was no gentler in his reply. He had met with Petit and Agostini; the latter had just returned from a trip to Amsterdam to review the production facilities and capacities of the film studios. He was reassured by what he saw there. His experience in cinema had taught him to produce more film than the final duration, to allow a decisive cutting in length at the end of production. Le Corbusier requested that Kalff authorize the studio to produce the surfeit footage, for only by these means could Agostini realize the intended goals.

In spite of the fact that only a month remained before the projected opening, and barely one quarter of the film had been completed, Le Corbusier retained his arrogance. In his second postscript, he added: "The moral of it all is this: it is Agostini who has my confidence, my orders. I have explained to him the spirit of the scenario; he understands. And it is he, naturally, who must transmit this to those undertaking the work at Geep Film. It can be no other way!"[38] One can only appreciate Kalff's patience and diplomacy.

In early April, with a trial of the musical score already scheduled, Kalff still hoped "that everything was moving along because the fatal date is rapidly approaching."[39] The opening reception for the exposition directorship and the international press was planned for 22 April; the following day a second presentation would be held with the Commissariat Géneral de l'Exposition, represented by M. Moens de Ferning, and his wife in attendance. Le Corbusier's presence was requested. By this time, even Le Corbusier showed signs of nervousness.

I repeat my absolute wish to have the map of the roads of Europe removed from the end of the Poème électronique and replaced with an element of the Kast film: the map of Europe drawn with the kids in charcoal on the roof of the Marseilles block, and limiting that to a quick glimpse in which I don't appear personally *(very important)! There must be in this film a moment where my hands hold the charcoal drawing.*[40]

He then crossed out what he had written; having checked, it would have been too complicated. (In the final *Poème électronique*, there were to be no maps at all, other than Le Corbusier's own urban plans.) The phrase "I am counting on you" appears twice.

The pavilion opened, in fact, on 22 April but closed immediately thereafter. The film was not completed to satisfaction and required further efforts to synchronize the two projectors. In addition, the music tapes were not properly geared to the scheme of loudspeakers, and Xenakis was troubled by the poor sound quality of his piece caused by re-recording.[41] A considerable amount of scrambling still would be required to make everything function smoothly. As the pavilion program was nearly completely automated—one of its demonstrative points—there could be no half-way measures, no half-successes. Everything had to perform perfectly or not at all.[42] No correspondence records the problems and their solutions, and the work was probably all undertaken on site, guided by Johan Jansen. On 2 May, the pavilion reopened to the general public and was a popular success, if rarely understood. In all, about a million visitors participated in the *Poème électronique*, and they did not appear to be disappointed by what they saw and heard. "It seems that after innumerable reports," wrote the architect, "our intention has been attained and the public has been greatly impressed."[43] In the succeeding months, tasks such as the paying of bills for the photographs and their reproduction rights—a misunderstanding that would try frayed nerves on all sides—and other matters of housekeeping required a certain degree of vigilance. Le Corbusier also wanted, and expected, Philips to pay for an excursion to Brussels for all those who had worked on producing the pavilion and the *Poème électronique*. Apparently, the Philips management did not leap at the opportunity, however. Untroubled by such trifles as lateness and electronic bugs, Le Corbusier sent the bill for certain photos with a sarcastic note:

I would like to point out to you, among other things, that I have never received a friendly word from your directors for the enormous effort that I have expended on your Pavilion. This is called, purely and simply, human ingratitude.

When an architect makes a building, the client treats him to a lamb chop or steak, and makes a little smile, even if this must be a [forced] smile. The publicity for Philips around the Poème électronique is enormous and no one at Philips can doubt the immense difficulty it was necessary to conquer in order to win.[44]

Ever the diplomat, Kalff again smoothed the waters:

> As for the ingratitude with which you reproach the directors of our committee, we do not at all agree with you. One must not forget that your scenario caused a number of bad reactions among the members of our managerial board and that any unanimous demonstration of enthusiasm on their part was impossible. [Because of] the delay in the inauguration of the pavilion and the slow mounting of our presentations, which were still incomplete at the opening and which are at this moment still in the course of perfecting, we have not had the opportunity to organize a stirring inauguration [for the collaborators].[45]

Kalff acknowledged and addressed the problem of the sound quality of Xenakis's work, and added, somewhat kindly considering the architect's accusation:

> Public opinion has changed considerably since the 22nd of April and the success of the Poème électronique with the general public, and above all the esteem shown by architects, musicians and artists for our realization has now convinced our administration that what we made is a success, and worth the effort and considerable funding invested in it. The small smile of which you speak will become a great smile, and not in the least forced. . . . "[T]he client," which in reality is a very complex organization, would like to offer you a monumental lamb chop the day that you come to Brussels, if possible during the month of September; with Xenakis and Petit and Agostini if they would like, and with some other members of our administration and Mr. Spaens from Brussels.[46]

The days set for consuming the lambchops were 29 and 30 September, and a full round of dinners and presentations were planned. Presumably, a good time was had by all—and the *Poème électronique* performed properly before its creators.

The images that Le Corbusier assembled to manifest these feelings and ideas were selected from various sources including the collections of the Musée de l'Homme, Musée Pédagogique, and even the American Library (which provided photographs of German concentration camps, American scientists at work, and the skyscrapers of New York). They ranged in subject from prehistoric figures to tribal sculpture to human settlements, science, and technology [figure 4-8]. Le Corbusier's earliest sketchbook jottings read as the abstract and concrete products of a stream of consciousness. Almost none of the sketchbook entries pictured the building and most are text notations rather than drawings. The sequence of color effects took more space than the constituent parts of the film. Le Corbusier assigned the final construction of the

4-8
Image from the *écran* of the *Poème électronique*, Sequence 1, "Genesis." This African tribal sculpture had appeared in Le Corbusier's 1925 book *The Decorative Arts of Today*.
[Philips]

film to his collaborators; he himself specified sources for the images. He linked items within the poem, and established tangential relationships between them. No clear comprehensive pattern was apparent, however, and the sketchbooks record a play of free associations, a quasi-surrealistic cascade of idea fragments that as a mosaic came to constitute the totality of the *Poème électronique* and its story of humankind.[47]

While the montage of images that comprise the film are completely Le Corbusier's own, the ideas that underlay *Poème électronique* owe a considerable debt to the work of André Malraux. Correspondence ascertains that a number of the photographs used in the film came directly from Malraux's books on sculpture, but the impact is far greater than several purloined art works. In Part One of *The Voices of Silence*, Malraux described the fundamental changes to the appreciation of art occasioned by the institution of the museum and the advent of mechanical reproduction. In the museum and in the photographic book, new relationships were established among the works of art, and between the work and the spectator.[48] Drastic changes in scale affected by photography and printing created a parity between the miniature and the colossus; through enlargement the tiny gesture of the historical object rivaled the facture of the modern painting or sculpture, fundamentally modifying the intention behind the original.[49] "Indeed," Malraux declared, "reproduction (like the art of fiction, which subdues reality to the imagination) has created what might be called 'fictitious' arts, by systematically falsifying the scale of objects; by presenting oriental seals the same size as the decorative reliefs on pillars, and amulets like statues."[50]

Sculpture—historically to be read from all positions under all environmental conditions—was captured in the biased lighting and view of the photographer, superimposing a reading that through continuous reproduction became synonymous with the sculpture itself. In addition, a fragment such as a head or torso acquired an identity as a complete work, further distorting the sense and purpose of the original. More central to the Philips project, Malraux demonstrated that through the curator's or author's juxtaposition of objects, there arose new affinities among works that may not be obvious when each resided in its own context—this created a Museum without Walls.[51]

Malraux himself ended the first part of his study with a discussion of the motion picture. Within the framework of his thesis, the motion picture achieved the status of an art form only when "emancipated from the limitations of the theater," that is, when it had triumphed over real time and space—both concerns of twentieth-century painting and sculpture. Real time was challenged through "cutting," using jumps, lapses, and freezes. Camera position and the focal length of the lens, on the other hand, manipulated space using close-up or distant views.[52] By this judicious choice of camera lens and movement "the producer *chooses*, instead of having [the area shown on the screen] imposed on him."[53] It was with just this juxtaposition and manipulation of images that Le Corbusier developed the scenario for the electronic poem, creating his own Museum without Walls by drawing upon the history of art and culture according to his own construct and for his own purpose. This, too, reflected Malraux's belief that "our culture is not built up of earlier cultures reconciled with each other, but of irreconcilable fragments of the past."[54]

The earliest entry for the project in Le Corbusier's sketchbooks, dating from 25 September 1956 reads:

> Sound // *low // medium // high*
> rhythm // *of stifling slackness // of breakneck speeds //*
> *frightfully spaced // quivering // horribly fast*
> color // *in the red // the yellow // the blue // the violet*
> *or better: // 2 Sound and Rhythm // 1 Light and color // make a* grid *S sound*
> = *R rhythm. C color // and pull at the strings.*
> *3/* volumes // *Euclid // the Dolomites // the Pyramids*
> Surfaces // *the sea the tide // the clouds // the wheat-fields //*
> *the tundra // the desert*
> lines // *slow trace // lightning- fast [trace] //*
> *(crocodile) //*
> *snake // lightning*
> [Entry 700][55]

The first study thus outlined both the abstract program for bonding the media of sound, color, image, and the physical elements of the volumes, with some suggestion of the connections between them. For example, the formal association of the snake and crocodile with lightning, and thus with speed, transcended the simple syntactical listing of components. Three days later, Le Corbusier focused on abstraction, but terminated with:

> *(sky/sea/prairie) // bird/fish/sheep // vulture / shark / rhinoceros //*
> *airplane /cruiser / tank // parachutes / swimmer //*
> *family gathering flowers horse of the Apocalypse //*
> *Jesus on the waves / lying in the hay // storm lightning / naval battle typhoon //*
> *euphoria*
> *attack.* [701]

Here, Le Corbusier outlines the dichotomies that will recur in the *Poème*: nature and culture, the tranquil and the tumultuous, apocalypse and salvation. Following thereafter are entries investigating the participant subjects in finer detail:

Application to: OBJECTS // a) inert = OB-in // b) mobile =
OB-mob
OB-in
crystals // houses palaces and temples statues pictures //
skeleton: death's-head man horse bird fish // immobile
mechanics: wheel, gears, etc. // star // inert atom
OB-mob natural // artificial
ocean dunes desert sky Vesuvius flood delta mobile
mechanics cannon // automobile // screw // stellar
gravitation atom fission
grain seed plant flower fruit (in accelerated [motion] tree worm snake crab
spider grasshopper dragonfly
cat dog horse elephant seal // hand trunk [703]

The following entry follows a different tack: the dichotomy of good and evil, mixing Laurel and Hardy, Fernandel, and Walt Disney with Hitler.[56] The concern for syntax reappears; how to present these materials?

Means
Light. / atmosphere // circumscribed points // streaks //
fixed or moving zones
Color / the harmonics of the prism // a) successive
b) simultaneous // parallel, vertical, horizontal //
the complementaries 2 by two a) successive b) simultaneous
1/2 + 1/2 spherical
Light white object on black // black [object] [on] white //
gradual movement from one to the other

Two months later, in the middle of December, his thoughts turned to the folly of man—battlefields and concentration camps—played against the antics of the comedian.[57] 3 February 1957 the conclusion became ghastly:

finish with catastrophic war volcano Typhoons
Buchenwald = important and not with optimistic
apotheosis [827]

But the vision of the apocalypse as the conclusion to the visual poem—and to humankind—was tempered in Le Corbusier's vision from 28 (?) February:

peace // slowed down / blossoming flower seed // this sequence = next to
last = new spring
nebulous worlds sun moon
baby's smile life of savages [833]

then:

light / sound / images
red lights in the corners blue - red - violet, yellow
roar silence roar // "think about it" - huge roar "think about"
"Spanish suave *Italian* prima donna - *nightingale //*
"The Electronic Poem is finished. This way to the Exit. This way to the Exit"
images *of horror from Buchenwald and war // rhythmic reversal of the sequence I // one [834]*

A visit to the Musée de l'Homme on 8 March 1957 and a review of several books produced a detailed listing from which Petit would presumably derive the images used in the final form of the eight minutes of film:

Various races Belgian Congo Pygmies
Greenland Eskimos Spanish haymakers (Pierre Verger)
Bali Dancers [(Pierre Verger)] Bavilean statue (Herdeg)
Skull. from the chapel of the Saint New Zealand
Tattooed\man Bucher Chinese junk (Verger) Funerary urn
Colombia
Herdeg Head Hawaii Herdeg.
research easter Island Head.
[research] planispheres
Greek ruins Nelly's pictures // Sphinx [ruins]. [836]

These studies suggested rather than established the explicit imagery to be bilaterally thrust upon the pavilion's visitors. The notebook studies provided only a lexicon from which Le Corbusier would select and configure, using a grammar still to be defined. Certain themes emerge and reemerge, for example, comic figures like Chaplin played against catastrophe (in a manner recalling the works of Samuel Beckett), human invention, tribal art and sophisticated technology, science gone amuck, and order and chaos. The relationship among the parts was always difficult to discern with precision, however. If, as he claimed, Le Corbusier already had a firm conception of the *Poème électronique* in his head, the sketchbook entries bear little witness to it. The semantic dimension of the film accompanied the development of its syntax, and the manner powerfully informed its meaning.

The visual production of the *Poème électronique* comprised seven sections of roughly a minute each, although they varied slightly in duration. These constituent sequences were distinguished from one another with a short return to darkness, but were difficult to identify without extended study.

Genesis, 0" to 60"
Matter and spirit, 61" to 120"
From darkness to dawn, 121" to 204"
Manmade gods, 205" to 240"

How time molds civilization, 241″ to 300″
Harmony, 301″ to 360″
To all mankind, 361″ to 480″

The definitive manuscript for the scenario, dated 27 November 1957, is retained in the archives of the Fondation Le Corbusier, although it was modified to some degree during the course of image acquisition and filming.[58] For the most part, Petit and Agostini were able to faithfully translate Le Corbusier's ideas into cinema, although there was considerable room allowed for their own creative additions. The major deviation from the 27 November *minutage* was the elimination of the film within a film originally proposed. Instead, the final sequence ends optimistically, returning to the human figure, the child, and the maternal.

The *Poème électronique* opens with parallel texts in French, English, and Dutch drawn in white against a black ground.[59] The ambiant Xenakis score *Concrète P-H* fills the space as the aural partner to the introduction by Philips of their role in the production about to begin, a new art form in which audio and visual electronics allow the audience to experience sounds and images as they have never before.

The poem begins in darkness, in blackness, with moments of silence before the gong announces the Varèse score. Recall, however, that the projected film—the *écran*—is but one component of the visual spectacle. Accompanying it were the *tri-trous*, or stencilled projections; the *volumes*, i.e., the mathematical and female figure invisible during most of the presentation; and most influentially, the *ambiances*, or colored lighting effects, created by filtering projected lamps concealed behind the parapet walls. These colored ambiances effectively tinted the black areas of the film and gave them color.

In the beginning is "Genesis," Sequence 1, which lasts for exactly one minute [figure 4-9]. From the darkness rise horizontal bands of gray light suggesting the dawn. Lamps mounted within the concrete shell and electronically modulated to simulate flickering stars punctuate the tinted dusk. An image of a bull's head set left against a black ground emerges from the darkness accompanied by the sounds of the Varèse score. To the right, a matador is counterpoised. The positions reverse as the give-and-take of animal and human in combat set the theme of oppositions that permeates the film. The Bull occupies the full screen. Then the Matador, full screen. The world will be in a conflict, partly the product of human technology, part the product of human hubris and frailty. The head of a Greek statue fades in center screen; darkness. Slowly the supine body of a woman, face down, is discernible in film clip produced for the *Poème*. (Most of the images, it should be remembered, are still photos animated through cinematic techniques.) She lifts her head with a smile, turns to the camera, and startled, recoils with a look of terror. The scene recalls classic surrealistic cinema, in particular Luis Buñuel's *L'Age d'Or* (1930). Each of the segments averages about five seconds and gives the film, to this point, a regular rhythm.

The woman's stare dissolves into a skull, which flashes, with variants; Sequence 2, "Matter and Spirit," commences [figure 4-10]. In contrast to "Genesis," Sequence 2 hurls its images at the audience in a rapid, staccato rhythm, each image remaining on the walls for an average of a single second. The skull implied by the woman's stare gives way to a scallop shell seen from above, then rotated flat, magically metamorphosing into a skull once again. A photo of four men, probably scientists of some sort, is spot-illuminated so that the faces appear in several variations, culminating in the revelation of the entire photo. In contrast to the tame pace of Sequence 1, Sequence 2 progresses quickly. The image of an African woman melts into a carving that zips to the upper right and then looks down upon the reclining female figure from a painting by Courbet. Quick flashes. Colors are dark for the most part, turning to blue. Heads appear and fade in the various corners of the screen; the head of a dinosaur skeleton emerges into center screen, then zips to the left; the lighting changes and the head is seen as a part of a complete skeleton. A monkey ponders, multiplies, and is seen in sets of two statements symmetrically disposed, multiplied into a crowd—like the human crowd that will reappear later in the spectacle.

Eyes and heads characterize the third section, "From Darkness to Dawn," which extends from 121″ to 204″ [figure 4-11]. The close-up of an owl pulls back to reveal the face of the bird. Splitting the screen doubles the image, which is then flopped to symmetrically duplicate the image about a horizontal axis. On a normal screen, this appears as a simple manipulation, but its effect would have been significantly augmented by the curvature of the pavilion's walls. Owls turn to a turkey head seen from the side; then two similar heads, one left, one right, turned toward each other. Then three fowl looking one way; then turning to look left. A new still shot, probably of the same woman encountered as she reclined earlier in the film, focuses on her eyes, which split and invert along a horizontal axis—like the owl eyes earlier in the sequence.

Tribal sculpture provides the subject of the next cluster of images. The camera first shows each of three carved figures in close-up, the heads of the statues shown sequentially in one image after the other. The camera then moves back to include more of each statue in a single frame. Finally, the three carvings in their totality appear together, now set against a white rather than a black background. The three variants of tribal art establish commonalities among cultures and link our past and our future.

The skeleton of the human hand is shown full-screen extending to the

4-9
Images from Sequence 1, "Genesis." *Poème électronique.*
[Philips]

4-10
Images from Sequence 2, "Matter and Spirit."
Poème électronique.
[Philips]

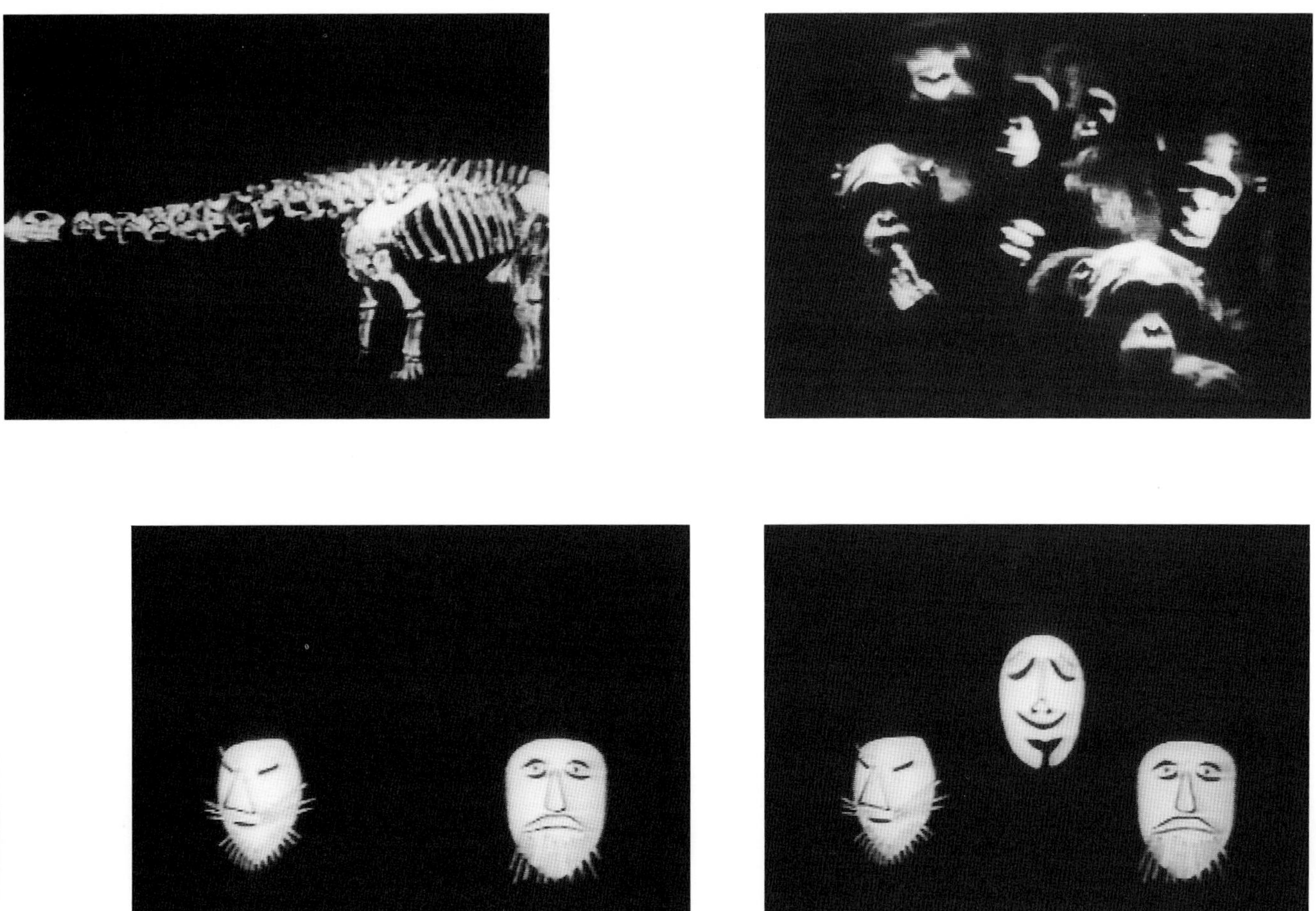

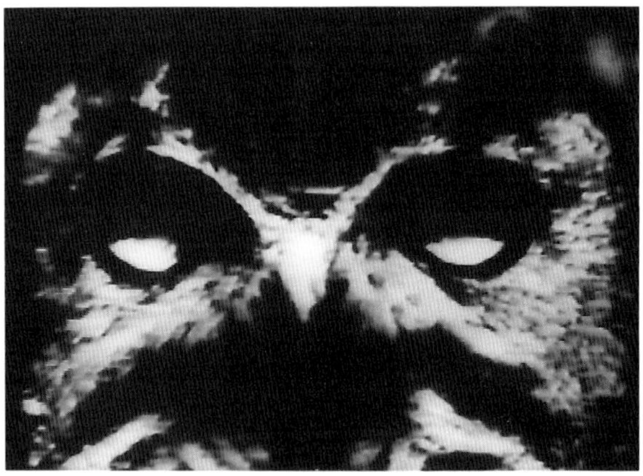

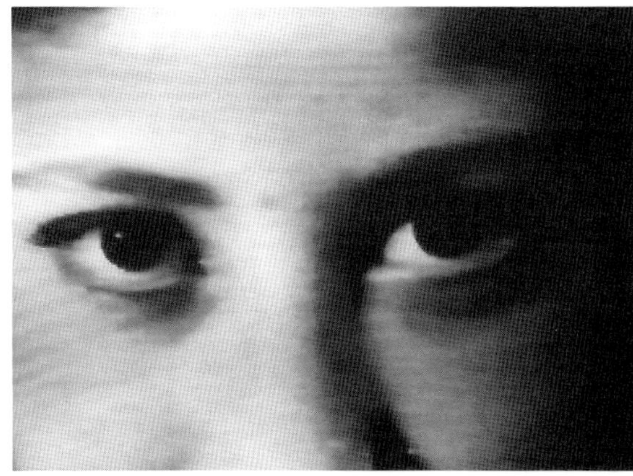

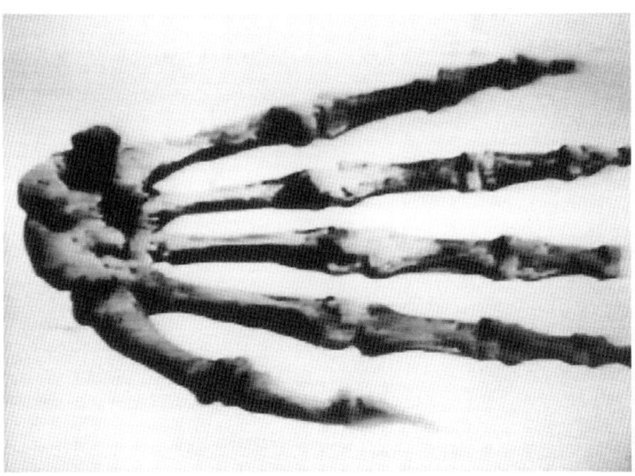

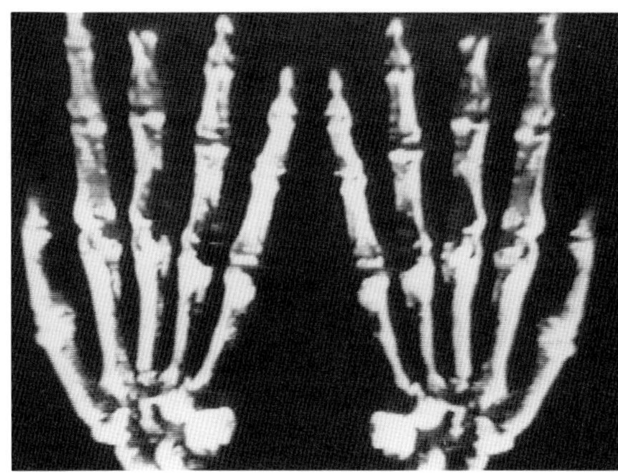

4-11
Images from Sequence 3, "From Darkness to Dawn." *Poème électronique*. [Philips]

right; then to the left; then it is doubled, and turned vertically. A quick flash of open trench graves from a concentration camp are harbingers of destruction. In his 1938 book, *Des canons, des munitions? merci! Des logis . . . s.v.p.*, Le Corbusier set his discussion of the development of urban Paris against the forces of war. Given the misery and ruin effected by the Great War and the war clouds threatening France in the late 1930s, Le Corbusier's concern was well founded. Humans, he tells the Brussels public, are condemned to repeat their mistakes. It will be no different in the future than it has been in the past: *Le Poème électronique* was at root a cautionary tale. Toy soldiers suggest how the minds of men are molded. The rapid alternation of the graves, toy soldiers and tanks, cowboys and Indians, a prisoner of a concentration camp, and guns, depict the roots and fruits of war. Giotto's *Deposition*; Christ as Man of Sorrows; medieval sculpture from Chartres and other churches suggest the religious dimension; the calm of the Buddha. While the images are thematically grouped, there is no attempt to assemble them in strict chronological order. True history is not chronological. From the blackness, the ambiance turns to regal purple with Our Lady of Sorrows. Fade.

"Manmade Gods," Sequence 4, begins with second 205, and again materializes from darkness [figure 4-12]. Gradually, the colors change from pink to blue to yellow, articulated by vertical stripes. At the twelfth second of this sequence a stone head from Easter Island emerges from the abstract chromatic ambiance. Whether by accident or design, the voice which utters something like "ooga" links the human voice with the human constructions. Four cubes follow the heads, spaced on bottom screen, suggesting the work of humans on the planet and geometry as primary order. Close-up of the Sphinx. A tribal carving of a head is doubled within the frame right and left; organic vegetal shapes are superimposed with a jump. The hands of the Buddha. The suspended "volumes" of the woman and the mathematical object are lit alternatively with ultraviolet light. This is a short sequence of only 35 seconds.

"How Time Molds Civilization," Sequence 5, commences with complete white light rather than darkness, and is held for five seconds [figure 4-13]. Here the fact of industry and power becomes a symbol for what humankind has wrought, these ten seconds set against a green background. A worker at an atomic plant is sequentially spotlighted to feature the face or the machinery; light animates the single image. The pace quickens, returning to one-second flashes. The crowd. Technology accompanies people, individuals alternate with their loss of identity within the crowd. A telescope operator suggests the dimension of the heavens to follow in another sequence. The Crowd. A surgeon; a miner; cement workers in India; the cow in India; the farm horse. What people can do; what animals can do. Comedy dissolves into technological threat: Charlot (Charlie Chaplin) looks to his left from the up-

4-12
Images from Sequence 4, "Manmade Gods." *Poème électronique.* [Philips]

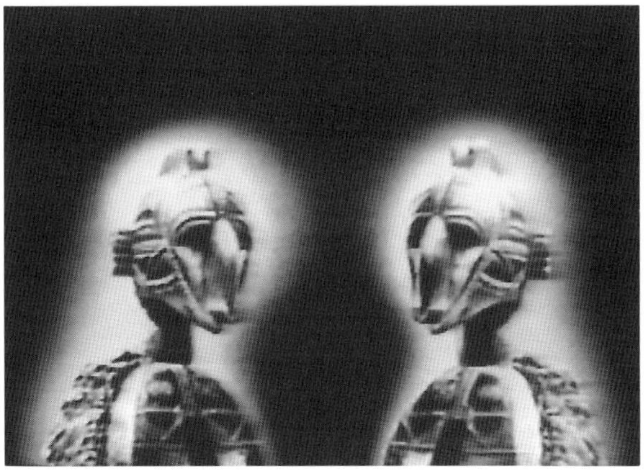

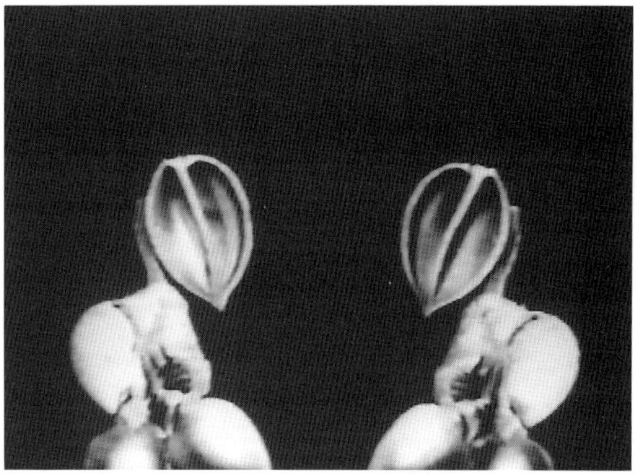

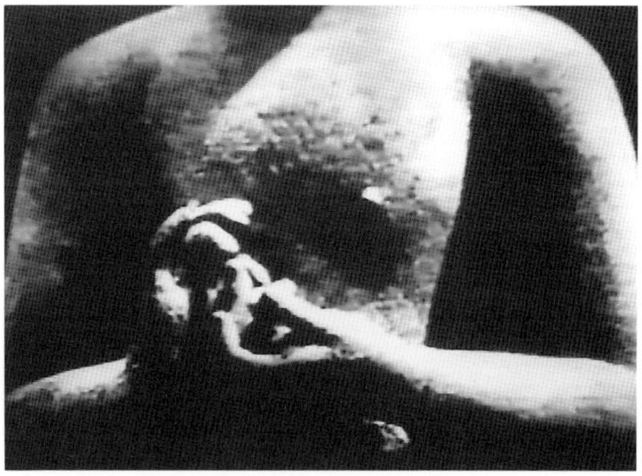

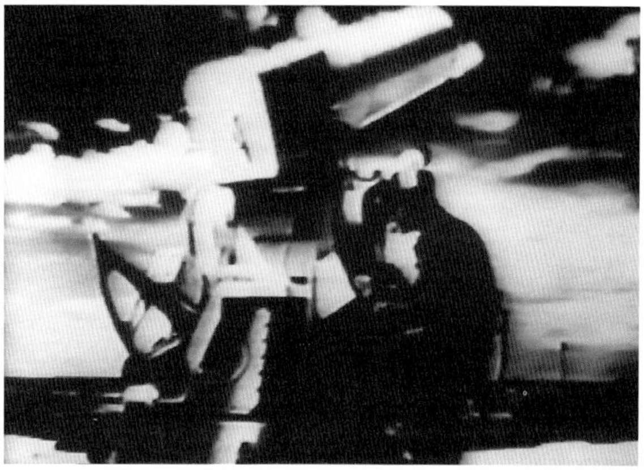

4-13
Images from Sequence 5, "How Time Molds Civilization." *Poème électronique.*
[Philips]

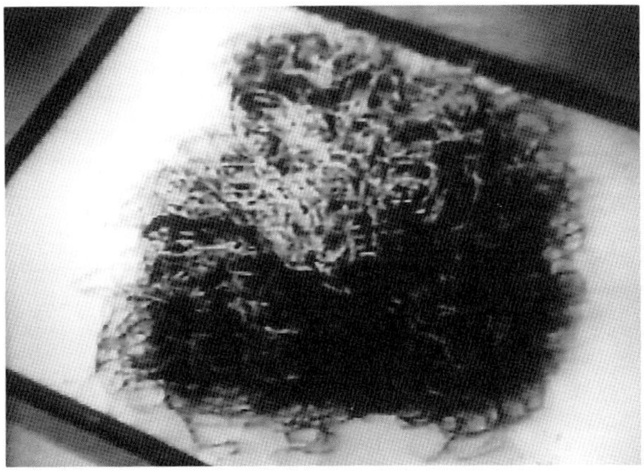

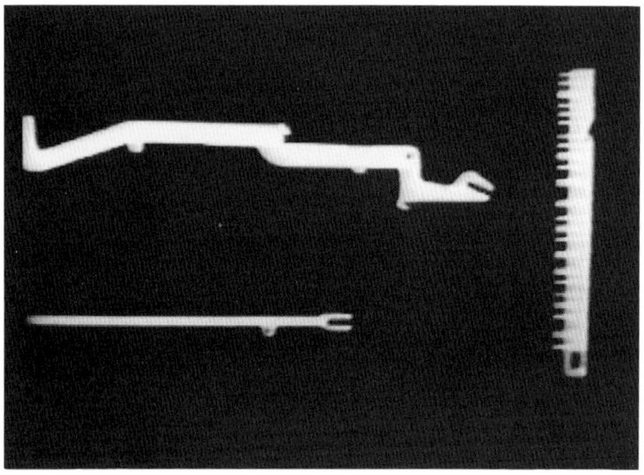

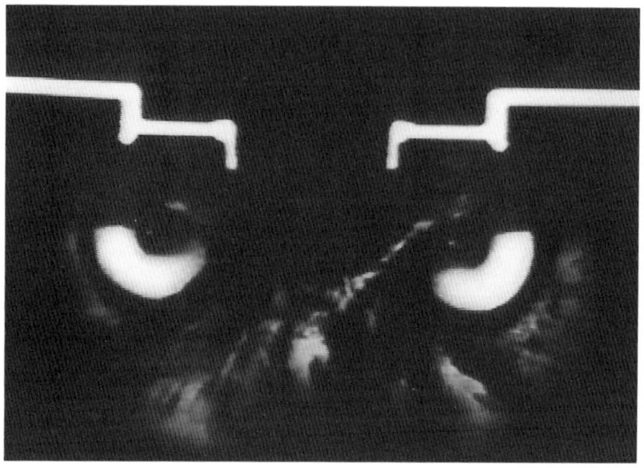

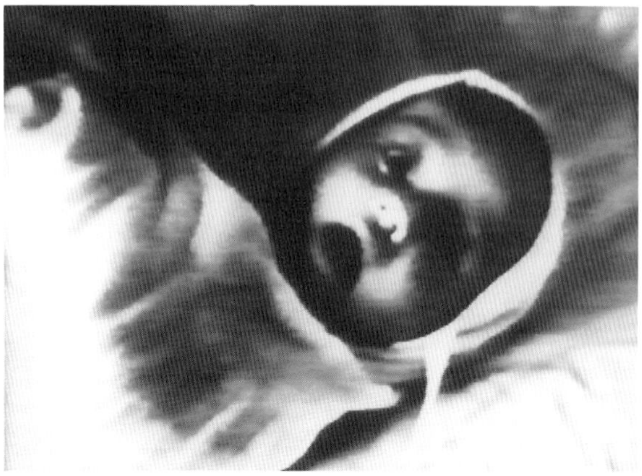

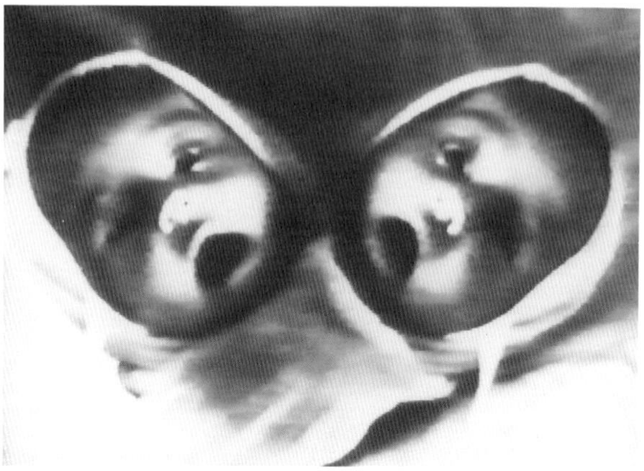

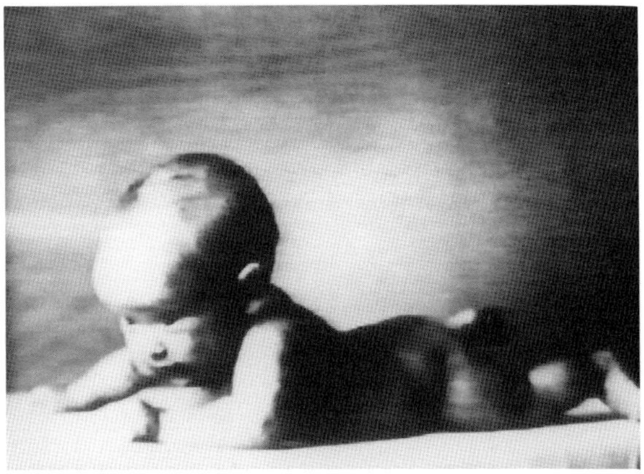

4-14
Images from Sequence 6, "Harmony." *Poème électronique*.
[Philips]

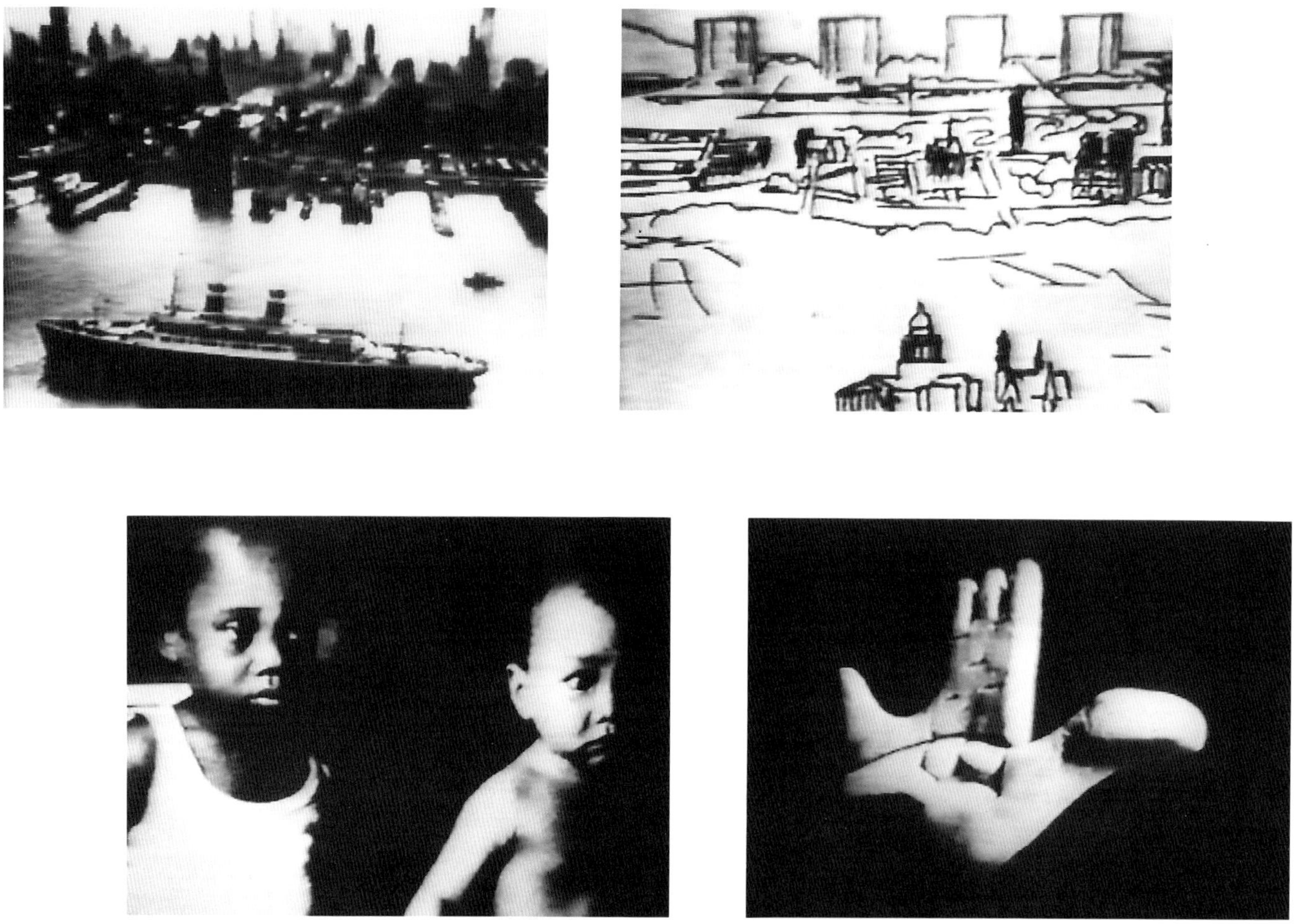

4-15
Images from Sequence 7,
"To All Mankind."
Poème électronique.
[Philips]

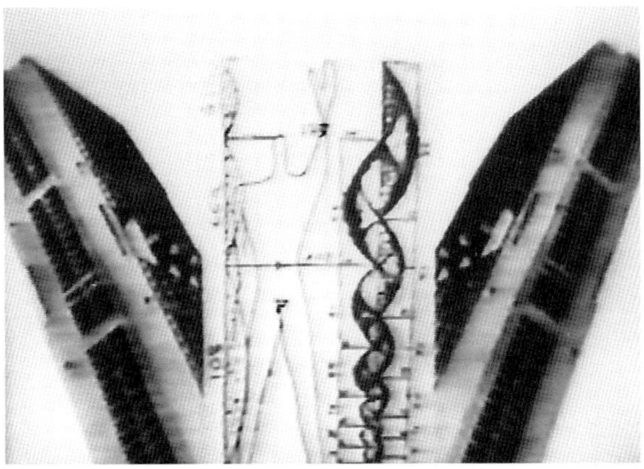

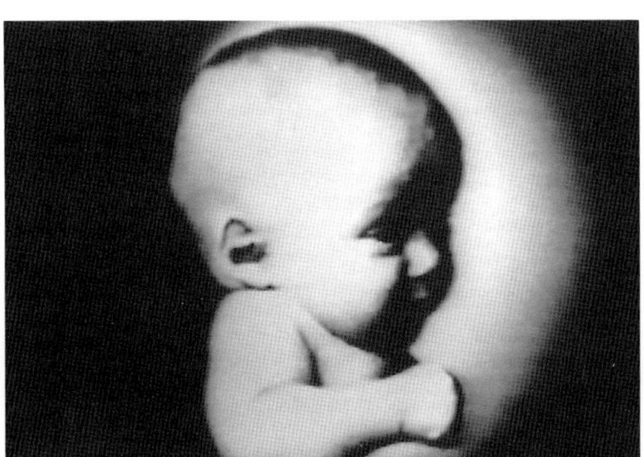

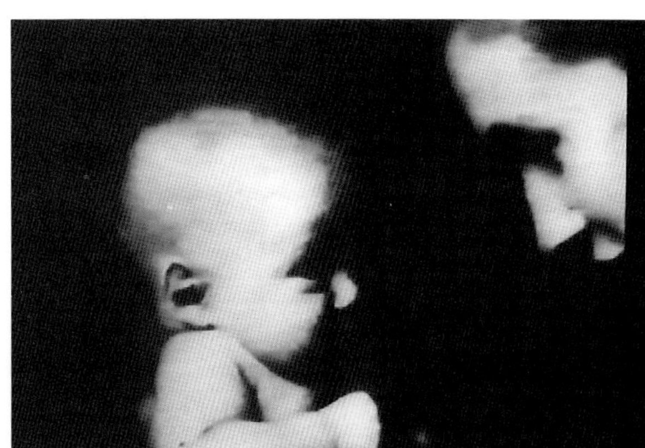

per right corner of the screen: his visage dissolves into a bomber, then a rocket, then radar antennae. The sequence inverts, with the rocket and the bomber. Flash of the faces of apprehensive children as a mass staring to the left, set against a yellow field. The antennae reappear and above them shines the moon, center screen. The scale of inquiry has become galactic. The clouds of the stratosphere are somehow foreboding in spite of their beauty. The clouds metamorphose into the mushroom clouds of nuclear destruction, and are doubled and tripled, and heightened with red light for increased effect. The children continue to stare apprehensively; the red accompanies them. Curiously, the musical score does not reflect the visual cataclysm, instead maintaining its autonomous phrasing of sounds and acoustic space. The third of the cinematic sequences concludes the segment. A rectangular flat box of some threadlike mass is set askew to the film frame. It is animated and pulses ambiguously, somewhere between tumbleweed blown by the wind and a heart extracted but still palpitating. Is this tangled mass our civilization, disordered and out of control? The starlike lights mounted on the top of the pavilion's concrete shells are illuminated; it is the Day of Judgment.

Sequence 6, "Harmony," suggests a more optimistic vision, with organic and mechanical order as the principle elements [figure 4-14]. The first six images, set in blue, are presented rhythmically at five seconds each, and illustrate the open ironwork of the Eiffel Tower—in positive alternating with negative—and the mechanics of machines. These fragments are treated as abstract shapes that suggest simultaneously Fernand Léger's *Ballet mécanique* (1924) and the early films of Man Ray, as well as his Ray-o-grams: photos made by laying real objects upon photographic paper and exposing them to light. The machine parts appear as white shapes on a black ground, and in their positioning, repetition, and animation they create their own mechanical ballet. At times, they share the screen with human beings. The miners reappear, as does the telescope worker. Perhaps this is the benign side of technology and its human interface. Chaplin, then Chaplin with an abstract machine shape; Laurel and Hardy, and then accompanied by the abstract shape. Homo Ludens? Does the presence of comedians suggest comic relief, or do they reveal the foolishness of human play with technology and perhaps the ultimate absurdity of life?[60] The cultural historian Elie Faure, in *L'Esprit Nouveau*, characterized Chaplin in 1921: "And I see where, like a painter, a geometer, a musician, he enters the empire of poets conquering."[61] The scale changes at the speed of light. The rapid fire of astronomical images of solar eclipses, nebulae, and galaxies becomes the darkness. In the last ten seconds two lovers first sit and then embrace; babies appear set against a pink interior. Each shot ends with the diagonal sweep dissolve of the windshield wiper.

In *Foundations of Modern Art*, Ozenfant had claimed that "the needs of an epoch equal the eternal needs plus the needs of tomorrow." In this final sequence Le Corbusier pictures the epoch—chaos, poverty, the metropolis, and children—its wants and its hopes.[62] The optimism is an antidote for the sequence on war and destruction. The last segment, "To All Mankind," approaches the almost requisite publicity that one has come to expect at such presentations [figure 4-15]. But in the case of the *Poème électronique*, the advertisement was not for Philips but for the architect. Here architecture offers the hope for humanity, and the greatest hope is from Le Corbusier himself. True, there are some images of the skyscrapers of New York, a disheveled city with no apparent order. But the drawings and models flashing on the yellow walls—the projects for Paris, the skyscrapers for Algiers, the housing blocks at Nantes and Marseilles, the High Court at Chandigarh—all offer the architect's latest projects or most powerful urban visions. For the most part, these are simple photographs or drawings, although their point is made through comparisons. A rough stone wall, piled without mortar suggests, through juxtaposition, the chaos of New York. In contrast, the careful joining of an ashlar stone wall is linked to the plans by Le Corbusier for urban sites. The drawing of Le Corbusier's proportional system, the Modulor, is accompanied by a human voice. Its center-screen image shrinks as it is diagonally flanked with mirrored images of the model of the Algiers skyscraper project (1938).

Although the only real architecture shown is Le Corbusier's, he returns to the human presence to conclude the *Poème*. A white baby's hand rests on a dark-skinned knee. The camera retreats to show both children, some small hope that all peoples can exist in peace. Le Corbusier's Monument to the Open Hand offers help and comfort to all—peace is possible [figures 4-16, 4-17]. In *Le Poème de l'angle droit*, Le Corbusier had described the open hand as a gesture which binds human beings:

> *It is open because*
> *all is present available*
> *knowable*
> *Open to receive*
> *Open also that others*
> *might come and take . . .*
>
> *Full hand I received*
> *full hand I now give.*[63]

People follow: an old woman, a child, people at the dinner table; a man sleeping in the street; a dirt road, the ruts of transport deeply etched in its surface. A baby; another; and a third [figure 4-18]. The fourth photograph, and the final image of the film, remains a baby, but this time with the partial face of the maternal presence looking from screen right.

Overleaf:

4-16
The Monument to the Open Hand. Le Corbusier. From Sequence 7, "To All Mankind." *Poème électronique.* [Lennart Olson]

4-17
The Obus Plan for Algiers. Le Corbusier. From Sequence 7, "To All Mankind." *Poème électronique.* [Lennart Olson]

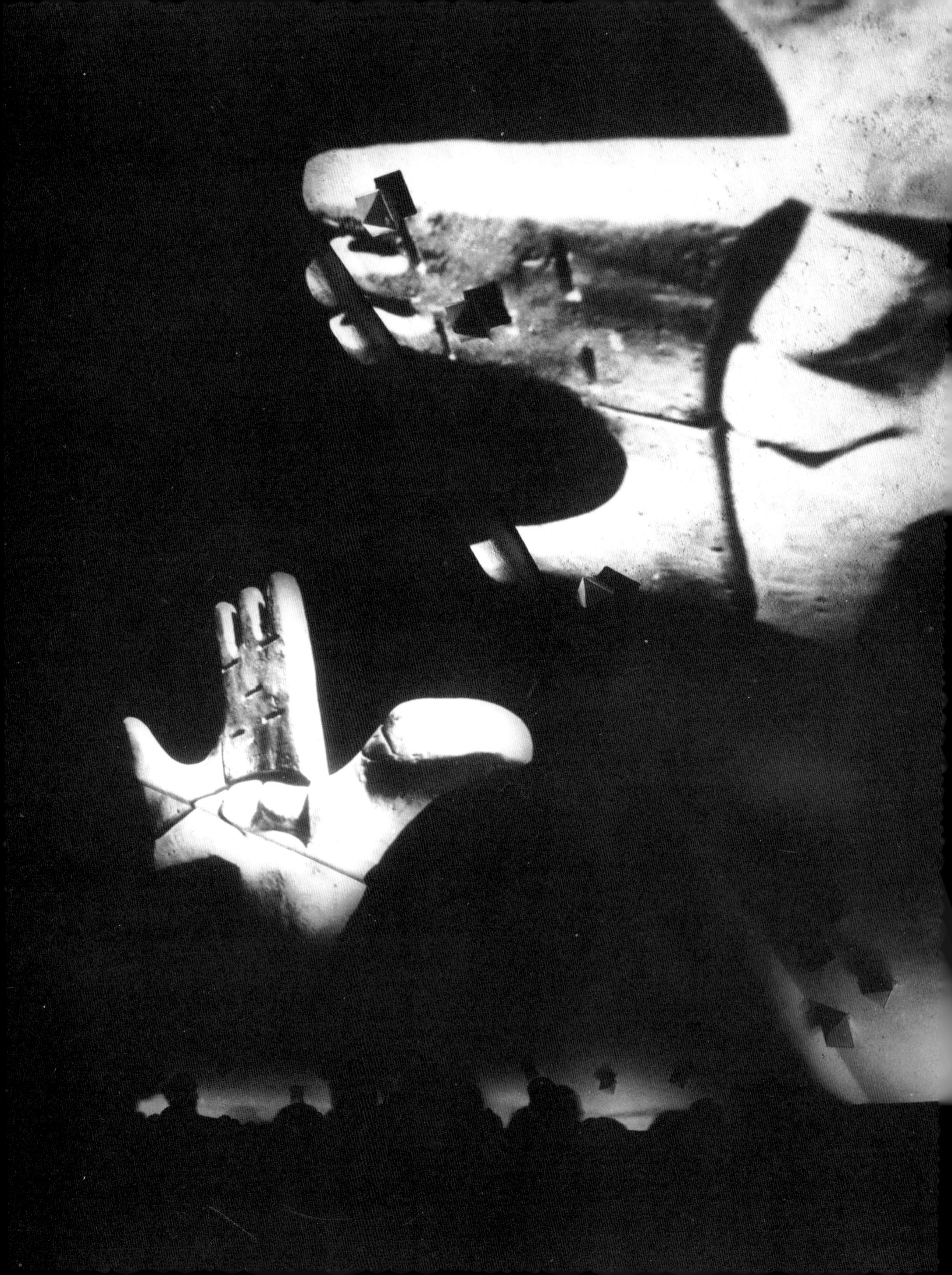

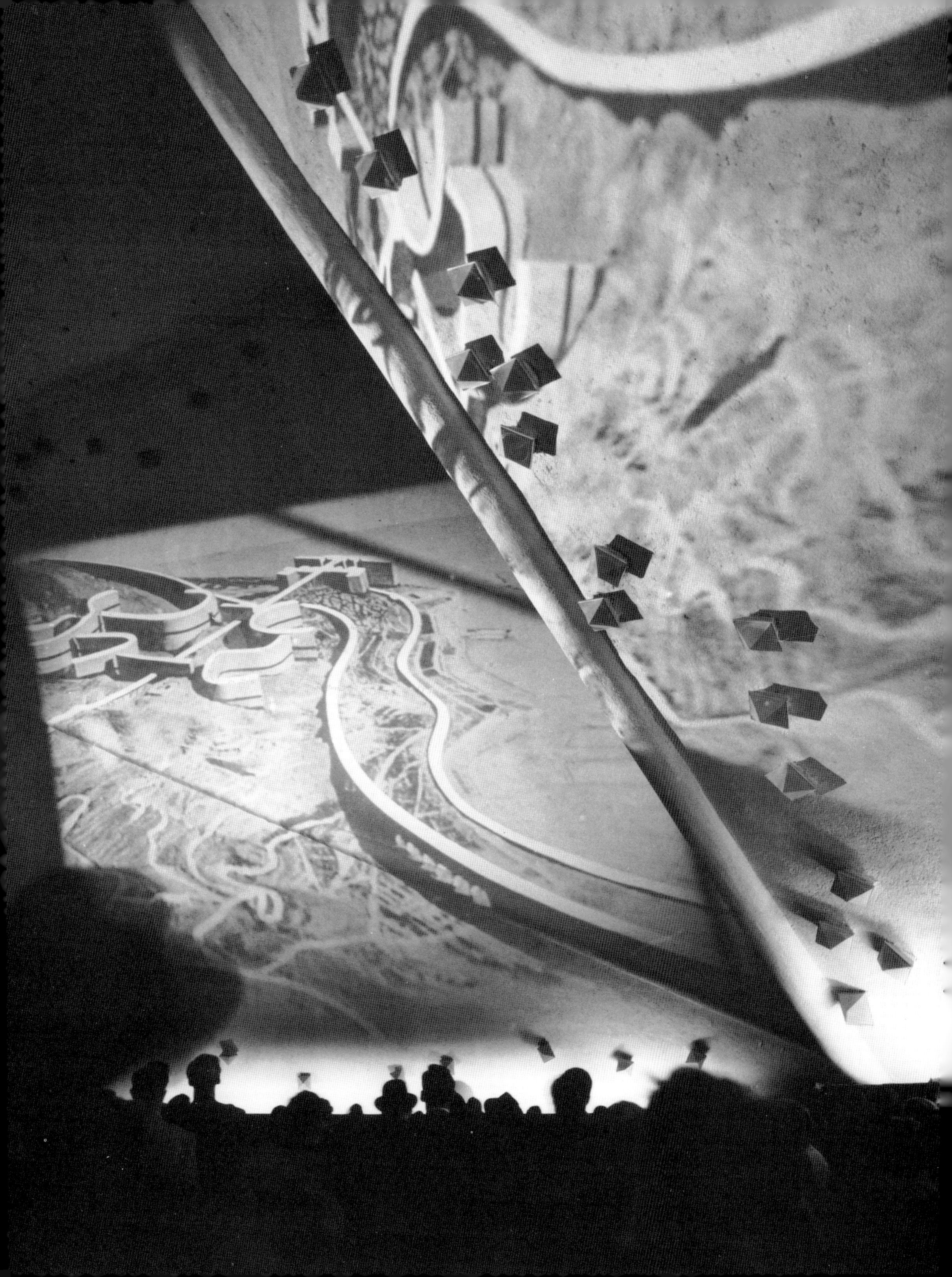

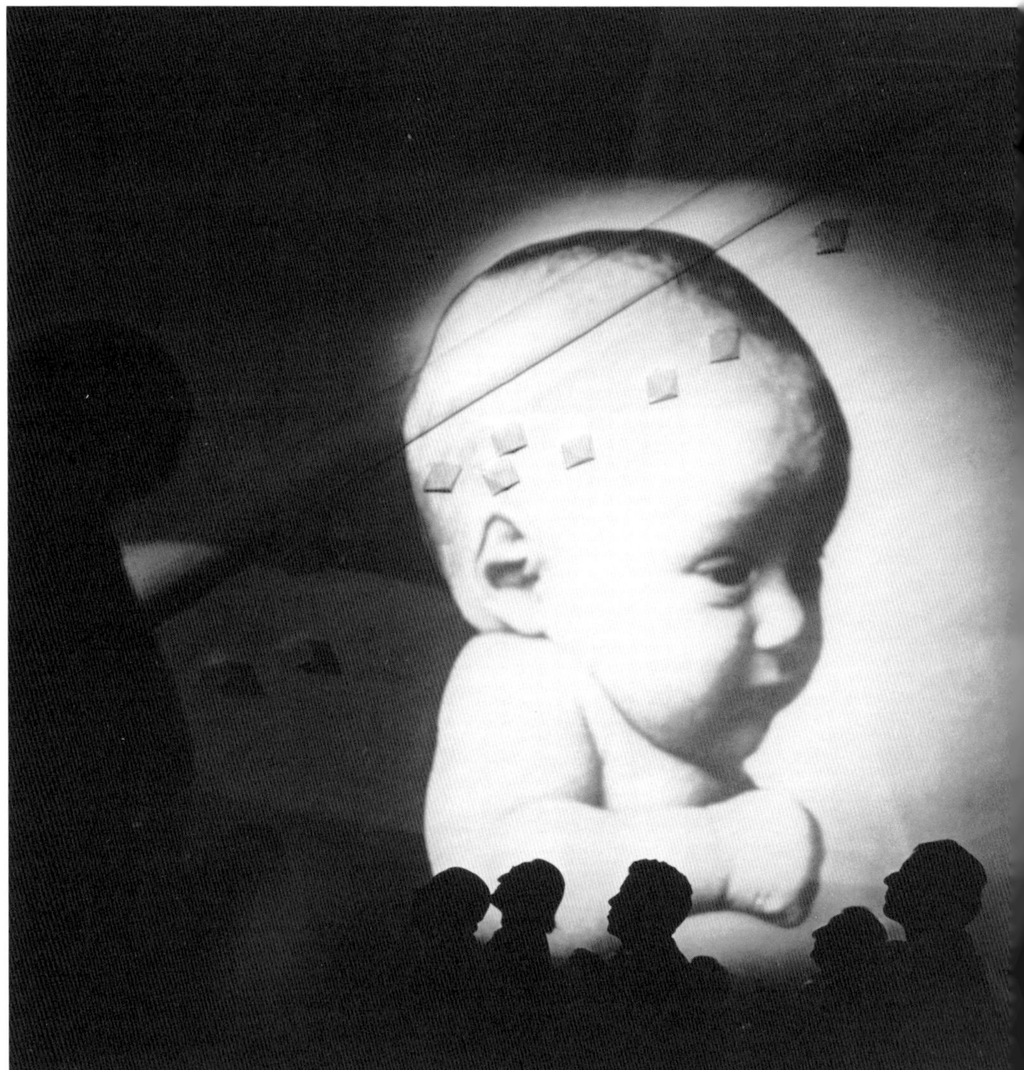

4-18
Image from Sequence 7, "To All Mankind." *Poème électronique.* [Philips]

The total time elapsed during this whirlwind journey was only eight minutes, but the audience left exhausted, if not befuddled. We must recall that in 1958, even sophisticated film audiences were not accustomed to the jump cuts that would characterize the work of Jean-Luc Godard, for example in *Breathless* (1959), or more recently the staple of MTV, where fractional-second editing replaces content, and syntax supersedes semantics or becomes semantics.[64] The *Poème électronique*—and this was Agostini's achievement—linked the content to the technique [figures 4-19 through 4-25].

Given that virtually everything within the *Poème* was a static image, one could question why they were not presented in slides rather than in film, and why not in color rather than in black-and-white. Since no documentation supports my conclusions, I must make some assumptions here. To begin, most of the source material for the film was monochrome, taken from books and photographic prints. Certainly color film existed in 1958; Kodachrome had been available to cinema professionals since the 1930s and to the public since after the war. But the colors tended to be soft, and it could be that under the high-intensity projection necessary within the pavilion, the colors would rapidly fade. Perhaps contemporary slide projectors were incapable of the power and throw required for the distance, although this is doubtful. Le Corbusier, in a sketchbook note, reminds himself that intensity is linked to the square of the distance.[65]

Philips was a technical, sound, and projection company; it was not in the photographic business. A black-and-white film was easier and far cheaper to produce than one in color, and the multiplication of images within single frames was far more easily achieved without "seams" at the point of collage. In addition, the beams from two projectors crossing in space would have distorted the colors of both, a problem that hampered the production of the *tri-trous*. And perhaps most of all, the task of synchronizing two slide projectors—and the inherent risk of jamming—would have been monumental. At least the relationships between the images of a film were fixed. All in all, and given the very stringent time requirements, black-and-white was probably all that could be hoped for. This was not a negative constraint, however, just a limitation. After all, the photographs that comprised the film were themselves in black-and-white, and by separating the colored ambiances and projections from the film, Philips was able to showcase more of its products. Kalff was clever in realizing, and exploiting, the property of light in which color is additive. More colors could be created by mixing a smaller number of light sources.

It was Philippe Agostini who ultimately shaped the sequences described on the *minutage* into a coherently incoherent piece of cinema. While Le Corbusier had established the primary temporal relations between the di-

4-19
Mask from Sequence 2, "Matter and Spirit." *Poème électronique.* [Philips]

4-20
Masks from Sequence 2, "Matter and Spirit." *Poème électronique.* [Philips]

4-21
Tribal sculpture, Sequence 2, "Matter and Spirit." *Poème électronique.* [Philips]

Plate 1
A postcard from the fair: "Europe Avenue: Pavilion of Philips, further on Tunisia and Morocco."
[*Philips*]

Plate II
View from the terrace: The Philips Pavilion; in front, the Mathematical Object
[*Getty Center*]

Plate IIIa,b
Two pages from the definitive *minutage* showing corrections made in the course of installation. Gouache on mimeograph.
[*Getty Center*]

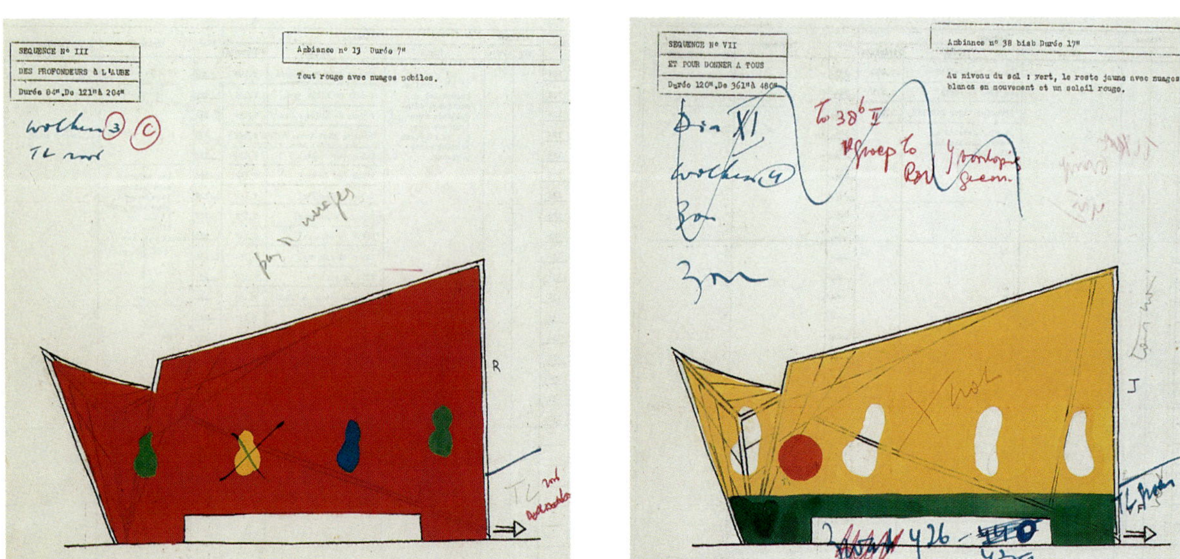

Plate IV
A violet *ambiance*; above is the geometric *figure*. The barrier screens the projection and lighting equipment.
[*Getty Center*]

Following pages:

Plate V
Diagrams showing the full presentation of colors and projections. Each increment is one second.
[*Mark Schwettmann*]

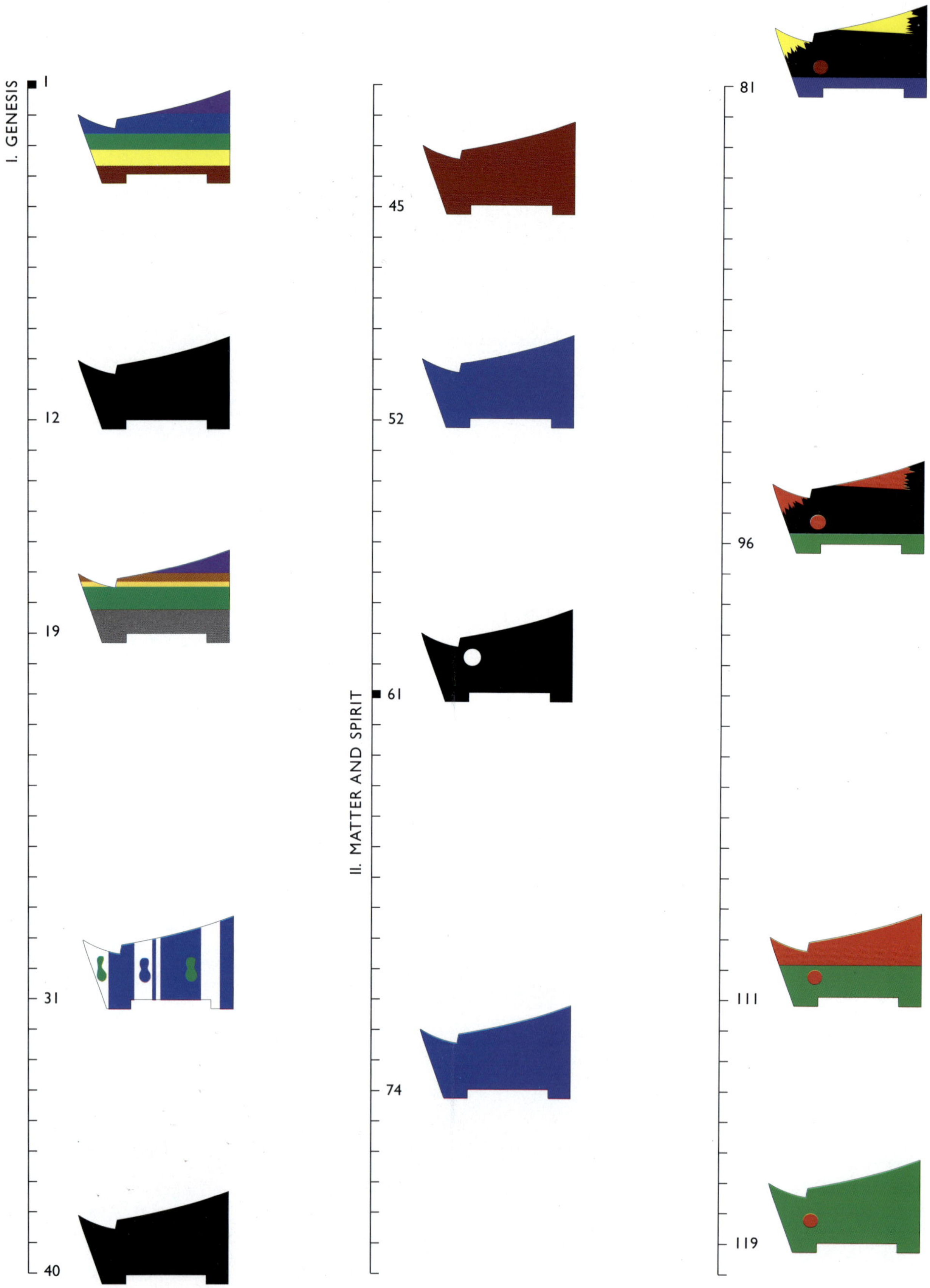

I. GENESIS
1
12
19
31
40
45
52
II. MATTER AND SPIRIT
61
74
81
96
111
119

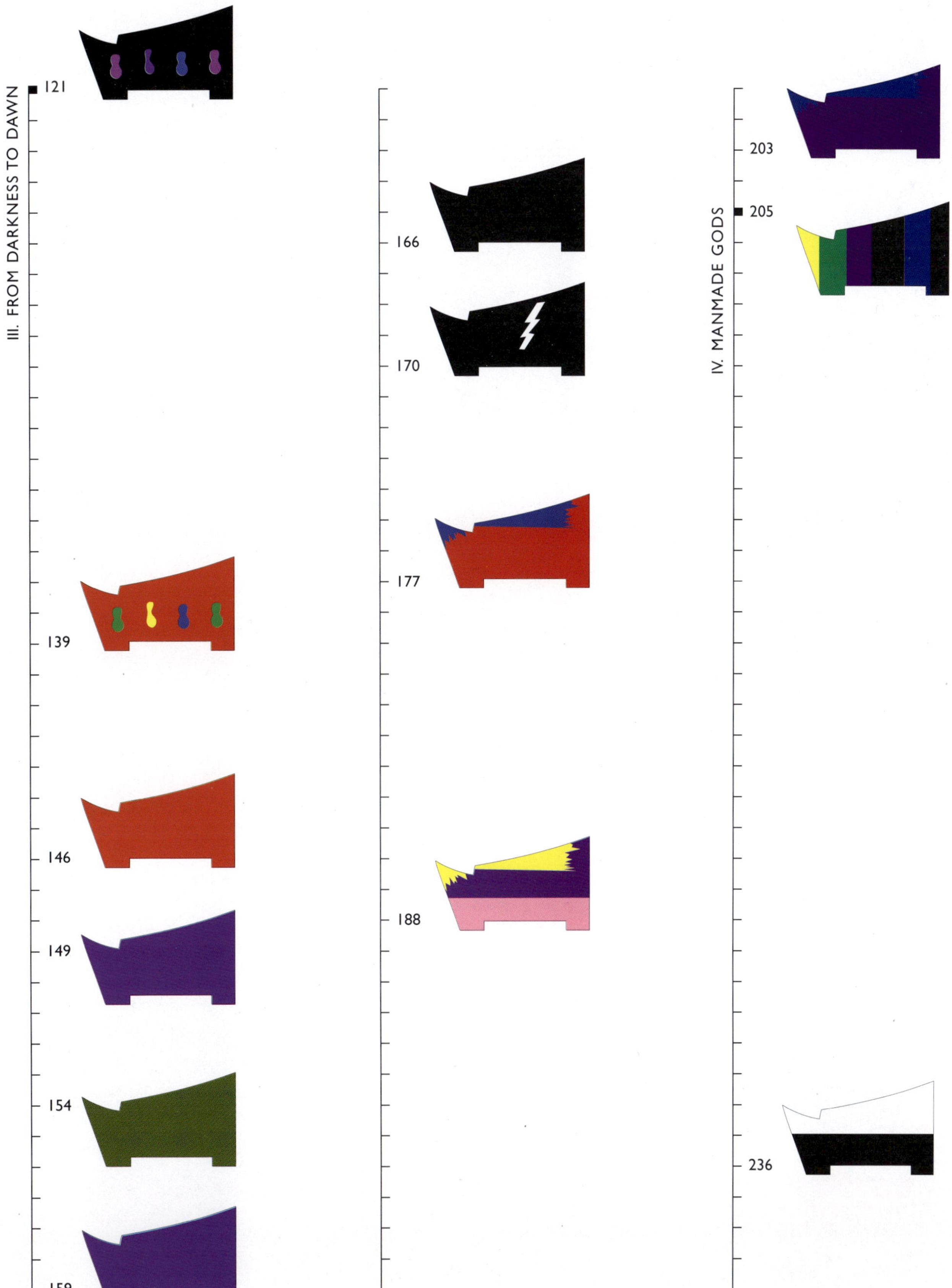
III. FROM DARKNESS TO DAWN
121
139
146
149
154
159
166
170
177
188
IV. MANMADE GODS
203
205
236

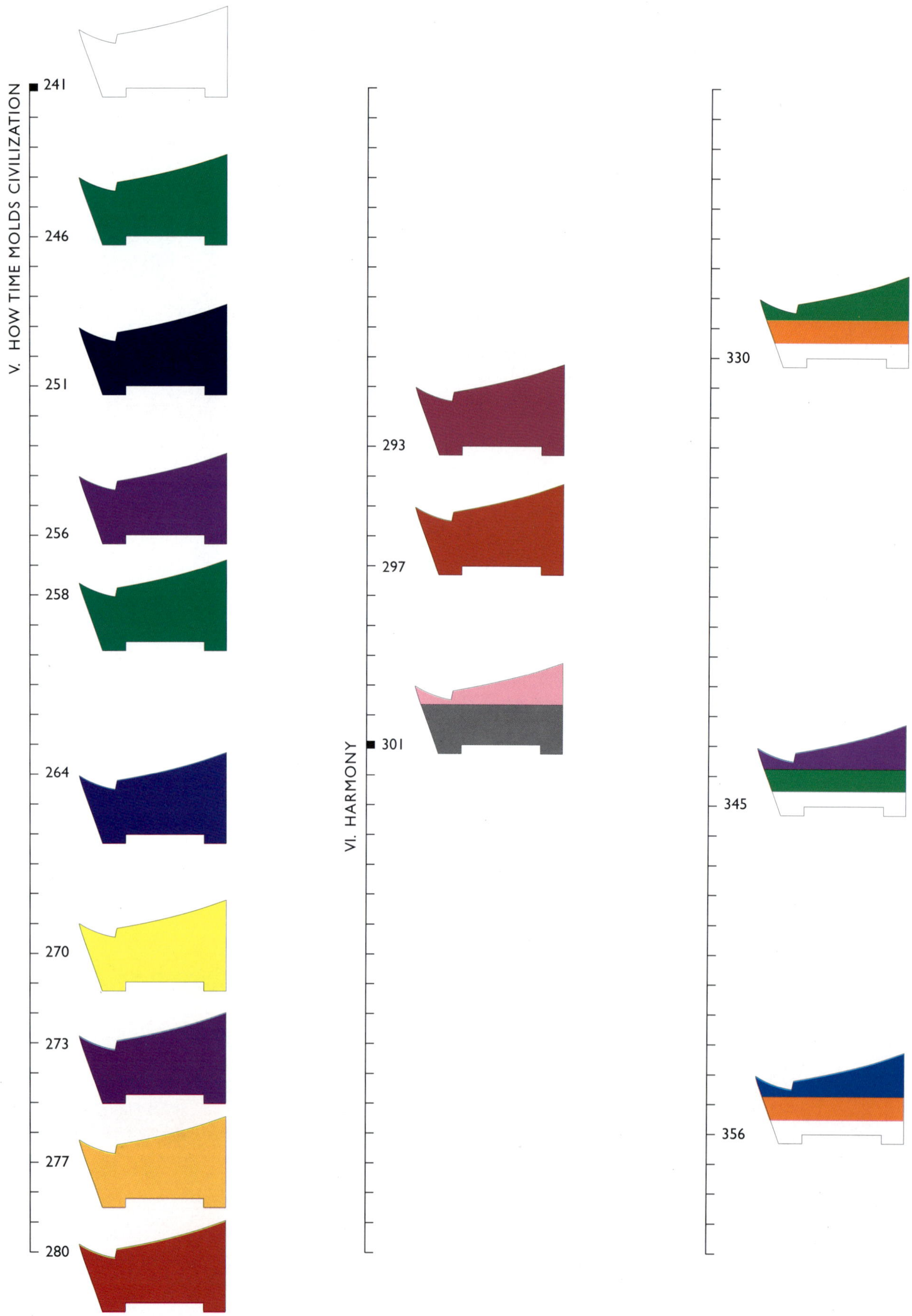
V. HOW TIME MOLDS CIVILIZATION
241
246
251
256
258
264
270
273
277
280
VI. HARMONY
293
297
301
330
345
356

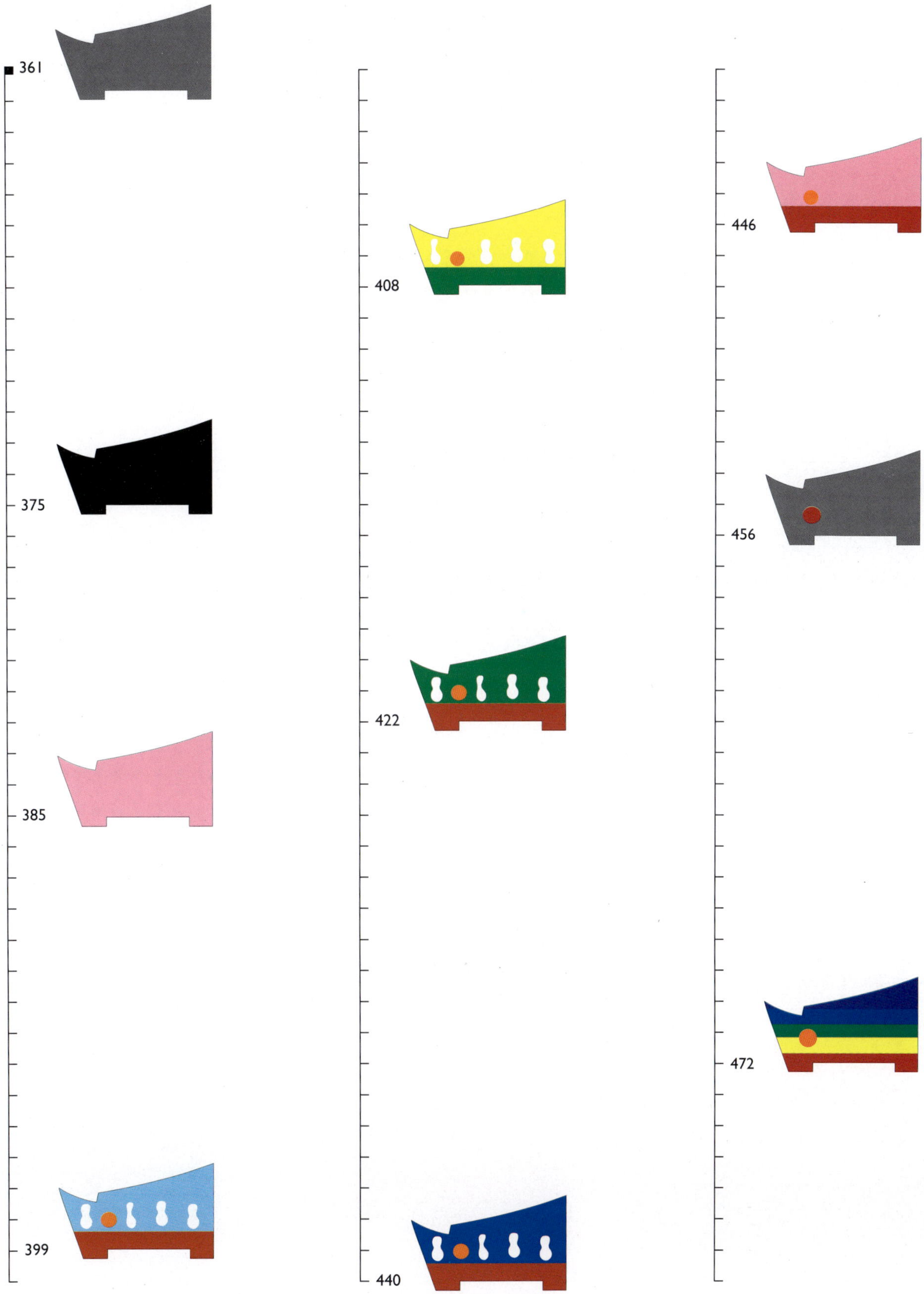
361
375
385
399
408
422
440
446
456
472

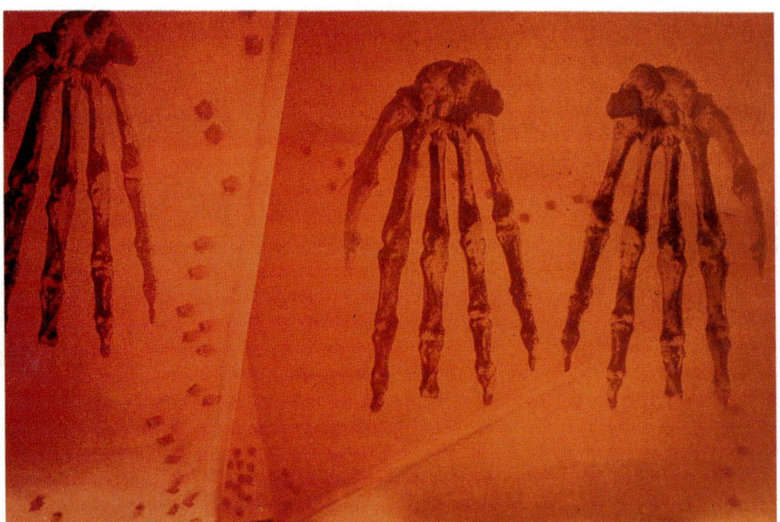

Plate VI
A pink *ambiance*.
[*Getty Center*]

Plate VII
Skeletal hands projected in light; the loudspeakers are distributed along the intersection of shells and over their surfaces.
[*Getty Center*]

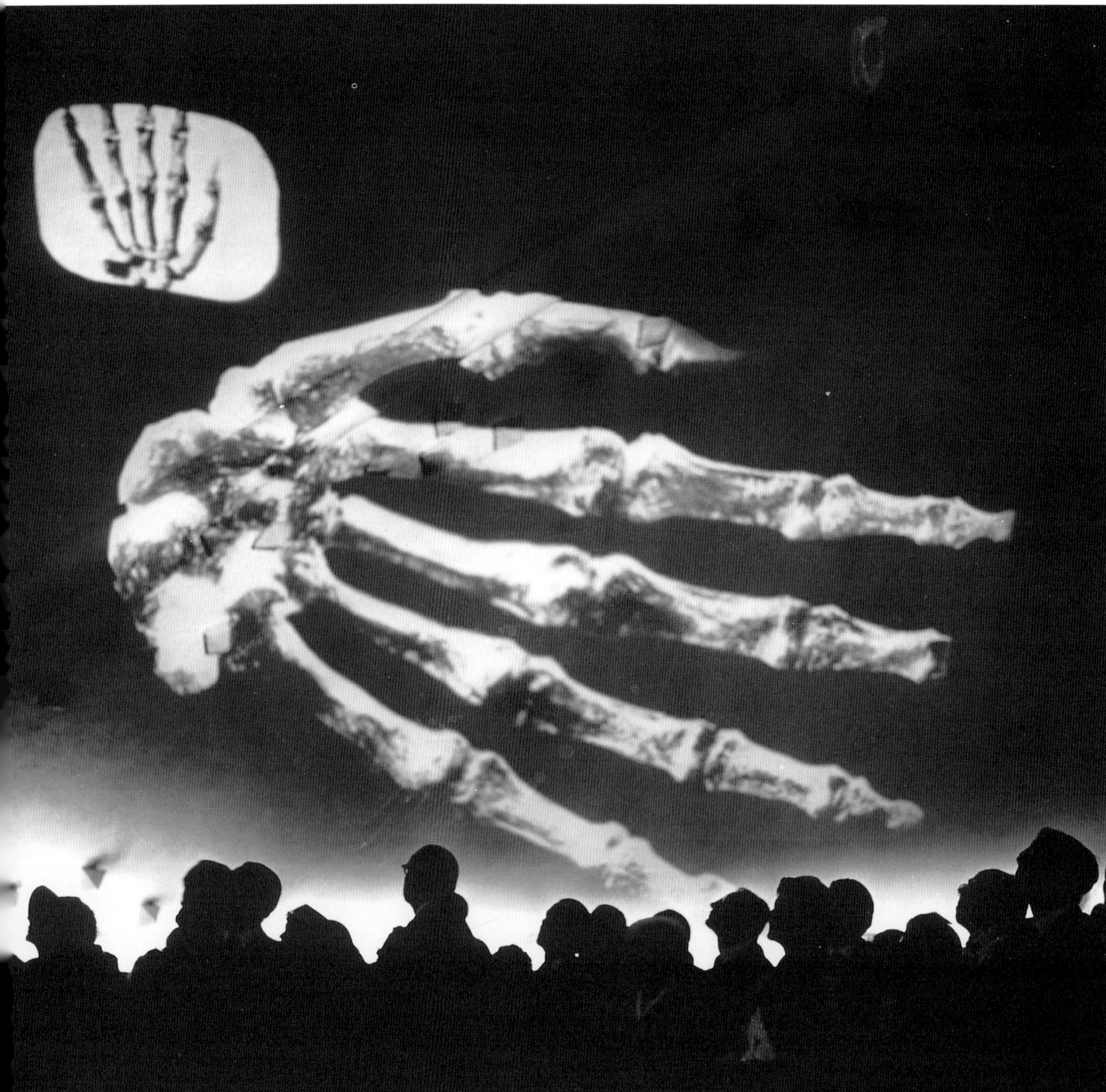

4-22
Skeletal hand, Sequence 3, "From Darkness to Dawn." The smaller image is a *tri-trous* projection. *Poème électronique.* [Philips]

4-23
Head of Christ, Sequence 3, "From Darkness to Dawn." *Poème électronique.* [Philips]

4-24
Nuclear cloud, Sequence 5, "How Time Molds Civilization." *Poème électronique.* [Philips]

4-25
The child as hope, Sequence 7, "To All Mankind." *Poème électronique.* [Philips]

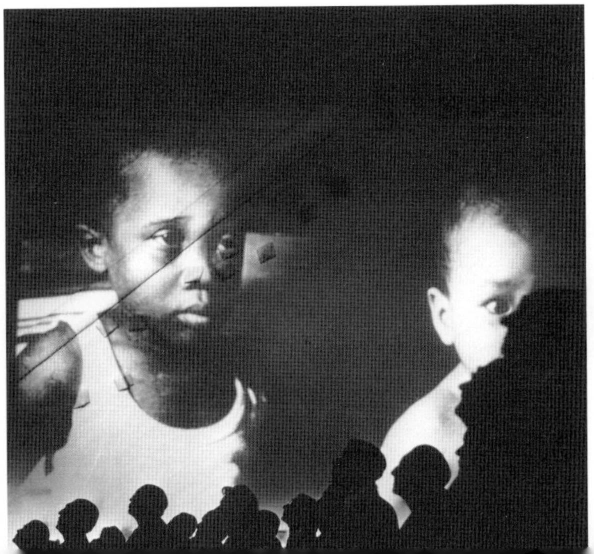

verse elements of the film, their disposition and framing were left in the hands of the film-maker. Agostini created drama within the frame by playing centered, often inversely duplicated, images against objects banished to the extremities of the screen. He thus manipulated the composition within the frame as well as the rhythm of the linkage from frame to frame.

Le Corbusier's familiarity with cinema probably dated from at least the mid-1920s, when his friend Fernand Léger was immersed in his experiments with film. For Léger, "Cinema is the age of the machine. Theater is the age of the horse. . . . Speed was the law of the world," a law that linked aviation and cinema as true expressions of the twentieth century.[66] Léger understood that cinema was essentially an art of fragments assembled to form a new unity. As such it bore only a passing resemblance to reality. Léger's *Ballet mécanique* of 1924 used the medium of cinema as its real content, expressed as the rhythm of mechanical movement generated by the rapid passage of the film's frames. The repetitive swing of the pendulum and the cycle of machine parts were countered by the rhythmical movement of the film: repeating the same sequence, animating static shapes, and playing on the persistence of vision and afterimage [figures 4-26]. The commonplace objects in the film appeared in purist as well as cubist painting. And like their role as elements of a painting, in the *Ballet mécanique* they were the objects rather than the subjects. The subject was cinema.

Le Corbusier's association with the periodical *Plans* from 1930–33 brought him into contact with artists such as the director René Clair and the composer Arthur Honegger, both of whom wrote on cinema. At that moment sound was challenging the conception of the motion picture, suggesting that in the talking film sound could lead rather than follow the story line. For Honegger, the fragmented nature of film instigated a rethinking of music in accord with the montage of images. "One day," he wrote, "without doubt, music will inspire the film."[67] As early as 1931, René Clair raised an issue still relevant twenty-five years later at the time of the Varèse-Le Corbusier collaboration: "If the use of sound must be limited to those normal sounds corresponding exactly to the images which the eye sees on the screen, the enrichment is quite poor, because the mind is quite quick to hear internally that which the senses do not perceive."[68] Le Corbusier's library contained few books on cinema, all dating from the 1940s. An issue of *Rond-Point* from 1945 (found in Le Corbusier's library) proposed the state of the motion picture at 100 years, or "Some Scenes from the Future in the Year 2000."[69] Accompanying images of space travel and discussion of space travel were proposals for a multisensory cinema engaging the senses of smell, touch (through vibration), and sound. The illustrations to Lo Duca's small volume

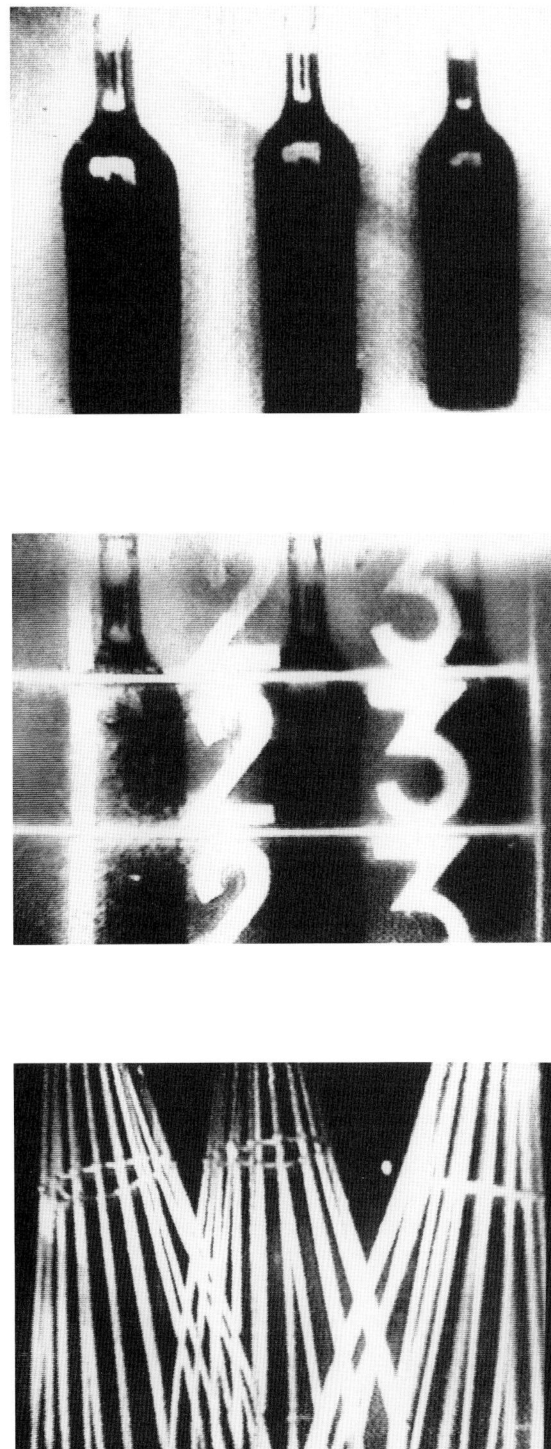

4-26a,b,c
Stills from *Ballet mécanique.*
Fernand Léger, 1924.

4-27
Illustration of cinematic framing.
[*Technique du cinéma*, 1943]

on film-making, also in Le Corbusier's library, presents framing and montage suggestive of Russian cinema [figure 4-27]. Le Corbusier had met Sergei Eisenstein in Moscow in October 1928, which Jean-Louis Cohen described as a "major point in his trip [to the Soviet Union]."[70] The architect inscribed a copy of *The Decorative Art of Today* to Eisenstein, expressing his affinity with the director: "It seems quite clear to me that I think like Mr. Eisenstein when he creates a film."[71] "Cinema and architecture are the only two contemporary arts," Le Corbusier told an interviewer in Moscow, and for his kinetic play of photographs in Brussels, he relied on the lessons that Eisenstein had taught.[72]

The ground rules for cinematic editing had been established in the late teens and 1920s.[73] Whether Eisenstein learned framing from the pictorial art of Asia or merely found the confirmation for his own ideas there, the Russian director believed that the position of the camera's eye in relation to the objects of its view created a visual syntax that was a major constituent in the resultant meaning of the film. Eisenstein wrote in "The Cinematographic Principle and the Ideogram," that in Asian painting a fragment of the object condenses the power of the totality; the fragment can be more evocative than the whole because so much is left to the imagination. Cinema can act in just this way: "Hewing out a piece of actuality with the ax of the lens."[74]

As the construction of the individual image established the significance of each shot, the relation between images established the significance of the film as a whole. Peter Wollen relates Eisenstein's construct, and cites his training in engineering as the basis for seeking the basic cinematic unit. "Science knows 'ions,' 'electrons,' and 'neutrons,'" the filmmaker wrote, "Let there be 'attraction' in art." From the sphere of industry and mechanics, from the practice of "assembling . . . machinery, pipes, machine tools," Eisenstein drew his analogy:

> This striking word is "montage" which means "assembling," and though it is not yet in vogue, it has every qualification to become fashionable. Very well! Let *units of impression combined into one whole* be expressed through a dual term, half-industrial and half-music-hall. Thus was the term "montage of attractions" coined.[75]

Experiments by Lev Kuleshov at the Khanzhankov Studios had graphically illustrated the power of montage to create a new reality. Among the illusions conjured by the cinematographer were the spatial juxtaposition of the Kremlin with the White House and the creation of a synthetic woman from cinematic bits.[76] These experiments vividly illustrated that meaning is not inherent within any individual view, but instead derives from the context established by the preceding and succeeding frames.

In the early years of Russian cinema, an internecine dispute on the theory of montage divided filmmakers into two camps. On one side, Vsevolod Pudovkin—whose work also appeared in *Plans*—saw montage as the linkage between shots, the entirety developing as a continuity.[77] Eisenstein, in contrast, saw the potential of collision as the model for juxtaposing scenes. In this respect, Eisenstein applied to filmmaking what Fernand Léger had concluded in regard to painting: "contrast = dissonance, and hence a maximum expressive effect."[78] There was no reason that any frame should "logically" follow another; the rhetoric of film created its own logic, and montage created its own meaning:

> The science of shocks and their "montage" in relation to these concepts should suggest their form. Content, as I see it, is a series of connecting shocks arranged in a certain sequence and directed at the audience. . . . All this material must be arranged and organized in relation to principles which would lead to the desired reaction in correct proportion.[79]

The director's ultimate intention was to create a synthesis or "mental fusion" within the psyche of the film's viewer. The cinematic state operated on five levels—metric, rhythmic, tonal, overtonal, and intellectual; these were purely physiological, however, and were intended as the means to the psychological end.[80] Eisenstein utilized the overarching model drawn from physics because it still allowed him to use linkage, rephrased as a completely elastic collision:

> Pudovkin loudly defends an understanding of montage as *linkage* of pieces. Into a chain. Again "bricks." Bricks arranged in scenes to *expound* an idea! I confronted him with my viewpoint on montage as a *collision*. A view that from the collision of two factors *arises* a concept. . . . So montage is conflict. As the basis of every art is conflict (an "imagist" transformation of the dialectical principle).[81]

Although Eisenstein acknowledged that his source of images was the real world, through dissection and reconfiguration the new reality was his alone: "I take photographs of reality and then cut them up so as to produce emotions . . . I get away from realism by going to reality."[82] The *Poème électronique* evoked its reality using precisely such means, even if its source images were still and were once removed from the everyday world.[83]

Le Corbusier believed that color affected the viewer both psychologically and physiologically. Like color, the rapid flicker of images created sensations greater than the sum of their constituent parts. The German psychologist Hugo Münsterberger had turned his attention to film shortly after his arrival in the United States to become the head of the department of psychology at Harvard. Münsterberger was fascinated by the new world of perception created by cinema, and studios were pleased to have their business of entertainment regarded as worthy of academic enquiry. In his classic study from 1916, *The Film: A Psychological Study*, Münsterberger first investigated the way in which image and reality were perceived by the deceiving mechanism of the motion picture camera. Of course, there was no motion in any of the frames, but in their combination, "*the motion which [the viewer] sees appears to be a true motion, and yet it is created by his own mind.*"[84] Thus, the motion in film—and the world it conveyed—were just a psychological construct.

While Münsterberger missed the mark when he predicted that sound would ruin the purity of the cinematic project, he has been proven correct in his assessment that montage is the most powerful of cinema's tools. He realized that cinema possessed the power to move both forward and backward (the flashback or cut) and to move inward and outward (the close-up):

> The detail which is being watched has suddenly become the whole content of the performance, and everything which our mind wants to disregard has been suddenly banished from our sight and has disappeared. The events without have become obedient to the demands of our consciousness. In the language of the photoplay producers, it is a "close-up." *The close-up has objectified in our world of perception our mental act of attention and by it has furnished art with a means which far transcends any theater stage.*[85]

Later in the essay, Münsterberger concluded that "the close-up heightens the vividness of that on which our mind is concentrated."[86]

By assembling through montage fragments of our visual and psychic world, Le Corbusier, Agostini, and Petit could lead the visitor to the Philips production into a worldview that he or she had never witnessed. It was an edited world based on a moral vision of the apocalypse that technology might bring. It also offered a way out of the malaise: through planning and architecture, or more specifically, planning and architecture by Le Corbusier. This world was first the product of the photographer, and thereafter of the filmmaker.

Le Corbusier had earlier used the worlds revealed by the camera in his photomural for the Swiss Pavilion in Paris of 1931. He selected microscopic patterns from nature, arranging them in a black-and-white mosaic of photographic panels mounted on the curving wall. He continued these investigations with collage in the exhibitions on the city and dwelling displayed within the 1937 Pavillon des Temps Nouveaux, and in 1938, with the cover image for *Des canons, des munitions? Merci! Des logis . . . s.v.p.* Le Corbusier's use of photomontage allowed him to mix realism and abstraction, creating new pictorial ensembles by combining in a single credible panel radically differing scales and points of view. The photomontage drew upon the *papiers collés* developed by Picasso and Braque, who had incorporated real elements such as newspapers and decorative papers into their paintings. Le Corbusier, on the other hand, tended to use the technique for propagandistic purposes, although these were mild in comparison to the tough anti-fascist works by the German John Heartfield.

Not incidentally, *montage* describes in French both the assembly of images and the cutting of films. Artists in both media employ essentially the same technique for essentially the same purpose. But photomontage derives its power from the novel juxtaposition of heterogenous images in space, while cinema juxtaposes images in time. More like writing than theater, the film evokes in the mind of the viewer a new reality through temporal dispositions. In this sense, the *Poème électronique* was literary and well termed a "poem."

The film that constituted the electronic poem was essentially an animation of still images through cinema. There was no motion but that created by cinematography. By rotating objects, dollying, or moving in or out on the subjects, by flashing from positive to negative images, the relationship between the viewer and the object was radically altered. This effect was difficult to attain in 1958 without the varying focal length zoom lenses of today or computer manipulation. As an image grew on the curving walls of the pavilion, the effect was that of the object's coming toward the viewer, gradually dominating the space. The possible psychological disturbance caused by an irregularly shaped space was mitigated by collaging repeated iterations of an image, distributing illuminated images—set against a black background—around the space. The peculiarity of the architecture dissolved into the cataclysm of image and color. Two sets of projection equipment displayed identical presentations, although the varying, asymmetrical form of the pavilion destroyed any of the formality and rigidity that usually characterizes symmetrical imagery.

In formulating his ideas for the *Poème électronique*, Le Corbusier listed the fairly abstract phenomenon of rhythm with the more tangible categories of color, sound, and light. His interest in rhythm may have developed from his brother Albert Jeanneret's work as an instructor of "La Rythmique" (Eurhythmics), based on the teachings of Émile Jaques-Dalcroze. A Swiss who had begun his work in body movement in 1905, perhaps with inspiration from Isadora Duncan, Dalcroze believed in the correlation between body and mind. To effect a healthy harmony between the two, one worked with the growing child who moved from the immediate haptic sphere to a world in which the sense of rhythmic integration was lost. "The consciousness of duration," wrote Albert Jeanneret in an article introducing La Rhythmique in *L'Esprit Nouveau*, "and its relation to place" must be established in the child.[87] In

1911 Dalcroze established an institute for the study of rhythm in the garden city of Hellerau, Germany, with buildings designed by Heinrich Tessenow. The theater there, white and clean, witnessed spectacles that integrated movement, word, and music: for example, the 1913 production of Gluck's *Orfeo* designed by Adolphe Appia.[88] When the war approached, Dalcroze relocated to Geneva and continued to proselytize for his work in Germany, Austria, Scandinavia, and Spain.[89]

Although trained as a musician and composer, Albert Jeanneret worked as an instructor of La Rythmique, and his advertisement for courses—accompanied by a sketch of dancing figures clad in tunics—appeared in the early issues of *L'Esprit Nouveau* [figure 4-28]. While the description of the method is rather vague, the advertisement claims that "the 'rhythmatician' will be armed for pursuing the important field of musical studies, because he will acquire the essential foundation." This was critical, as Jeanneret explained in 1920: "France, we are certain, will not be late in generalizing a method appropriate to the Latin mind, and one that makes concrete the natural tendencies of the race toward measure and rhythm."[90]

Not only did rhythm affect body movement and musical composition, however; it was fundamental to the arts as a whole. "We hold the belief," Jeanneret wrote, "that it is with the study of body movement that we must begin, independently of all freedom of expression; it is to the study of the play of values, or relations, of combinations, and in a word, of *abstraction, emotion fundamental to the artist*, that it must be linked."[91] Here, a connection to film can be discerned.

Albert Jeanneret contributed to *L'Esprit Nouveau* on a fairly regular basis during the early 1920s. In a review essay of Charles Koechlin's article "La Musique, plaisir de l'esprit, ou jouissance sensuelle," Jeanneret questioned whether music was an art or a science. He proposed that it was both, and that to produce music required both a mind and a heart. "[Music] is a science by the same laws of physics on which it is based. Acoustics, intonation, timbre, dynamics, etc., all of these: the givens that intelligence values. . . . Form: the frame assigned to an idea. Just that, no mystery."[92] But it is also an art.

The cellular aspect of the motion picture, according to Arthur Honegger, would precipitate a major rethinking of music and its image. Music had at first covered the sound of the projection apparatus, thereafter served as an accompaniment to the film. In the third stage, however, it was the nature and rhythm of the montage that would suggest the character of the film—and perhaps vice versa.[93] Honegger cited Mickey Mouse cartoons as leading exponents of this phenomenon; Sergei Prokofiev's collaboration with Eisenstein on *Alexander Nevski* provides a more sophisticated example. Honegger's

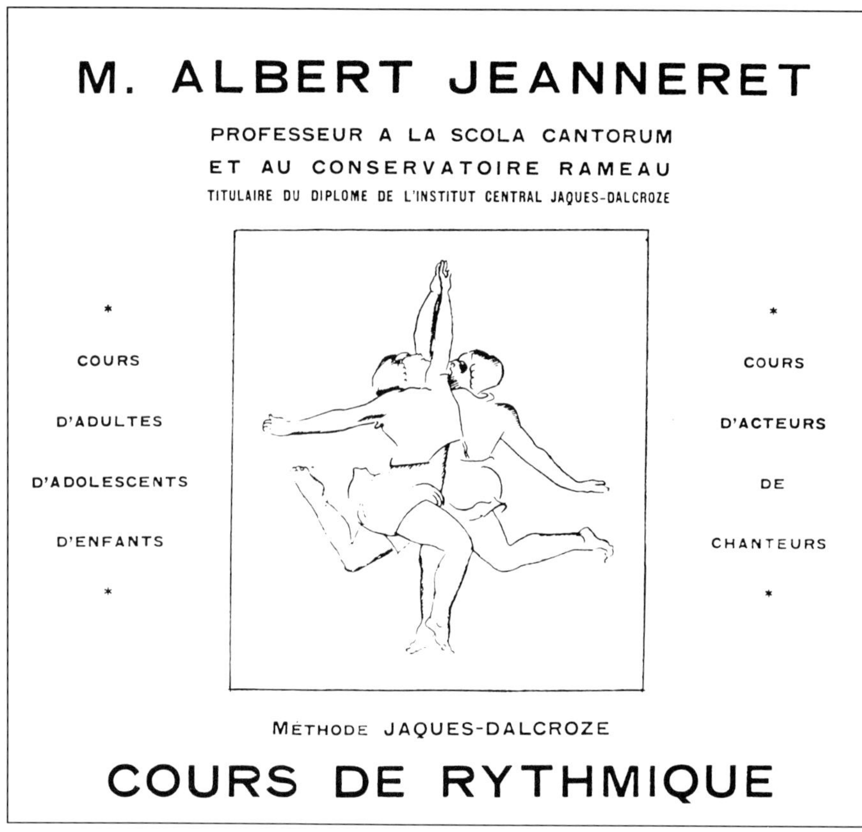

4-28
Albert Jeanneret's advertisement for his course in Eurhythmics. [*L'Esprit Nouveau*, 1920]

1924 *Mouvements symphoniques* was more commonly known as *Pacific 231*, a classification of steam locomotive. Honegger deplored such a literal reading of the work, claiming that "in *Pacific* I pursued an abstract and quite ideal notion by giving the impression of a mathematical acceleration of the rhythm, then of a movement that stabilizes itself on its own."[94] Although the composer claimed that the title was added whimsically, and after the fact, the associations have endured. In 1946, the piece became the basis of a film—about steam locomotives and the railroad.

Varèse had also extended the definition of rhythm, noting that it was "too often confused with metrics":

> Cadence, or the regular succession of beats and accents, has little to do with the rhythm of a composition. Rhythm is the element of music that gives life to the work and holds it together. It is the element of stability, the generator of form. In my own works, for instance, rhythm derives from the simultaneous interplay of unrelated elements at calculated, but not regular, time-lapses. This corresponds more nearly to the definition of rhythm in physics and philosophy as a succession of alternative and opposite or correlative states.[95]

If Le Corbusier had established that the house could be defined as a machine for living, his older brother also established a connection with the machine in relation to music, and beyond that, to the cinema. Reviewing the year 1921 in music, he opened with the ultimate modern metaphor:

> The work par excellence of our time is the machine: the machine to fly, the machine to record, the machine to reproduce, by sound and by image, the machine with which to talk. The very condition of the machine is movement. The machine succeeds in moving regularly, with security. Well regulated, it becomes sensitive to and produces movement at all degrees of speeds and delay. The sprinter slows down to regain his speed, in order to run. The cinema slows down in order to analyze, and speeds up to create intensity by a rapid succession of images.[96]

The cinema constituted the means to educate the eyes of our times, but the products to that point had not truly exploited the potentials of the medium, particularly those of speed and rhythm: "Ridiculous example, excellent method." Rhythm provided the means for structuring visual poetry: "Film frees the dream from the laws of gravity."[97] Rhythm was central to the creation of the *Poème électronique*, and like its architecture, challenged the same laws of gravity.

In the interval between the reality of film image and the abstraction of illuminated colored washes of the *ambiances* were the *tri-trous*: "At certain parts of the wall around the 'écran,' brilliantly colored beams of light or figures in black and white are occasionally projected. This is done by means of a film strip each frame of which is entirely opaque except for three holes ('trois trous') in which the colours or figures are introduced."[98] This technique allowed bands of colors and certain shapes to be projected [figures 4-29, 4-30, 4-31]. In addition, flat shapes such as a "red sun, a moon, stars and clouds"[99] possessed the luminous intensity to override both the ambiance and the *écrans*, creating a projection that pulsated within the space. Le Corbusier had discussed with Kalff the device of the *tri-trous* only in November 1957, but had stressed its importance. Kalff explained the technique as "projected images, mostly black and white, but sometimes with simple colored filters, appearing and disappearing on three places on each side of the pavilion"[100] [figure 4-32].

A note in Le Corbusier's sketchbooks, later used as the basis for discussion at a meeting with Agostini and Petit, established the conditions for executing the *tri-trous* and the way in which the technical problems of color addition and light intensity could be overcome. To some degree the constraints imposed by the characteristics of projection determined the positioning of the *tri-trous* in relation to the film images:

> *If one desires sharp projection, the ambiances must be of a dark color.*
>
> *Therefore, you should use images only when ambiances are dark:*
>
> *red*
> *blue*
> *green*
> *The tri-trous are possible only on a dark ground.*
>
> Therefore, *for each tri-trou, find an area that already possesses a shadow.*
>
> *Therefore, no [film] image on the screen during the bright ambiances, because the six 16mm projectors for the tri-trous can be located behind the barrier at the position at which each can be effective.*[101]

The technology required for these effects was not particularly advanced—projectors and "stencil" slides—but coordinating the two sets of projection equipment tested the limits of the available automation [figure 4-33, 4-34]. Because the two projections were thrown from opposite sides of the pavilion, their intersections would have distorted their hues, as colored light is additive. Kalff and his collaborators solved the problem by using intensities as low as possible, avoiding large areas of white, which would have wiped out any other chroma, and by choosing "suitable combinations of colors."[102] By

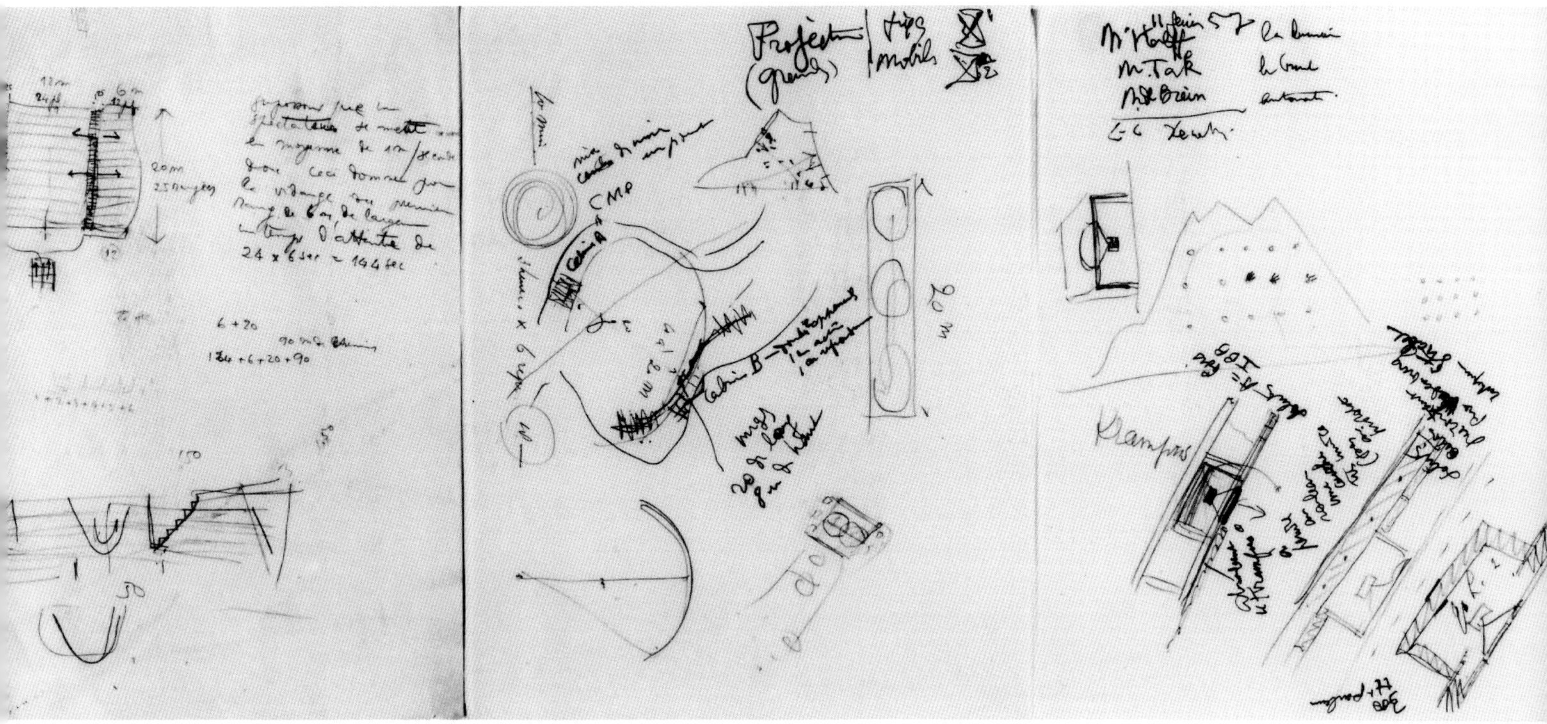

4-29

Sketchbook studies for projections and sound. 11 February 1957? Le Corbusier. [FLC; ©1996 ARS, New York / SPADEM, Paris]

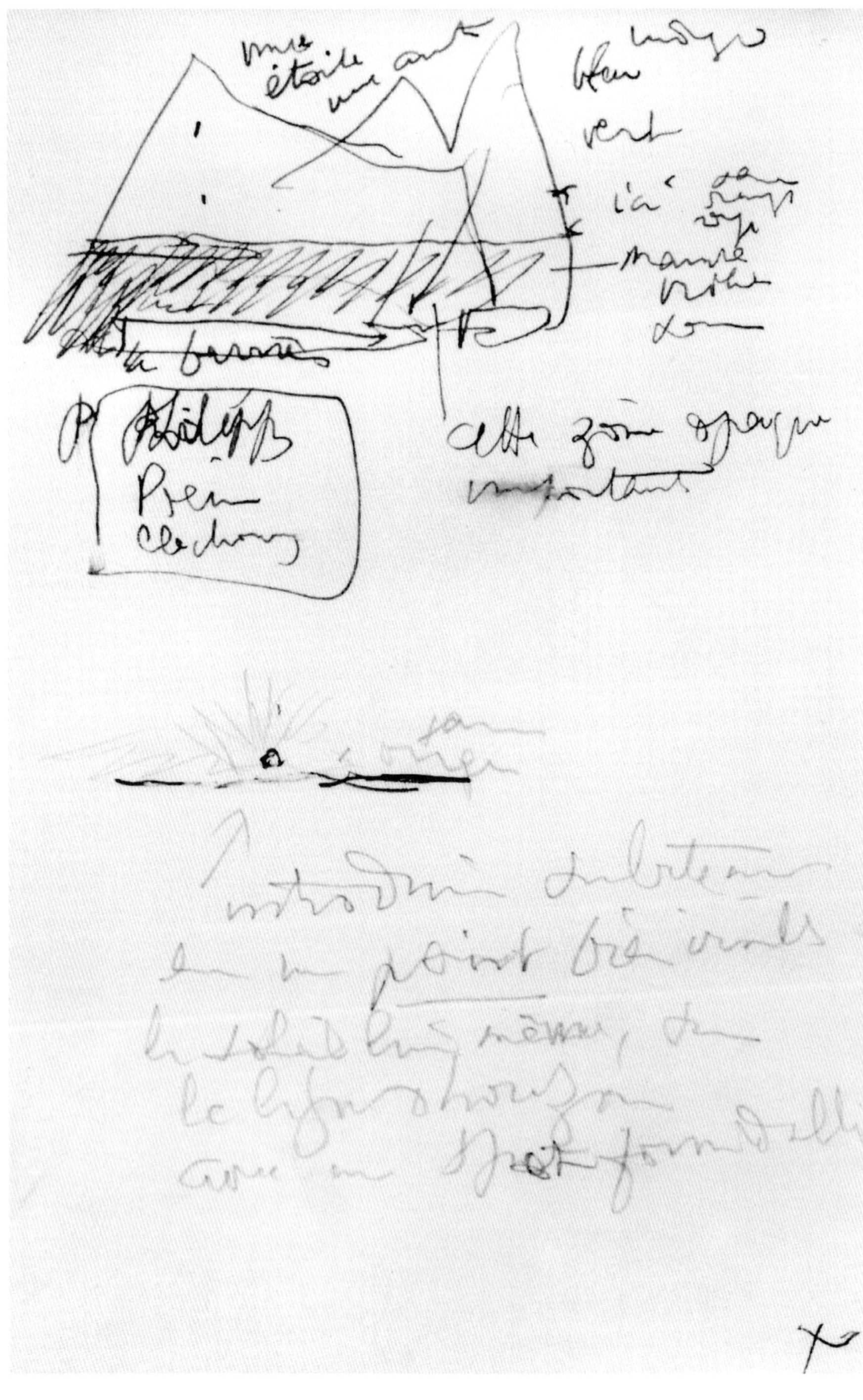

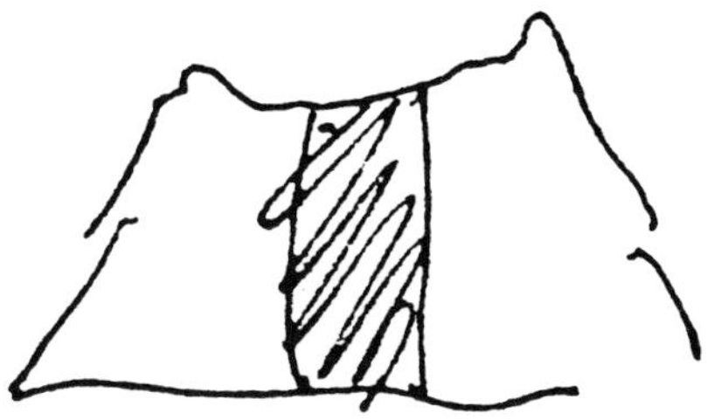

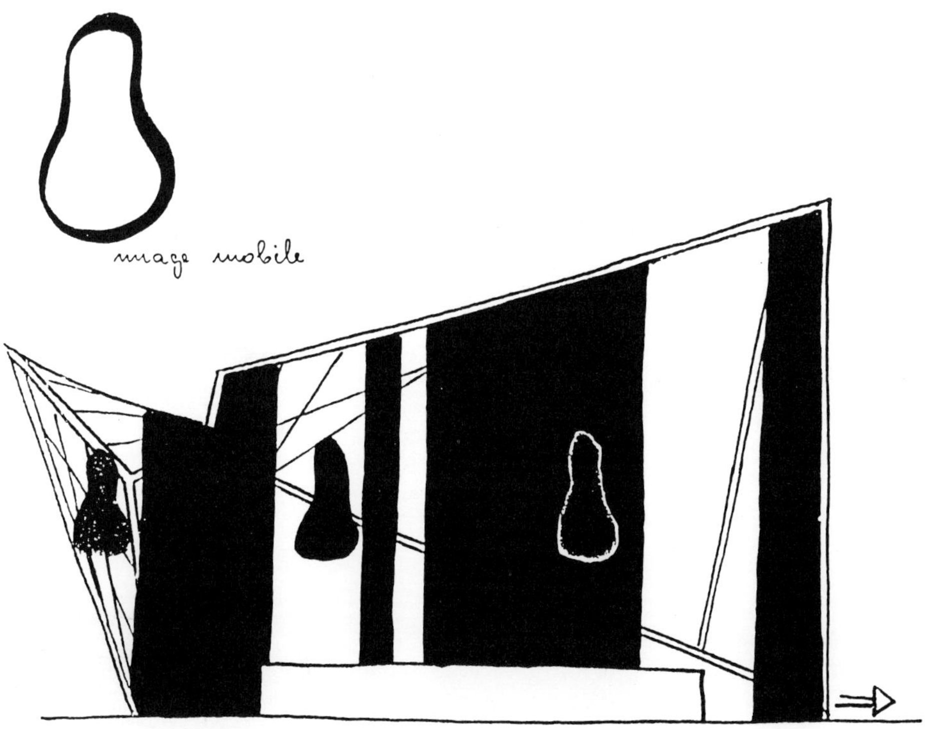

4-30
Sketchbook study for the colored illumination.
Le Corbusier notes: "Suddenly, introduce the sun itself as a [precise?] dot on the horizon line with an intense spotlight."
[FLC; ©1996 ARS, New York / SPADEM, Paris]

4-31
Sketchbook studies for the light projections. (*tri-trou*).
Le Corbusier.
[FLC; ©1996 ARS, New York / SPADEM, Paris]

4-32
Ambiance from Sequence 1, "Genesis." Alternating bands of black and white with projected images of "clouds." Gouache colored scenario page.
Le Corbusier.
[Philips]

employing noncomplementary colors, the possibility of inadvertently creating white light was reduced.

The list of equipment finally used included four cinema projectors, two for the film and two for the *tri-trous*[103] [figure 4-35]. Mirrors directed the light beams of the *tri-trous* to their desired destinations. Four projection lanterns, with rotating disks equipped with colored filters activated by servomotors, paired with fluorescent lights, created the ambiances. An additional two projectors contributed the cloud effects, and a final pair projected the sun and moon. Six more spotlights—near the peaks of the structure—added to the colored ambiances, as directed light sources. Four lightbulbs fluoresced the volumes of the mathematical object and the female figure suspended from the peaks. Fifty "stars" were scattered over the upper limits of the space, and by switching them on and off, a twinkling effect resulted. In addition to these projectors and lights, forty groups of five fluorescent lamps in white, red, yellow, green, and blue, mounted behind the barrier, could be regulated to create the illusion of a bright horizon sunrise. All the controls for the lamps, like the sound system, were coded onto magnetic tape. In addition to the aesthetic lighting, emergency illumination and secluded bluish white fluorescents were mounted on the underside of the barrier wall to demark the limits of the space and to aid the visitors in entering and exiting.[104]

Painted glass plates, now in the Special Collections of the Getty Center, may be the remnants of an early attempt to project either distinct shapes or the more atmospheric *ambiances* [figure 4-36]. Arranged as an arc, the colored shapes could have rotated sequentially before the projector lamp, spraying them as sharp [or diffused, depending on the lens] tints on the pavilion walls. No documentation records either their intended use or their maker, and they remain a bit of a mystery. In any event, this technology was discarded.

One might question the role of the *tri-trous* projections within the overall scheme of the electronic poem. Without doubt, the film provided the most lucid statement of the architect's narrative, rooted as most of its images were in the recognizable world. The colored ambiances tinted the visitors' perceptions of the moving images and their changing psychological dispositions. At a perceptual level the *tri-trous* projections were just another type of image: abstract, flat, intense in light and hue. Their power to override the narrative of the *écran* images suggests ties with Le Corbusier's mural paintings, which developed as an extension of his easel painting.

Early in his architectural career, Le Corbusier tended to treat the wall as an integral unit, painted only with solid colors. In fact, he took issue with Dutch artists such as van Doesburg, whose painting animated the wall surface by fracturing it into several shapes of varying perceived depths. By the late 1930s, however, Le Corbusier's attitude toward the wall had changed. Perhaps

4-33
Partial section through the pavilion showing the parapet wall used to screen the equipment, the projectors, and low-frequency loudspeakers. [*Philips Technical Review*]

4-34
A view of the equipment behind the parapet wall. The large floodlights were the primary vehicle for creating the *ambiances*. [Philips]

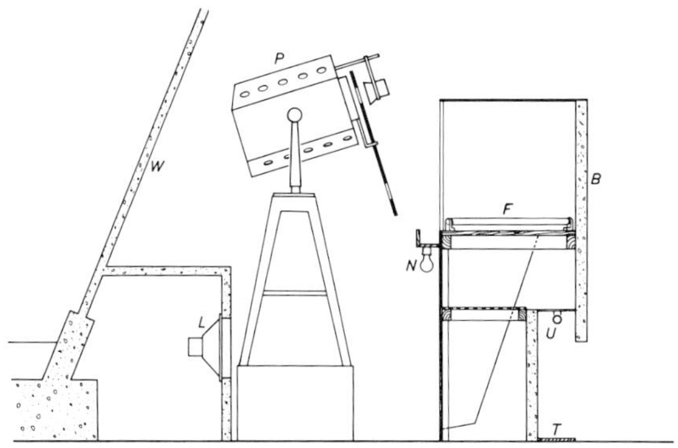

P
W
B
F
N
L
U
T

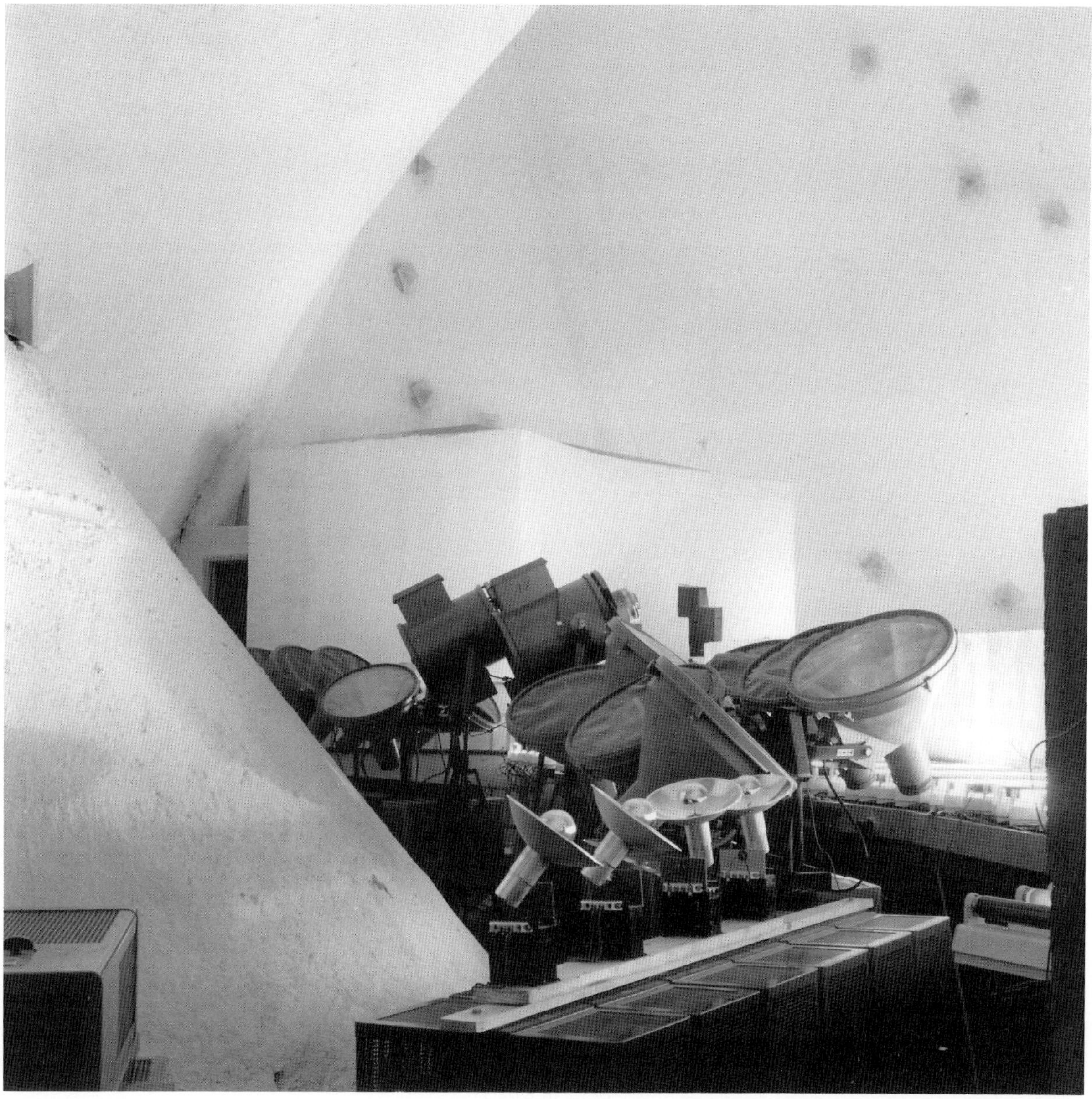

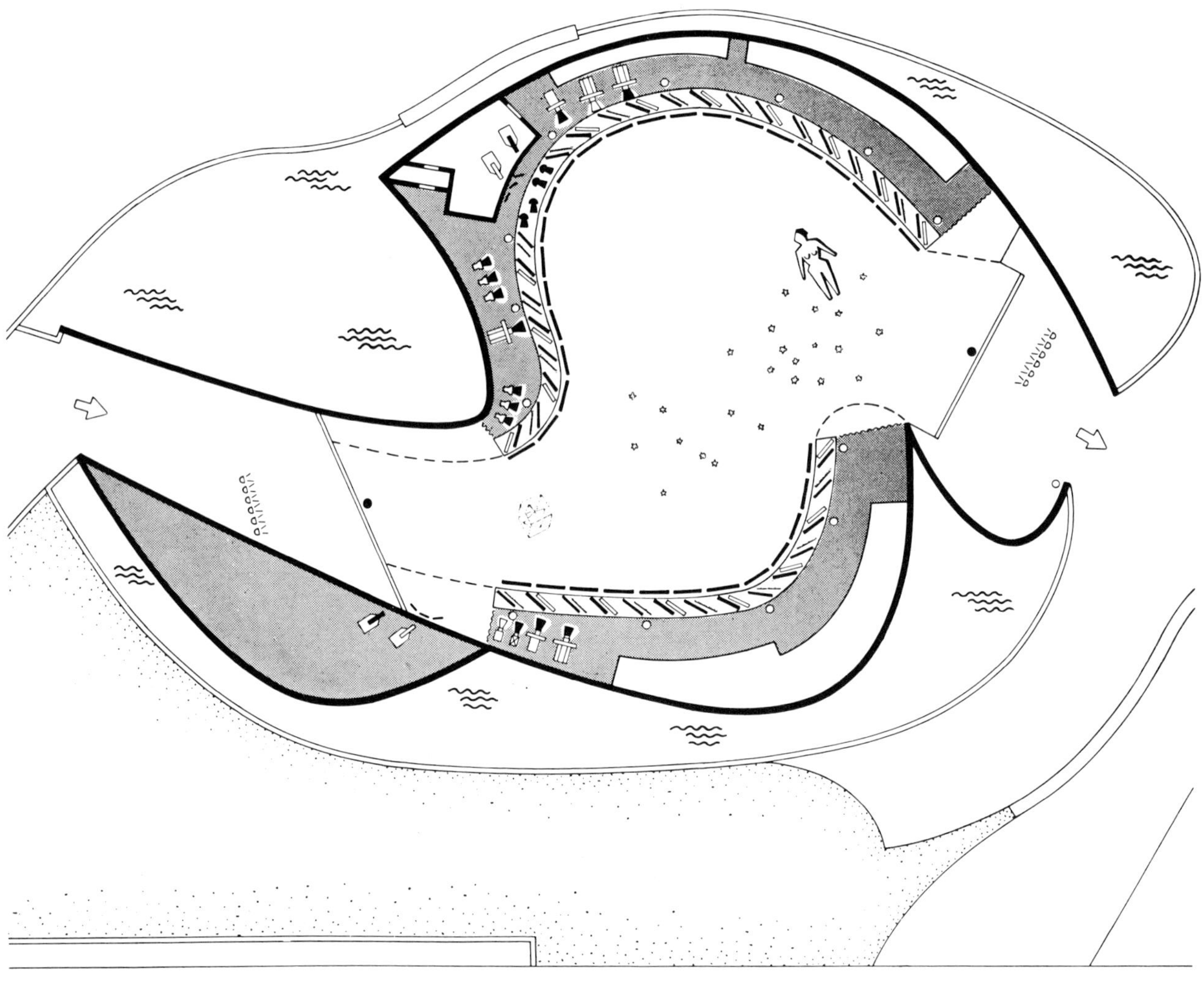

4-35
Plan of the pavilion showing the location of all light and sound equipment.
[*Philips Technical Review*]

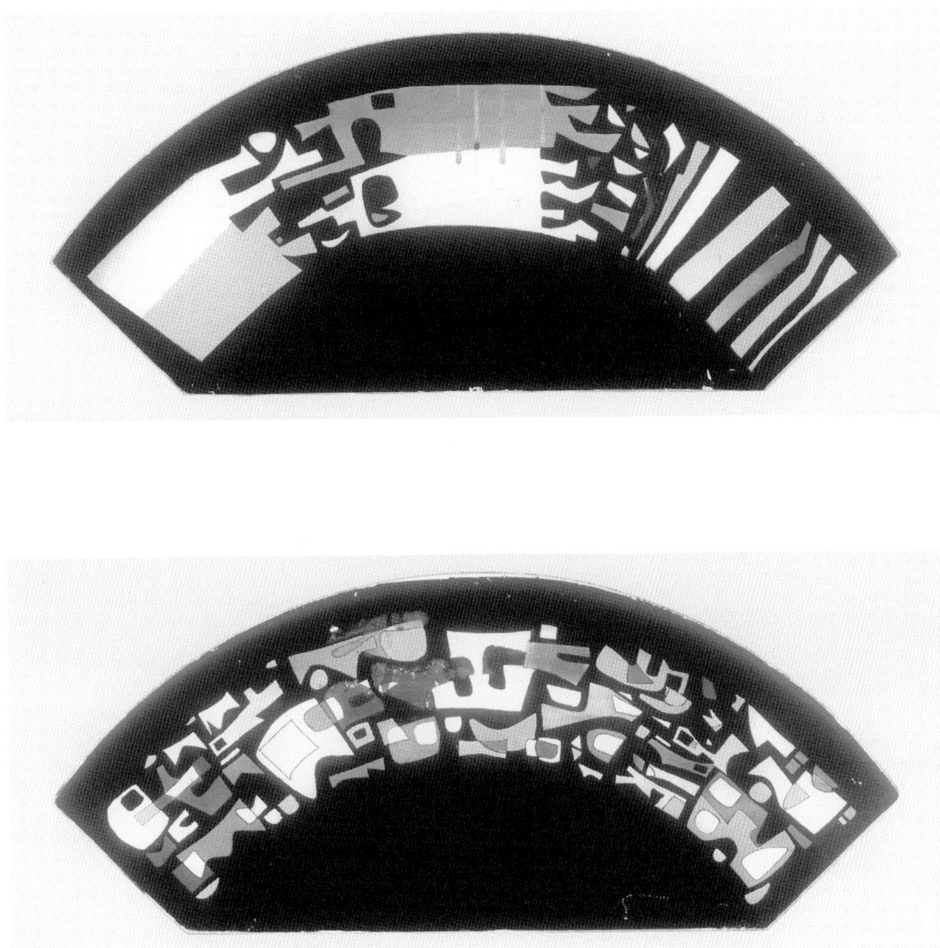

4-36
Painted glass plates which may be a study for the colored *ambiances*. Origin and source unknown. [Getty Center]

due to his developing ideas of a synthesis of the arts, and the role that painting could play within architecture, he executed a series of mural paintings for villas in the south of France (one of them being the house designed by Eileen Gray and Jean Badovici at Roquebrune).

The question of the wall had occupied Le Corbusier's colleague Fernand Léger by the early 1930s. Even the architect himself had admitted that "an entirely white house is a pot of cream"; color must have a presence in modern architecture, but of what sort? "My architect friends," Léger wrote, "we should be able to get together about the wall. You want to forget that painters are put into this world in order to destroy dead surfaces, to make them livable, to spare us from overly extreme architectural positions."[105] Léger chided the architects: "But who is responsible for dispensing color? We are. You've tried to do it, you have made planes seem to be in motion through the addition of tones. That is no longer the question. Color demands more than that, it is for us to activate color, working in close relationship with you."[106]

Painted several years later, Le Corbusier's murals suggest that the architect and the painter were by that time in complete accord. Le Corbusier noted that "if the place is suitable, call in a painter and ask him to register here his plastic thoughts and at one blow open all the doors to the deep land of dreams—especially there, where there is no real depth."[107] Le Corbusier would not paint just any walls, however, as he explained a decade after the fact: "The walls chosen to receive nine large paintings were precisely the most dull, the most insignificant. In this way, the beautiful walls have remained and those which were neutral have become interesting; and the sparkling white of the former brings out the colors of the latter." He added that "the full range of [perceptual] deformations [of the wall] until collapse or dissolution is available."[108] With ten years, the projections—both static and animated—would efface almost completely the racked planes of the Philips Pavilion.

Complementing the film and *tri-trous* projections were the defined or diffused colored illuminations "intended to conjure a particular atmosphere. They were produced by projecting colored light on the pavilion walls; long bands of color accentuated the peculiar shape of the building."[109] The *ambiances* shared a role of importance equal to that of the image in producing the intended emotional state. They not only tinted the walls, and thus the film, with varying colors, but in shifting direction, intensity, and gradation, the illuminations actually conjured varying spatial perceptions of the interior itself. The effects had precedents in several of Le Corbusier's buildings, principal among them La Tourette and Ronchamp. At the monastery's chapel, light "canons" illuminate the subsidiary altars whose ceilings and walls have been painted in saturated colors. The interior of Ronchamp, on the other hand, demands punctuations of light and color to animate its darkness. To create the desired effects at Brussels, banks of fluorescent lamps were paired with projected images.

The ambiances could be divided in two general groups. The first provided an overall chromatic tint to the interior to affect the psychological mood of the viewer. Absolute darkness could be included in this category, and it was effectively used to signal pauses in the sequences or to return the film images to a pristine black-and-white. There is some evidence, however, that Le Corbusier conceived the interior of the pavilion as a celestial space. Indications for the ambiances noted a dawn or a sunset and included references to *tri-trous* projections that recall the sun, moon, and clouds. In addition, the lightbulbs on the shell's apex twinkled like stars [Plate IV]. Given these abstract yet realistic references, the poem can be read as images from the human world overlaid on a celestial surface, or as the terrestrial world set within the nebulous.

For Le Corbusier, color was a basic aspect and requirement of human life. He agreed with Léger that "[c]olor is a human need like water and fire."[110] *L'Esprit Nouveau*, which Le Corbusier co-edited with Amédée Ozenfant in the early 1920s, published several articles on color theory, among them a rather technical essay, "La Lumière, La Couleur et La Forme," in which Charles Henry proposed a "psycho-physical" relation between color and human perception.[111] Early in his career, the architect acknowledged the "danger" of color: too dynamic, too sensual, too filled with memories, too close to the interior sources of emotions.[112] To counter these subjective tendencies, Le Corbusier applied objective methods, that is to say "a color scale of definite and constant expression, and one logic of use."[113] But this stance changed over time, particularly at the end of the 1930s when color received an increased prominence in his painting. To increase the architectural presence of the canvas Pavillon des Temps Nouveaux, for example, Le Corbusier used planes of saturated colors—purple on the walls, gold on the ceiling, green and blue walls: "The sun, from humble gravel, made a beautiful beach of yellow ochre."[114]

Given the psychological as well as physiological properties of color, memory greatly influences chromatic perception and links phenomena in the real world with the purely physical aspects of chroma. Early on, for example, Le Corbusier associated blue with elements of the environment such as the sea and the sky, and more metaphorically, to the elongated line of the horizon. In his early paintings Le Corbusier avoided the primary colors of the Dutch de Stijl artists and sought instead colors he believed more in accord with the visual experience of nature.[115]

In 1931, the Swiss wallpaper firm Salubra had commissioned Le Corbusier to develop a "keyboard" of harmonic colors.[116] Organized along a "horizon," each color of the suite was named after a natural phenomenon or element: "sky, space, sand, landscape, velvet, wall," underscoring the important connection to the real world. These ideas evolved with time, however, so that the colors Le Corbusier applied to his buildings during the 1950s were far purer in hue than those of paintings from his purist period. In the ambiances of the Philips Pavilion, however, he again associated color and astronomical and natural phenomena such as dawn, night, stars, and clouds. At times the ambiances reinforced accidentally the black-and-white images on the wall, at other times they offered an emotional counterpoint. Like the patches of color detached from the drawn subjects of his later paintings, colored light and projected image within the pavilion floated somewhat independently of one another—not upon a flat picture plane, but within a twisting space. The synthesis that resulted issued from fortuitous congruence rather than design, although it was the architecture of the pavilion that ultimately joined together the many layers of the *Poème électronique* as an entity.

Rhythm and montage were critical ingredients for creating the poem's cinematic component; concerns for light and color were necessary for fusing the elements into a synthesized entity. Light and color became the most advanced media for theater design late in the nineteenth century, as the stage underwent significant reforms. Under the sway of the Wagnerian opera, the young Adolphe Appia (1862–1928) had proposed that movement on stage should be the child of the music. Rather than overwhelming the viewer with elaborate staging that diminished the presence of both the actor and the music, the stage design should be simplified and rendered ethereal. Within this context, the "actor's body itself [could] be seen to occupy and require space rhythmically—that is, to be engaged in active and living movement—which could be perceived and measured in terms of the static objects around him."[117] Eurhythmics was one means by which this could be achieved; light was a second.

Appia, whose work Le Corbusier had studied while still living in La Chaux-de-Fonds,[118] introduced himself to Emile Jaques-Dalcroze by letter, believing that the two men shared attitudes and knowledge which could be mutually beneficial. Richard Beacham notes that only under Appia's influence did Dalcroze realize that "what had begun as training in musical sensitivity had vast implications for theatrical reform, an area which, initially, had held little interest for Dalcroze."[119] Sensing its potential power, Appia insisted: "Eurhythmics is not an art form—I want to shout that from the rooftops—but a path toward art."[120] "Light, music, and movement could be carefully coordinated and integrated to create subtle and expressive effects which previously had been virtually unimagined, much less achieved."[121] Color passed from the perceptual to the psychological.

As Albert Jeanneret had tried to explain to the readers of *L'Esprit Nouveau* in 1920, the theater in Hellerau suggested a completely new generation of productions. What architects and plastic artists could learn from Dalcroze and Eurhythmics was an understanding of the body in space and all that it implied. Appia explored this interaction in his productions and his theoretical writings. In 1912, he wrote:

> Although in the spoken drama the actor may not find the equivalent of the art to which eurhythmics has introduced him, he will discover a common and essential element! *Space*. The discipline of rhythm will make him particularly sensitive to the dimensions of space, which correspond to the infinite variations of musical sequences.[122]

He bemoaned that "daylight is not at our command,"[123] although it can be modulated by physical means. Stage lighting, on the other hand, is at the beck and call of the designer, although our knowledge of illumination is rather crude when compared to our understanding of music [figure 4-37, 4-38]. "We must therefore study light with the help, and under the control, of music!"[124]

In theater past, light had been used to render the actors and the sets visible. Now, light would emerge as a medium in itself, powerful in that it lacked any inherent iconography:

> Light, no longer constrained to illuminate the painted canvas, can spread out into space, filling it with living color and infinite variations of an ever-changing atmosphere. It will no longer exclusively serve the scenic artist but . . . the dramatist![125]

Awash in the embrace of synesthesia, Appia was convinced a new form of theatrical spectacle would result:

> Then the all powerful light, subject to music, will join in the *mise en scène*—light which gives us plasticity, which fills the space with highlights and moving shadows, which flows in calm rays or bursts forth in colorful, vibrant flashes. The human bodies bathed in this vivifying atmosphere will recognize it and salute *Music in Space*.[126]

In many ways—its charge for the unification of the arts, the power of light and color, the distribution of music and light in space—Appia's 1911 call could be read as program for the *Poème électronique* avant la lettre.[127] Le Corbusier, as co-editor of *L'Esprit Nouveau* and as Albert Jeanneret's brother, knew of the Dalcroze-Appia collaboration almost from its inception in the 1910s. In addition, in Léger's 1924 essay, "The Spectacle," the artist declared the traditional dead. To resurrect its corpse it was necessary "to make *the ob-*

4-37
Stage design for Act III, *Iphigénie à Aulis* by Glück. Charcoal and white chalk on blue paper. 1926. Adolphe Appia. [Cabinet des dessins, Musée d'art et d'histoire, Ville de Genève]

4-38
Rhythmic space. Overture. Charcoal and white chalk on beige paper. 1909. Adolphe Appia. [Cabinet des dessins, Musée d'art et d'histoire, Ville de Genève]

jects act . . . it is the image alone which acts on the spectator, and he submits to it, is moved, and conquered."[128]

The painter Wassily Kandinsky regarded color as a central element of painting, although he acknowledged that form will affect the perception of color as color will affect the reading of form. Underlying the particular variants, however, were general properties that could be assigned the various chroma. To Kandinsky: "The power of profound meaning is found in blue, and first in its physical movement (1) of retreat from the spectator, (2) of turning in upon its own center. The inclination of blue to depth is so strong that its inner appeal is stronger when its shade is deeper. Blue is the typical heavenly color."[129] On the other hand, "Orange is like a man, convinced of his own powers. Its note is that of the angelus, or of an old violin."[130] Violet is "sad and ailing"; "gray is produced by a mixture of green and red, a spiritual blend of passivity and glowing warmth."[131] "A red sky," he wrote, "suggests to us sunset, or fire, and has a consequent effect upon us—either splendour or menace."[132] But he cautions us that no color possesses any meaning implicitly, and that the human reading is greatly modulated by the relation of other objects to the color.[133] Naturalistic rendering of objects or a picture will bias the interpretation towards naturalism; a composition of abstract forms will lead the viewer to treat the color on its own terms.

The same was true of the colored ambiances at the Philips Pavilion [Plates IV, V, VI]. When the colors seemed to reinforce the content of the film—yellow and red with a nuclear holocaust, for example—a sympathetic vibration resulted, and the colors were read naturalistically. When any obvious connection occurred between the colors—like the music—and the film images, the audience reaction must have been to treat chroma as an abstraction. But regardless of when and how it appeared, the color was never neutral. As Virginia Woolf had written "[i]t must be remembered that when bright colours like blue and yellow mix themselves in our eyes, some of it rubs off on our thoughts."[134] Like Varèse's music, color—and light—became concrete.

5. ORGANIZED SOUND

From the very origins of the project for the *Poème électronique*, both Philips and Le Corbusier had considered sound a central ingredient. The increased demand for consumer products after the conclusion of the war included systems for reproducing sound within the home. The brittle 78 rpm record had given way to the 33 rpm extended play album, and the emerging field of graphic design (as opposed to the older field of commercial art) had invented the designed record jacket. Since recorded sound appeared to the eye as only an unfathomable pattern of spiraling grooves, the protective album covering was cast as a marketing product that correlated the visual and the aural. In time, packaging became such a concern in its own right that recordings could be sold almost on their jacket designs alone.[1]

The association of record packaging to the form of the Philips Pavilion is not entirely fatuous. Introducing the company's latest high-fidelity sound and image systems, the pavilion ultimately functioned as a giant speaker enclosure and screen for projection and illumination. At the outset of the project, the company sought an original score that would demand the range and talents of a major symphonic orchestra, while showcasing their equipment's ability to re-create the aura of the concert hall using electronic sound alone. That Benjamin Britten was suggested in early correspondence is not surprising.[2] European, although not Dutch, Britten would neatly complement the selection of the French architect and the Russian-born sculptor Ossip Zadkine, illustrating the international reach of Philips products. Britten's music was frequently performed and extremely well received in Holland during the postwar period. While progressive, Britten retained sufficient references to the majesty and tonalities of the grand symphonic tradition to embrace rather than repel the general audience visiting the pavilion.

Le Corbusier brought any ideas about collaboration with Britten to an abrupt end when he firmly stated that it must be Edgard Varèse who would create the score for the *Poème électronique*[3] [figure 5-1]. In some ways, this was a curious choice; Le Corbusier had met Varèse only in passing, apparently in 1935 on his first trip to New York, where Varèse had been living since 1915. According to the architect's recollection, he had not seen Varèse for about twenty-five years.[4] But Varèse remained very much in Le Corbusier's mind nonetheless. In 1951, the architect noted in his sketchbook to contact Varèse about composing the score to a film on the Marseilles apartment block, although nothing seems to have come of it.[5] Le Corbusier wrote the composer in 1954 to seek his assistance in creating an electronic carillon for the proposed campanile at Ronchamp. In his letter, he proposed a score of collaged sounds that in some ways prefigured the score of the electronic poem:

> *I have recently invented the term "tape music." By this I intend that musical elements will be recorded one after the other on wire or tape, thereby func-*

5-1
Edgard Varèse and J. W. de Bruyn in Eindhoven. [Philips]

> *tioning as a library of sounds. The broadcast director [then] edits these segments of tape music.*
>
> *In this particular case, I imagine . . . beautiful fragments of ancient liturgical music set against a background of modern music, disrupted by violent or impersonal bursts . . . of modern music.*[6]

Although this proposed collaboration also evaporated, and the scope of the campanile was reduced from the gridded steel cage shown in preliminary drawings to a few wooden posts supporting the bells, the incident suggests that in the 1950s Varèse remained Le Corbusier's composer of choice.

Le Corbusier was not known to have appreciated music of the modern variety, and it may have been, as Iannis Xenakis has suggested, that Le Corbusier was more interested in Varèse's ideas about music than in their actual sounds.[7] Despite any possible lack of sympathy on the architect's part, music—or more specifically sound—was to be a central ingredient of the electronic poem, as Le Corbusier noted:

> *Sounds, noises, unlimited power. . . . A new creation opening all before it. . . . Immediately I thought of Varèse, with whom I had had no dealings for nearly 25 years. My feeling about this was so strong that I was forced to say I would not undertake this task except on the condition that Varèse should create the music. They said: "We have our own musicians; we have our own composers." I said: "You must take it or leave it, as you wish. And moreover, I also make it a condition that Varèse shall be accorded a remuneration worthy of him."*[8]

Varèse replied enthusiastically to Le Corbusier's offer to participate, realizing the resources that might be placed at his disposal and the role the commission could have in allowing him to implement long-held ideas about music. First in June, and then again anxiously in July 1956, he wrote Le Corbusier, but received no response.[9] Only in October did the architect reply, excusing his tardiness by citing vacation time and job interviews. There was no need for concern on the composer's part, however, as negotiations with Philips were proceeding:

> *[The Philips representatives] arrived in Paris Friday, 5 October. We have had one very intense meeting; not everything is yet settled. But I am sending you this note to tell you that we are on track, galloping, and I will keep you posted as soon as something definite comes through.*[10]

The day before the presentation of the pavilion's preliminary design in October 1956, Le Corbusier wrote an extended letter to Varèse, explaining in great detail the terms of the contract negotiated with Philips. The fee had been settled (although Philips had at first offered only one-tenth the sum Le Corbusier thought appropriate for Varèse's art); the question of artistic rights would warrant further, amicable, discussion. Kalff requested that the architect review the matter with Varèse and establish a "gentlemen's agreement" in the form of a letter contract. In closing, Le Corbusier suggested that the collaborators accept the existing technology offered by Philips rather than proposing anything new.[11] The exigencies of the time framework and the budget precluded unbridled fantasy; creation using existing resources, therefore, would constitute their approach to the invention of the electronic poem. Varèse, quite typically, replied immediately, and agreed to the conditions of the contract, but like Le Corbusier himself, he cautioned a close reading of its terminology. In a postscript he noted his interest in seeing photos of the preliminary scheme.[12]

The birth of the composition was not without its pains, and like Le Corbusier, Varèse was chronically late providing the Philips staff with any suitable evidence that the composition was actually progressing. Even from the start, Philips never really supported him, and it is to Le Corbusier's lasting credit that he so steadfastly stuck by his original choice of collaborator. Kalff wrote to Le Corbusier nervously at the end of 1956:

> The information we have received until now about Mr. Varèse's work is not very comforting. It appears that more and more Mr. Varèse is concentrating on *musique concrète* and that he is thus avoiding all the traditional instruments and their compositions. This is just the thing that we are trying to avoid. Naturally, we wish to respect your desire to collaborate with Mr. Varèse but we always reserve the [final] decision on this subject until we have heard the most recent works of Mr. Varèse on records. Thus, we wish to make the decision on this subject after our meeting with Mr. Varèse and you in Paris.[13]

Xenakis replied almost immediately, trying to strengthen Varèse's position with Philips, and to suggest that Varèse did not work in an abstract mode alone:

> A meeting with you, Mr. Le Corbusier, and Mr. Varèse is, without doubt, indispensable. But permit me to establish that Mr. Varèse has never composed music which is solely "concrete." His last work, "Déserts," performed in Paris in December 1954, was written essentially for an orchestra of wind and percussion instruments, with only interventions of concrete music. And until now, he has always written for instruments. I myself think that music by Mr. Varèse is exactly the kind which would show to best advantage the electronic systems of the Philips Pavilion because it utilizes all the registers and all the dynamics of each instrument which are not usually "sent" through ordinary broadcasting. I think that Mr. Varèse, by the power and the quality of his instrumental music, is one of the best choices that you and Mr. Le Corbusier could make.[14]

Xenakis argued from both the head and the heart. Given the musical preconceptions of the Philips management—perhaps Kalff among them—rational argument would have little effect. That Le Corbusier had essentially blackmailed them into the choice of Varèse certainly did not help eliminate the friction. The situation did not substantially improve with time, and the artistic and corporate differences, which increased with the passing of deadlines, plagued the composition of the music and were never dissipated.

At the year's end, the composer traveled to Paris, meeting with Kalff and Le Corbusier on 20 December. On his return to New York, Varèse became anxious, and as the months passed, his disquiet grew. He wrote Le Corbusier toward the end of February, expressing his concern: "I have been without any direct news from Philips and I have no idea on what conditions they are considering associating me with the project. Mr. Xenakis tells me that they will certainly write me directly. I have heard nothing so far . . . "[15] To return to Europe in April without some assurance that the commission would materialize would be financially impossible. "This collaboration with you interests me so much," he closed, "I hope that everything will work out."[16] Le Corbusier replied to his colleague's letter immediately: everything is progressing, a contractor has been engaged, everything will work out as he had envisioned it. Trust in Kalff's ability and his objectivity.[17]

An April 1957 meeting in Eindhoven gave Le Corbusier and Xenakis a much better understanding of the Philips technology and its artistic potential. Xenakis wrote Varèse in June, outlining the state of the enterprise and the electronic possibilities that the composer would have at his disposal: about 300 loudspeakers distributed over the inside of the shell with the possibility of creating "routes of sound" around the space. Stereophonic effects would be facilitated by about ten magnetic command points which will regulate "a) the swarm of loudspeakers, b) the routes of sound, c) the lows alone or the mid-range or the high-pitches, d) special mixes, etc. I believe that until now no composer has ever had such electronic means at his disposal."[18] Le Corbusier's work on the "luminous poem" was quite advanced, but was still not defined, Xenakis admitted. There should be some concord between the sound and the light aspects of the poem, but these should not constrain the composer in any way—this was a fundamental condition of the collaboration. A sample presentation was planned for later in June or July, using the Varèse recording; the composer's presence would be necessary to understand the full capabilities proposed by Philips.

Some weeks later, Le Corbusier himself wrote Varèse and sent under separate cover twenty-seven sheets of the scenario for the *Poème électronique*. "You will not understand a bit of it, a thousand apologies," he explained. A thorough description of the imagery of the *Poème électronique* followed. "It sounds silly; but there is a way to make something out of it," Le Corbusier admitted and closed: "I repeat: the music does not coordinate with [the visuals]; each follows its own course and does not make contact until the silence at point 19 [so marked in the scenario, paired with complete whiteness]."[19]

The structure of the collaboration was not at all typical. In the operatic tradition, for example, the music could follow from the libretto, or both could be developed simultaneously with accommodations by both the librettist and the composer. In most cases, the set design follows long after, as the means to visually realize the production. But Le Corbusier set out the nature of the collaboration with Varèse based on a detachment. Each medium of the poem would be developed with almost complete autonomy. The correlation of the film, the ambiances, and the *tri-trous*, of course, was tightly governed; any relation between these visual parts and the music, however, was by chance encounter.

Varèse wrote to Kalff the following month, informing the Philips artistic director that he and his wife, Louise, were planning to arrive in Rotterdam on 2 September. He expressed his interest in meeting with the technicians and becoming familiar with the facilities, "even if Le Corbusier was on vacation." As he had just received a letter from Le Corbusier, he was still "assimilating" its contents. But he quoted what the architect had established as the relationship between the visual and aural scores, which de facto established their mutual working methods: "I have decided one thing—your music is free, my scenario is free . . . each follows its own way and will not make contact until the silence at point 19."[20] Varèse also mentioned to Kalff his article in *Commonweal* (December 1940), in which he had conceived of a technique that opposed sound and light. Thus, he was in complete accord with the working method that Le Corbusier proposed in Eindhoven.[21]

Kalff finally responded to Varèse's letter at the end of July, confirming the text of his earlier telegram: "YOU ARE WELCOME IN EINDHOVEN FROM SEPTEMBER SECOND." After a perfunctory rehearsal of travel and lodging arrangements, Kalff himself seems to become excited by the prospects offered the composer:

> These are effects of sound in space, therefore of movement, of direction, of reverberation and echoes, which until now have never been used in electronic installations other than to accentuate the realism of music in concert halls and theaters such as La Scala in Milan. If it were possible for you to utilize these effects, we believe that the demonstration will be still more interesting and novel than with traditional automatic reproduction.[22]

He suggested that Varèse spend his first month in Eindhoven examining the equipment and becoming familiar with its capacities; the creative work, in collaboration with Tak and S. L. de Bruin, would follow. The composition should be ready to test in the pavilion itself in February or March in order to perfect its distribution in the space.[23] Having received a copy of the letter from Kalff, Le Corbusier contacted Varèse and assured him that everything was proceeding well. He had been in Eindhoven on the 24 July for further demonstrations of sound, color, and image, and had greatly expanded his original vision.

> *I have to tell you that I am very pleased with my contact with Mr. Kalff and Mr. Tak, in particular (Mr. Tak = acoustics). They are great realists who respect inventors. As a result, I wish you good luck. I think that you will enjoy working with them in Eindhoven.*[24]

As noted above, Le Corbusier had established that roughly halfway into the *Poème électronique* there would be silence; beyond this parameter, Varèse was given a completely free hand. But even this minor gesture of concordance was not to be. "You will do as you wish," Le Corbusier wrote to him:

> *I shall leave you quite free, thinking of your music as a presence surrounding a man reading, for example, some book or tale and whose ear catches noises from outside (a barrel organ, a brass band marching past, a revolution coming along the street, up the staircase, smashing in the door).*[25]

With spasmodic starts and abrupt dead ends, accepting constraints and rejecting them, Varèse continued work on the composition throughout the autumn months in Eindhoven. He followed his own internal program, as Le Corbusier had followed his while creating the *minutage* of the poem's visual components.

Throughout most of the century and much of his career, Varèse remained an outsider to the musical establishment and his works were performed only sporadically. To propose this maverick to Philips was both a brilliant stroke in itself and a curious choice for an architect who tried to control virtually every aspect of the pavilions's creation—even at one remove. The nature of the collaboration was itself peculiar, with Le Corbusier in India several months each year from 1956 to 1958; Xenakis in Paris responsible for the pavilion's design; Petit also at work in the capital; Agostini, though based in Aix-en-Provence, on location in various places; and Varèse in New York and later in Eindhoven. In some ways, the work on the pavilion and its aural and visual scores staked out a new genre of collaboration: one based more on the fortuitous intersection of collaged elements than carefully correlated aspects of *son et lumière*, or sound and set design. Its conception of integral structure as unstructured independence prefigures developments in the modern arts and those later prompted by poststructuralist literary theory.

Throughout the century, composers had undermined the rigid structure of music, at times incorporating the world of real sounds ("concrete music") or introducing chance into the process of creation. In his score for *Parade* (1916), for example, Erik Satie mixed the sound of the typewriter with the instruments of the symphonic orchestra. Later in the century, in his bid to free musical composition from the binds of tradition, John Cage joined sounds taken from the world with chance organization. "A composer writes parts," Cage wrote, "but, leaving their relationship unfixed, he writes no score. Sound sources are at a multiplicity of points in space with respect to the audience so that each listener's experience is his own."[26] Thus, there is no single work, certainly no single experience, and the identity of the piece is constantly in flux: "Where people had felt the necessity to stick sounds together to make a continuity, we . . . felt the opposite necessity to get rid of the glue so that sounds would be themselves."[27]

Varèse's music for the *Poème électronique*, in contrast, was constructed on tape, and fixed in composition and time. Each performance, completely automated, would be exactly the same. In this respect, all notions of chance were eliminated. In other ways, however, the entire effect of the mixed media was generated by chance. Working in different locations, with few mutual contingencies except the duration of 480 seconds, the meeting of Le Corbusier's visual score and Varèse's organized sounds was almost coincidental. The electronic poem depended upon both design and accident, thus reflecting both traditional and modern ideas toward film and sound and the arts.

Varèse believed that art and science are inherently linked, and that "music has its place in the company of mathematics, geometry, and astronomy."[28] There were, in particular, at least three aspects of Varèse's character and oeuvre which would closely ally him to Le Corbusier's cause. The first was the composer's dissatisfaction with the limits imposed by the existing instruments of the orchestra; Varèse looked to technology for the production of new sounds and saw a connection between art and science. Second, from a very early date Varèse pursued the spatialization of sound and also felt hampered by the orchestra's fixed palette of instruments and the sounds that they make. Why, as Varèse had once stated, should a contemporary composer be limited to the tones of the string choir, which had reached its peak performance during seventeenth century in the works of composers such as Vivaldi? Varèse sought new sounds from old instruments, but also new sounds from new instruments.

The search for new sources led him to electronics, since sound itself was the product of controlled disturbances of the air. As an art-science, music was "the corporealization of the intelligence that is in sounds," a definition proposed by Joseph Marie Wronsky and Camille Durutte in their nineteenth-century treatise on harmony, and quoted by Varèse in a 1939 lecture.[29] The artist here turns to science for new means of expression in sound.

> Composers, like everyone else today, are delighted to use the many gadgets continually put on the market for our daily comfort. But when they hear sounds that no violins, no woodwind or percussion instruments of the orchestra can produce, it does not occur to them to demand those sounds of science. Yet science is even now equipped to give them everything they may require.
>
> Personally, for my conceptions, I need an entirely new medium of expression: a sound-*producing* (not a sound-*reproducing*) one. Today it is possible to build such a machine with only a certain amount of added research.[30]

The direct creation/recording of sounds allowed the composer to eliminate the intermediary performer, which guaranteed a more explicit expression of the composer's ideas in music. "Anyone," he wrote, "will be able to press a button to release the music exactly as the composer wrote it—exactly like opening a book."[31]

Varèse's search led him first to the siren, which could produce a full gradient of sounds rather than the discrete intervals of traditional Western music, and he was as much interested in the device for the spatial implications of its sounds as he was the range of its tones.[32] "My first physical attempt to give music greater freedom was by the use of sirens in several of my scores (*Americas* [composed 1918–21, premiered 1926] and *Ionization* [composed 1930–31, premiered 1933]), and I think it was these parabolic and hyperbolic trajectories of sound that made certain writers as far back as 1925 grasp my conception of music as moving in space."[33] Newly developed electronic equipment such as the Ondes Martenot, developed by Maurice Martenot, offered increased technical possibilities to Varèse, who began his inquiry into the electronic medium in 1927.[34] But it wasn't until 1954 when invited to work in the electronic studios of Radiodiffusion France, that he finally approached the goal for which he had struggled for four long decades—in this case, to create the tapes for the "organized sound" for *Déserts* (composed 1949–54, premiered 1954), which combined episodes for both orchestra and tape.[35] From that point forward, Varèse avoided the term music, perhaps as much for the expectation instilled in the audience as for his own definition of the art. His term "organized sound" suggested its own aural medium, free from melody, apparent rhythms and harmony.

> Sound and noise. There is no difference between sound and noise; noise is sound in the course of creation. Noise is due to a non-periodic vibration, or a vibration too complex in structure, or a duration too short to be analyzed or understood by the ear.[36]

While Varèse sought the new in music, he did not entirely discard the old. He respected the works of the past and viewed them as the products of their age. "Preceding ages show us that changes in art occur because societies and artists have new needs," he told his audience. "New aspirations emanate from every epoch. The artist, being always of his own time, is influenced by it and, in turn, is an influence. It is the artist who crystallizes his age—who fixes his age in history. Contrary to general notion, the artist is never ahead of his own time, but is simply the only one who is not way behind."[37] If the times provided the new means, the artist should utilize them or create an appropriate art form, even if the products seem "to some people more like a cactus than a rose."[38] In spite of his comments on the artist and the epoch, Varèse did not imply that the artist should remain content with the existing conditions:

> Because for so many years I crusaded for new instruments with what may have been fanatical zeal, I have been accused of desiring nothing less than the destruction of all musical instruments and even of performers. This is, to say the least—an exaggeration. Our new liberating medium—the electronic—is not meant to replace the old musical instruments, which composers, including myself, will continue to use. Electronics is an additive, not a destructive, factor in the art and science of music. It is because new instruments have been constantly added to the old ones that Western music has such a rich and varied patrimony.[39]

Of course, the new technology offered sounds that were not possible with acoustic instruments. For one, timbre could be rendered independent of volume so that "a whispered passage of a song can be made to sound above an orchestral tutti, although it still retains its 'whispered' characteristics."[40] In addition, there would be virtually no limit to the intervals between tones, since the machine need not coordinate with the tempered scale.

In a recent study, Jonathan W. Bernard proposed parallels between cubist thinking and Varèse's music.[41] These included the avoidance of repetition, escape from illusionist perspective, multiple viewpoints and a "geometrical" structure, and most significantly "*music as spatial* . . . as bodies of intelligent sounds freely moving in space." In a 1953 interview with Fred Grunfeld, Varèse resorted to a visual simile to explain his musical ideas:

> Visualize the changing projection of a geometrical figure on a plane, with both plane and figure moving in space, but each with its own arbitrary and varying speeds of translation and rotation. The instantaneous form of the projection is determined by the relative orientation between figure and the plane at the moment. By allowing both the figure and the plane to have motions of their own, one is enabled to paint a highly complex and seemingly unpredictable picture with the projection. In addition, the qualities can

> be further enhanced by letting the form of the geometrical figure vary as well as its speeds.[42]

That the musical idea could be conceived spatially only when translated into more graphic imagery also reflects Varèse's interest in the plastic arts of sculpture and painting, which he saw as more innovative than tradition-bound music.

Since—as Varèse had pronounced in his 1939 lecture "Music as an Art-Science"—the production of sound becomes synonymous with the reproduction of sound,[43] there is no loss of composer's intention with each new performance. By using electronic means, sound could also be distributed throughout the hall, thus freeing the production of music from the fixed position of the orchestra, which limited the sources of sound. The spatial possibilities for electronic music become endless. Henk Badings and J. W. de Bruyn, who had worked in the Philips Eindhoven laboratory, suggested the future possibilities for electronic music in an article from 1957–58; in many ways, the essay outlines the technical aspects employed by Varèse for the *Poème électronique*:

> The composer might go a step further along these lines [than positioning the loudspeakers throughout the space] by not only arranging for different parts of the overall tonal pattern of his score to reach listeners from different directions and distances—a generalization on the idea of the trumpeter behind the wings in "Fidelio"—but by also giving these parts a different spatial character.[44]

Varèse's *Poème électronique* and Le Corbusier's and Xenakis's pavilion would address just these issues, adding yet another dimension of visual imagery to the aural/architectural mix. Under Philips's aegis, Badings had produced—within a month—an electronic composition for the ballet *Cain and Abel*, which premiered in The Hague in 1956. Given the corporate sponsorship of this electronic precedent and its publication in the Philips house organ, which included a small recording as a sampler of sounds, the management's initial resistance to Varèse and his music seems curious.[45] In addition, Badings remained a more traditional composer using a written score although incorporating electronically generated sounds.[46]

Edgard Varèse's quest for spatialized music continued a tradition begun by several noted composers before and during the Renaissance. Creating music to dramatize the acoustic properties of specific spaces in the Western tradition extends back at least to the medieval period when the composers of chant are believed to have considered the reverberation time of stone churches in creating musical phrasing and duration. In Venice, Giovanni Gabrielli

(1557–1612) developed the play of antiphonic choirs specifically in reaction to the shape and lively acoustics of the basilica of San Marco. Built in the shape of a cross, each arm with balconies, the very form of the Venetian church suggested the alternation of voice that constituted antiphony.[47] Not only was it crucial for the two major groupings of musicians and singers to hear each other, but the blending of the two voices drew upon the reflection of sounds and their amplification within the basilica. The writing of the parts derived to a great degree from the configuration and reverberation of the church, and the music could not be performed in other venues without some adjustment.[48] In these works, architectural space was itself an active component—one could almost term it a performer—in the generation of music.

Almost two centuries later Hector Berlioz (1803–1869) furthered these experiments by manipulating not only traditional orchestration by combining "single instruments from dissimilar families (flute, harp, cello harmonics and a muted trombone)"[49]but also the location of orchestral choirs within the performance space. Varèse regarded Berlioz's music with the greatest respect, especially because the latter had been "passionately interested in *sound*."[50] Berlioz's major works, among them *Symphonie Fantastique* and *Harold in Italy*, are compositions known for their brilliant tonal coloration and their skillful weaving of thematic fragments based on various *idées fixes*.[51] Whether attuned to architectural spaces or slanted to basic showmanship, Berlioz also exploited the properties of the space when the situation demanded. In his oratorio *L'Enfance du Christ*, for example, a band of angels accompanied by flutes sings from the rear of balcony as an evocation of the heavenly choir. This new source of sound takes the audience by surprise, while its elevated position suggests the celestial source.

Of the Berlioz repertoire, however, it was his *Requiem: Great Mass for the Dead*, Opus 5, that most fully exploited the potential of placing sounds in space—and indicatively, Varèse conducted the work in New York in 1917 as his first major public performance in the United States, and as a tribute "to the fallen of all nations."[52] Conceived for the church of Saint-Louis des Invalides, the composition was originally commissioned to celebrate France's 1830 revolution. Due to a series of intrigues, both political and aesthetic, the piece was not performed until some years later, on 5 December 1837. For the premiere Saint-Louis was draped in black cloth and illuminated by thousands of candles; the members of the orchestra and choir counted in the hundreds. The church is configured as four-apsed Greek cross. Its form and resultant acoustics, both of which were familiar to Berlioz, contributed to the scope of the work and its sonic investigations.[53]

After the opening *Requiem et Kyrie* follows the apprehension of the Day of Judgment, *Dies Irae*, the Day of Wrath. And then the *Tuba mirum*, "that Awesome Horn":

> A trumpet, spreading a wondrous sound
>
> Through the graves of all lands,
> Will drive mankind before the throne.
> Death and Nature shall be astonished
> When all creation rises again
> To Answer the judge.[54]

The register of the music drops; the bass strings move rapidly upward like the winds of a gathering storm, culminating in the piercing entry of the brass. Rising voices, set in series, foretell the doom of those who have not repented. The brass choir had been divided and surreptitiously dispersed to positions behind the audience in each of the three apses as well as in front of them on stage (their positions are indicated on the score in terms of compass directions). The sounds are announced: one, two, three, four choirs blare forth, their notes playing upon one another, challenging, answering, and interweaving their tonal color; ultimately, eight sets of timpani join the tumult. The floor rumbles under the force of the low tones; the music can be felt in the body as much as heard through the ears. Berlioz himself qualified the *Requiem* as "a musical representation of the Last Judgment, prepared for with such deliberation and employing an exceptional combination of forces in a manner at the time unprecedented and not attempted since."[55]

While the Berlioz *Requiem* reached a culmination in volume, Le Corbusier's and Petit's electronically projected poem found a consummation through images and light. The mushroom-shaped cloud prefigured the final judgement in the *Poème électronique*; for Berlioz, the drum and the trumpet, the trombone and the tuba, announced the day of doom. And one might further extend the comparison between the electronic Mass and the aural *Requiem*. Each introduced humanity and its failings; each suggested the fate that may befall those without conscience of the consequences of their acts. Each attempted to resolve conflict through benediction: the *Requiem* through the beauty of the tenor voice over a soprano chorus in the *Sanctus*; the *Poème électronique*, more egocentrically perhaps, with the adoption of Le Corbusier's architectural and urban visions, or more universally in its optimistic hopes for the newborn child. In both these works, humankind moves from chaos and despair through cataclysm, ultimately emerging into the light of hope, as eternal rest for Berlioz or the promise of a better life for Le Corbusier.

There have been two basic schools of electronic music (although a hybrid of the two, which includes instrumental music, could be counted as a third). The first approach, which was championed by Pierre Schaeffer in the

1950s, sought "concrete" sounds taken from the quotidian world. These could be patched, spliced, distorted, and combined in such a manner that a form of structured sound, or music, would result. The second school sought the sonic possibilities of electronic equipment to generate sounds that do not otherwise exist. Edgard Varèse shifted between the two camps, as his ideas and his need took him. Early on he was fascinated by the Ondes Martenot's ability to synthesize sound and in a 1929 Paris performance of *Amériques* he used it to substitute for sirens, which were not available.[56] His quest for the new dimensions of sound eventually took him to electronics, and yet, in spite of these preoccupations, Varèse continued to rely on the human voice and even traditional instrumentation, perhaps as a means to temper the inherent abstraction of unfamiliar and ethereal sounds. Even in the *Poème électronique*—despite its reliance on synthesized sounds—the human voice, both male and female, alone and in groups, recalls more familiar forms of expression and provides a human presence within abstraction.

Of the media that comprised the Philips Pavilion's eight-minute spectacle, the Varèse *Poème électronique* has remained the most enduring fragment. No part of its making, from the commissioning of the composer to the creation of the work to its presentation and acceptance, was easy. The means by which it was created also challenged the notion of collaboration, since the composer worked with virtually no knowledge of what the architect was preparing as the visual score of projected images. Xenakis stated: "For years I have advocated a technique of opposition for sound and light. I am happy that we are in complete agreement."[57] Given the distance between architect and composer, and their respective schedules, they could follow no other route.

As early as 1936, Varèse had formulated a musical score that could be taken as the imaginary prototype for the *Poème électronique*:

> When new instruments will allow me to write music as I conceive it, the movement of the sound-masses, of shifting planes, will be clearly perceived in my work, taking the place of linear counterpoint. When these sound-masses collide, the phenomena of penetration or repulsion will seem to occur. Certain transmutations taking place on certain planes will seem to be projected onto other planes, moving at different speeds and at different angles. There will no longer be the old conception of melody or interplay of melodies. The entire work will be a melodic totality. The entire work will flow as a river flows.
>
> We have actually three dimensions in music: horizontal, vertical, and dynamic swelling or decreasing. I shall add a fourth, sound projection—that feeling that sound is leaving us with no hope of being reflected back, a feeling akin to that aroused by beams of light sent forth by a powerful search-

light—for the ear as for the eye, that sense of projection, or a journey into space.[58]

Two decades later, with only scraps of ideas in hand, Varèse traveled to Eindhoven to create his score, planning to work in the Netherlands for several months. While Varèse's organized sound for the *Poème électronique* drew upon concrete sounds in the world and reformed them through electronic processes, it also incorporated the human voice and sounds of completely electronic creation. As he later wrote: "Studio recordings were used, machine noises, transposed piano chords and bells, filtered recordings of choruses and soloists. . . . Oscillators were used to record sinusoidal sounds, and literally unheard of sounds were made by mixing and combining all these."[59]

Recording the 480 seconds of the poem became a seemingly endless task fraught with impasses. The friction present at the beginning of the project, when Le Corbusier first proposed Varèse, continued throughout. This hosility was not mitigated when the composer arrived in the Netherlands without a clearly defined plan and revealed himself to be unfamiliar with the sophisticated equipment now placed at his disposal. After decades of only dreaming about access to new sources of sound, he appeared puzzled by the possibility for realization. Thus, critical time was lost as Varèse revised his preconceptions and responded to the technical possibilities offered by the Philips labs. Since his work method appeared to be completely nonlinear, he was unable to inform Philips how much of the work had been completed at any given point. Like almost every other aspect of the production, accomplishments were measured in seconds and minutes. At a personal level, Tak and Varèse seemed to work on different planets. Varèse started from scratch again and again, with neither rhyme nor reason—or at least not as perceived by the Dutch technical staff. One anecdote describes Tak laughing almost uncontrollably while recounting an incident that had happened one morning. After several hours of unsuccessfully searching for a particular sound, the studio door opened with a squeak of the hinges. "I must have that!" exclaimed Varèse. To the far more methodical Tak this was an almost ludicrous means of composition.[60] Problems also lay at the managerial level, and it is difficult to discern Kalff's exact position vis-à-vis the sound score. In letters, he appears to represent Philips as a corporation; his own opinions are not apparent, however.

In September 1957, Varèse wrote to his future biographer, Odile Vivier: "Here, I am working. As you know, it goes slowly. The pace of the country: more than slow—and the locals are complacent. Having lived for a time in New York, and before that in Paris, the lag is bothersome" [figures 5-2 through 5-4].[61] As the composition progressed, the difficulties and resistance increased. In December, he wrote to Vivier again:

> There are, without doubt, going to be fireworks. Messrs. Philips and . . . do not like the fragment of the composition that X [according to Bart Lootsma, this is Tak] . . . has made them listen to. It is true that it was presented cursorily with completely inadequate means. Verdict: no melody, no harmony. These good sirs apparently would be happy to get rid of me, but I am not in the habit of quitting things. The first part is already recorded on three tracks. The second part is well on its way in sketches. I have plenty of time, [as] the sound equipment is far from being ready.[62]

After some four months of work, Varèse had captured barely two minutes of the eight-minute total required for the presentation [figures 5-5, 5-6, 5-7]. By the turn of the new year, the Philips committee had neither seen the first minute of film and lighting effects nor heard the first two minutes of the score. With the projected opening of the building only four months away, and aware of the complexities of fitting the building for sound, the directorate became extremely anxious. Even Kalff, who had remained a loyal supporter of all aspects of the project, was starting to get cold feet. His 13 December letter to Le Corbusier (which was forwarded to the architect in India), provided specific soil information and requests for modifications that are more or less normal for any project. But the final paragraph indicated that displeasure was turning to more than mild annoyance:

> We sincerely hope that it will be possible to see you here with Petit and Agostini and the first minute of the film. We would like, in your presence, to show our directors this first minute of film, accompanied by some lighting effects and a part of the music by Varèse. Mr. Philips is beginning to be a bit uneasy with our presentation, which could be so abstract and completely foreign to anything with which the public is familiar that they would miss its aim; which is to reinforce the prestige of the Philips corporation in the world. This is particularly true regarding the music of Varèse . . . and we would thus like to have the possibility, before the end of the year, to decide if it would not be possible to have a score by another composer, so that we have two possibilities instead of one. We hope that none of this is necessary; Mr. Tak is sufficiently optimistic on the subject of the novelty and the surprising effects of three-dimensional sound. But I feel I must inform you of the state of our spirit at Eindhoven in order for you to understand that we are forced to make these serious and definitive decisions with the least delay.[63]

In Paris, Xenakis intercepted the letter and cabled Le Corbusier in India. Xenakis agreed to Kalff's proposed adjustments to the design of the building and its immediate surroundings, and did his best to calm the storm that was brewing in Eindhoven. As a supporter of Varèse, whom he regarded highly,

5-2

Sketch for the *Poème électronique*.
Edgard Varèse.
[Philips]

This sketch appears to contain notes on the character of various "sound objects" to be pre-recorded, then later mixed together. Elements such as rhythm and tempo are indicated, for the most part with great precision, suggesting the Varèse knew exactly what he wanted as far as the basic musical impetus of the objects was concerned. In this sketch, there is virtually one metronome mark for every sound object. [What looks like the indication "any tempo" can be taken as a further indication of this, not as its contradiction, since the motion from stochastic (indeterminate) pulses to recurring (determinate) ones is an inherent dialectic of such music, i.e., the tempo may be momentarily stochastic; the concept that dictates its necessity as a part of a progression to order is not.]

"Reminder" indications: "high; sped up—same intonation—notice M.M." (metronome mark) are also present, and there are attempts at timbre indication: "new timbre = same for all of these." Some timbral descriptions are by external reference, there being no precise way to notate such sounds and, in any event, what matters in such a situation is a meaningful cognitive tag for the composer's imagination (notation exists in order to enable *other* people to play one's music—clearly of no concern here). Among them are "Timbales" (Fr. *timpani*) and "Pilons" (Fr. *steam-hammer, pile-driver*, or *battering ram*). Of particular interest is "Jazz—spurts," "Jet" and (if I read it correctly) "Extase" with an arrow pointing to Perc (percussion) and Trmb (trombone)—not exactly the instruments that other composers would think of in relation to that sentiment. "Guiro" is a Latin American gourdlike instrument with a serrated side, allowing it to be scraped to produce a dry, brittle sound; it is clearly audible on the tape.

Thus, Figure 5-2 seems more in the nature of a film storyboard than a traditional musical score—not surprising, given the media similarity between tape and film.
[Richard Felciano]

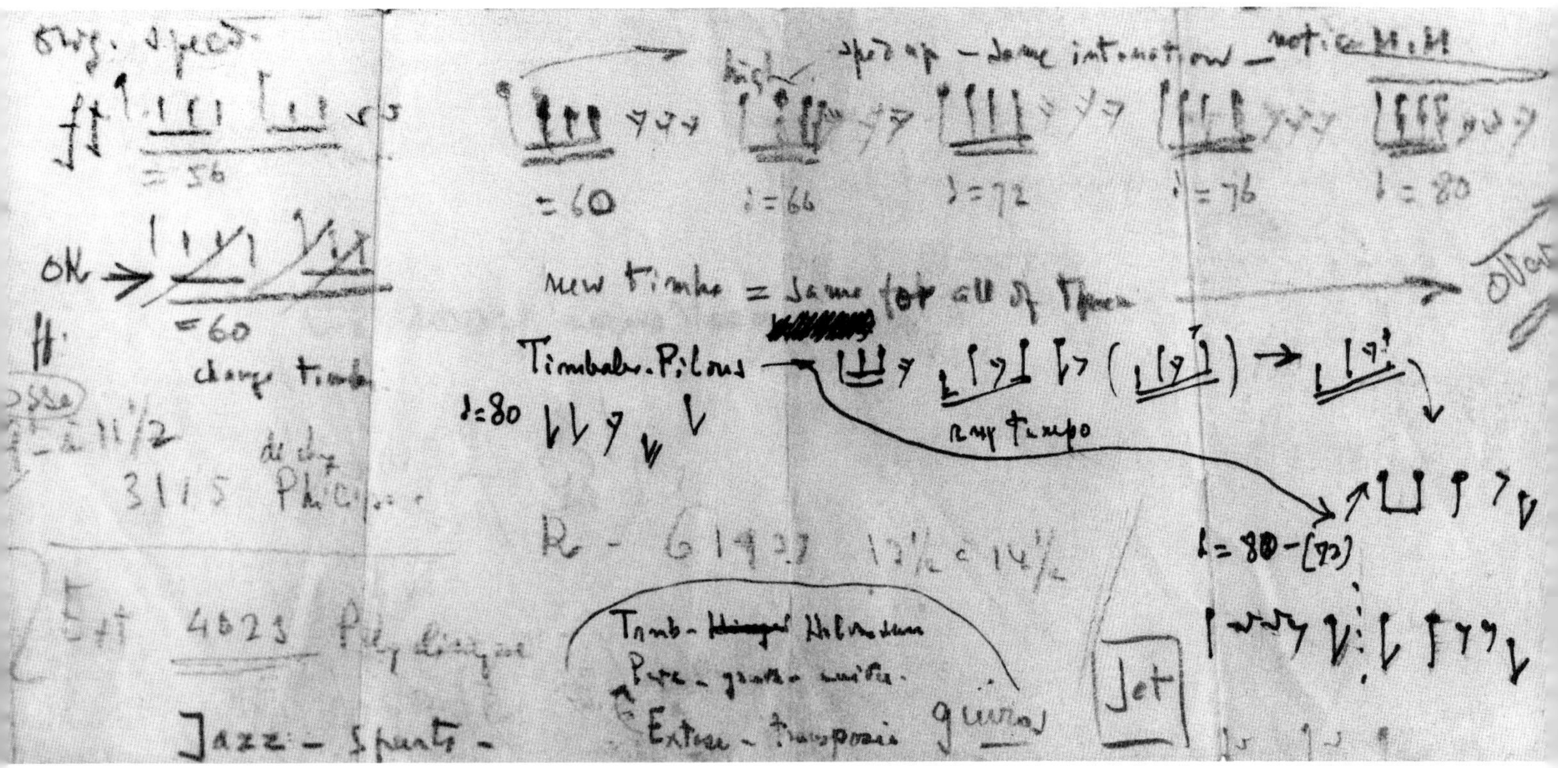
speed up — same intonation — notice M.M
new timbre = same for all of them
Timbales. Pilons
change timbre
Jazz — Sports —
Jet

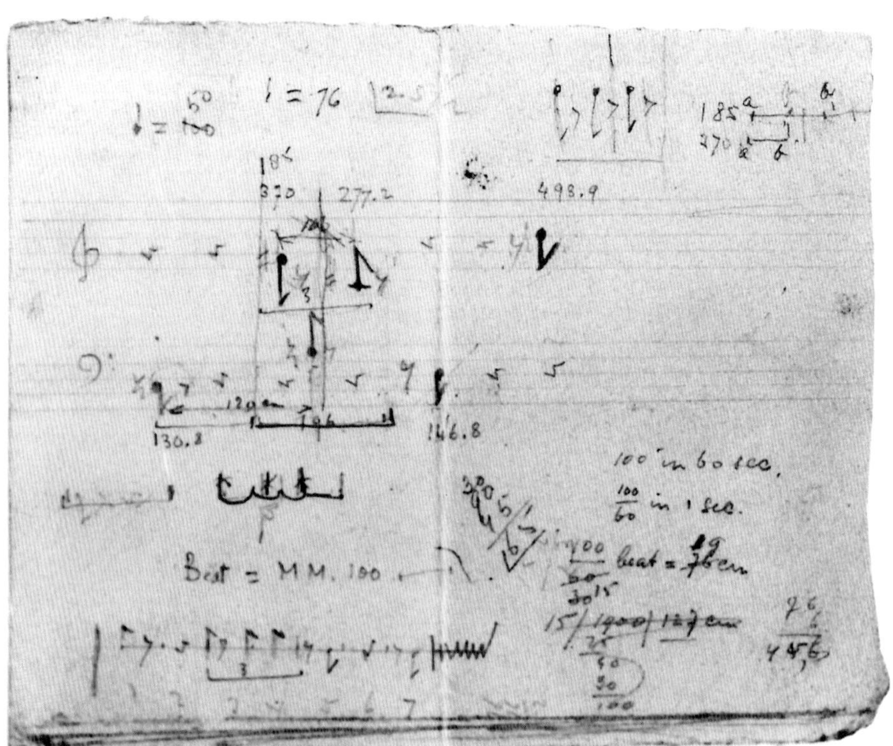

5-3
Sketch for the *Poème électronique* on scrap cardboard.
Edgard Varèse.
[Philips]
This is the most difficult of the existing fragments to decipher because it contains less actual information than the others. There are more tempi calculations, though these are precompositional and not attached to specific figures. "Beat = M.M. 100," meaning 100 beats per minute, is then verbalized "100 in 60 sec." and further converted to "100/60 in 1 sec." There seems also to be an attempt to convert beats (time) into length of tape in centimeters (space), indicated "cm"—an obvious necessity in planning such music.
[Richard Felciano]

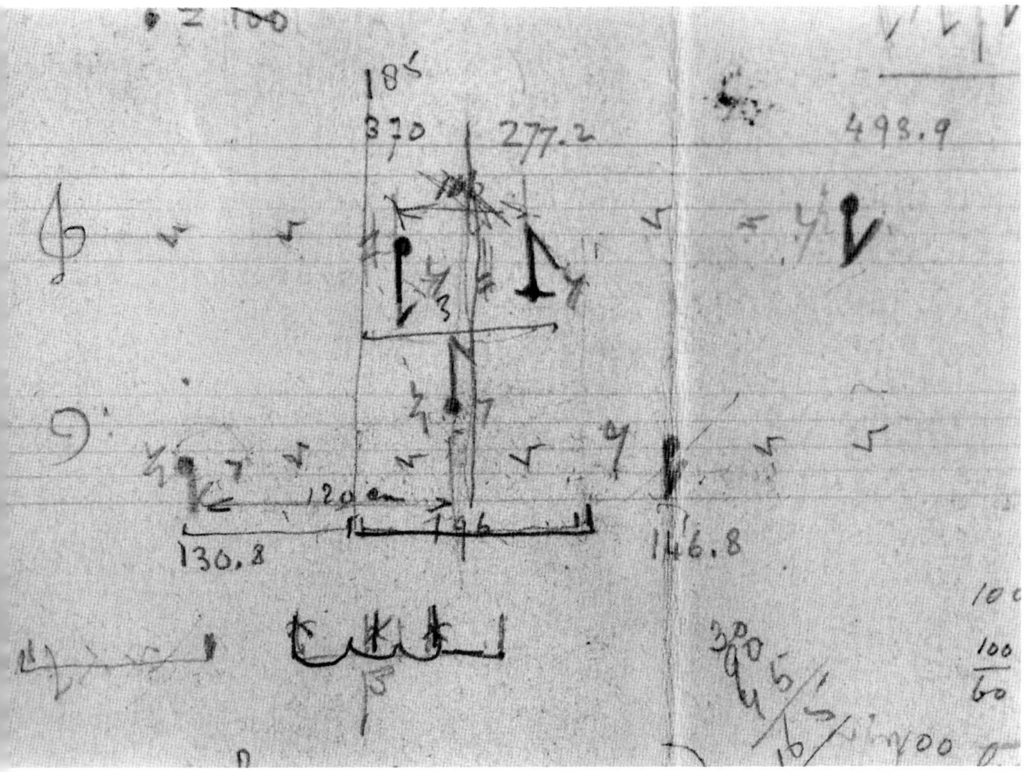

5-4
Sketch for the *Poème électronique*. Detail of Figure 5-3. Edgard Varèse. [Philips]

5-5

Sketch for the *Poème électronique*.
Edgard Varèse.
[Philips]

This drawing appears as a clear attempt to deal graphically with the shifting planes and sound masses of which Varèse often spoke. It may also indicate something about sound trajectories within the building (something like the building's paraboloid shape seems implied by the graphic shapes, but that may be happenstance or a parallel acoustical thought). What can be gleaned with reasonable certainty is an attempt to separate the evolutions of independent shapes and to provide some precise dynamic indications: "mf, p, fp, fff" (moderately loud; soft; loud then suddenly soft; very, very loud), some indications for dynamic change (crescendo-diminuendo), and "out" (a filmic indication). At the lower left is "cumulation amplification," an indication that seems to address mixing and channel distribution if not actual spatial routing. In the lower right is a rhythmic notation (fragment) with the tempo indication M.M. 112, followed by what looks like "Guiro" and "S. Dr." (snare drum). Immediately below is another M.M. indication, this time 80, "Tone clusters—give tempo" and then "Gong."

Running horizontally along the bottom is a more traditional type of staff (relatively speaking), presenting a timeline with the beats (probably one second apart), indicated by equidistant vertical slashes above the line. The line then easily serves to indicate the *point of entry* for the sound objects, though the objects themselves may have separate tempo indications. Below the staff is a crescendo-iminuendo with "wh. n." (white noise—also clearly audible in the piece), then a number of words that are indistinct, except for ending with "pilon" (pile driver) and "M.M. 80." The final five figures bear traditional string down- and up-bow symbols. [Richard Felciano]

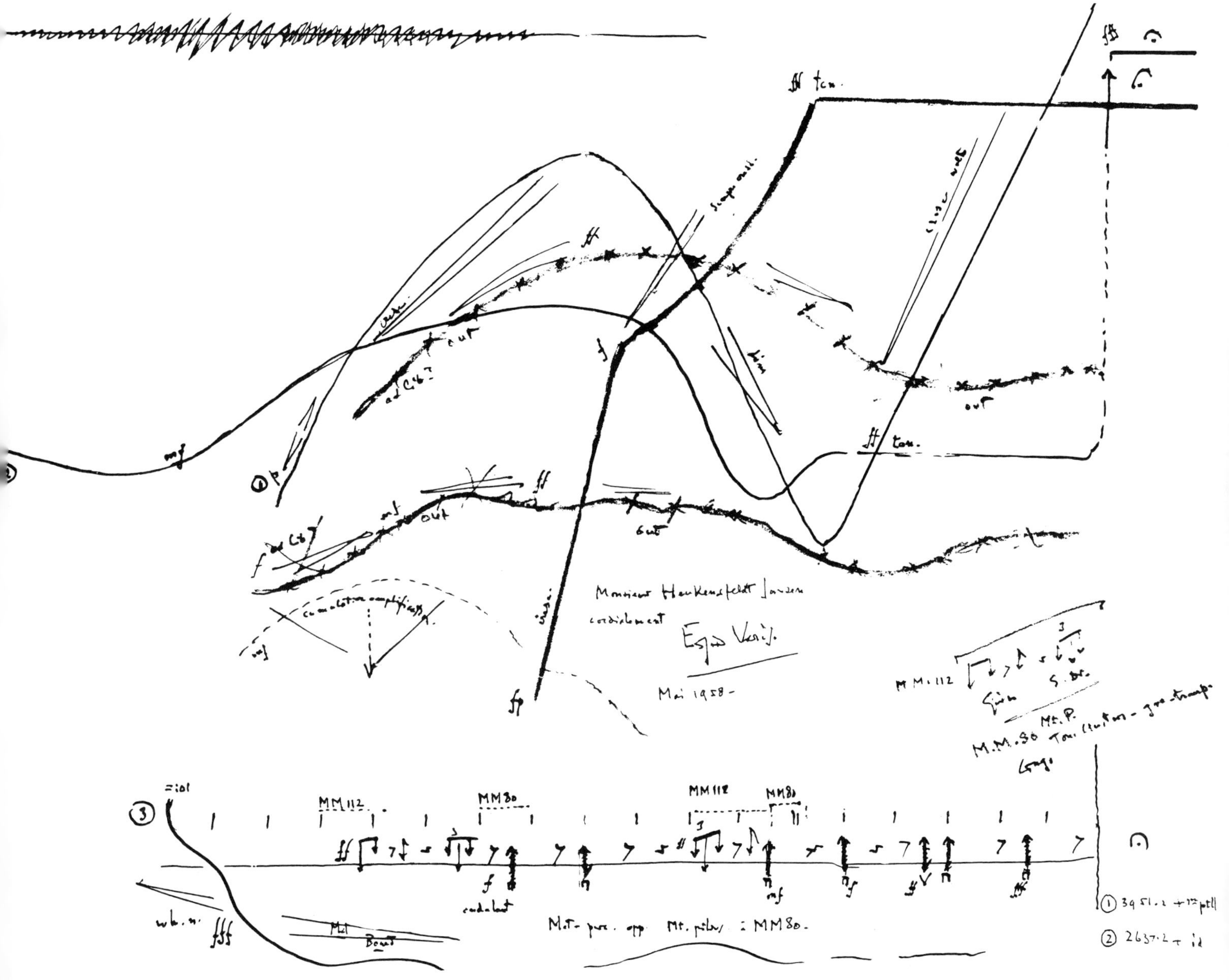
ff ton.
out
dim
Monsieur Heukensfeldt Jansen
cordialement
Edgar Varèse
Mai 1958 -
MM 112
MM 80

5-6

Sketch for the *Poème électronique*.
Edgard Varèse.
[Philips]

My guess is that this sketch noted some vocal material he planned to prerecord in order to have it available when constructing the piece. The score presents rhythms and dynamics but no pitches. The text may be a natural language (I am not able to identify it) or simply phonetic structures needed for the piece. In any event, it seems that little of this material was incorporated in the final piece, if for no other reason than that the prominent vocal quality displayed in the *Poème électronique* is that of the *solo* voice performing a kind of downward or upward sliding *moan*, processed by reverberation and perhaps filtering, and later, an elaborate solo vocal melisma–not the kind of rhythmic chanting implied by this sketch. Nevertheless, the page contains some additional clues to the type and precision of Varèse's acoustical imaging: "foule - confus" (crowd - confused); "sharp - biting"; "Pesante" (heavy); "O.K.!"; "nasal"; "très articulé" (very articulated); and "modulé" (literally "modulated" but meaning "carefully shaped dynamically"). The important deduction, it seems to me, is that the page represents not a score to the piece but a *scheme to produce a precompositional array of available material*, not all (or perhaps any) of which may have been used.

[Richard Felciano]

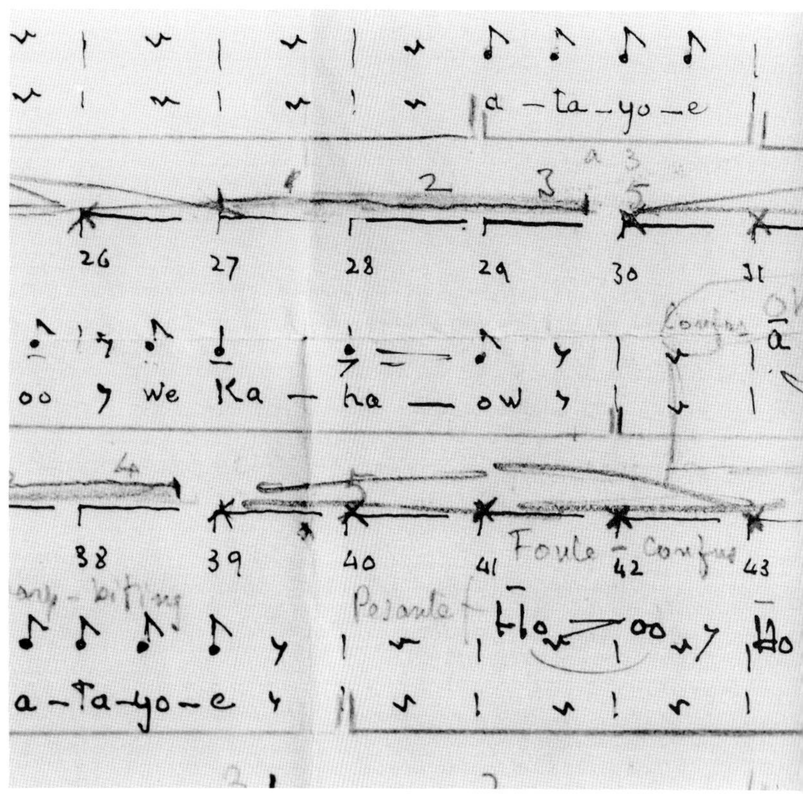

5-7
Sketch for the *Poème électronique*.
Edgard Varèse. Detail of Figure 5-6.
[Philips]

he tried to justify to Kalff the composer's presence on the pavilion collaborative and his music:

> You should not fear Varèse's music, he is the composer and it is the music that you must have. Your pavilion must attract attention by some avant garde "strangeness" and must create a scandal. . . . If on the site of this demonstration you will have a good little "Demonstration," fuddy-duddy, like those you can see in the trade shows of Paris (High Fidelity; Lighting, etc.) the impact will be reduced to that of a ball of cotton. The highly artistic novelty, which you categorize as abstract, is a characteristic of those works that endure. For a long time after the end of the exposition there will be talk of your Pavilion as a bold stroke in the imagination of the public.[64]

Perhaps it was a bit audacious for the young Xenakis to be lecturing an architect and accomplished engineer twenty-five years his senior on the place of the arts in society and their relation to corporate capitalism. Kalff was in a difficult position. He had gone out on a limb with the selection of the artistic team and with the inherent abstraction of the project. It should be remembered that until this time, commercial displays at world's expositions consisted primarily of items already in production or those planned for the future. To eliminate the product was, in the eyes of at least a number of the Philips directors, to erase the image of Philips in the mind of the public. Clearly, Kalff had conceived the pavilion at a much higher, far more abstract level, which was quite risky. Kalff received little aid from his team: his staff in Eindhoven could do little before Varèse, Petit, Agostini, Xenakis, and Le Corbusier did their work, and they were all late in meeting their deadlines.

Kalff's letter reached Le Corbusier in India; he immediately drafted a reply and sent it to Paris by special delivery to be typed and forwarded to Eindhoven. His letter reflected his high moral principles and his regard for architectural clients. It also revealed the architect's developed ego and suggested that all his previous accomplishments (and he can provide documents to prove it!) should excuse the delays on the current project.

> *I have your letter of 13 December. Well! You are in Eindhoven, an hour away from that* "little-great moral crisis" *which strikes every client at the decisive turn of a project.*
>
> Please calm down!
>
> *During these two years, I have worked* without stop, *in my head, on this electronic poem for Philips. It has been born completely on a few pieces of paper. For five years I worked on "The Poem of the Right Angle" for Editions Verve in Paris (a truly luxurious publication of only 200 copies). The publisher Teriade also had his moment of great worry. "The Poem of the Right Angle"*

is today regarded as a great success. I can give you a copy of a letter I received last month from Jean Rostand.

Here [in India], Chandigarh has become a pilgrimage destination for the entire world: the heads of state, the foreign delegations, tourists, students all come to Chandigarh. India, because of Nehru, has undertaken two great things: Chandigarh *and the* Bhakra Dam, *the largest in the world. The Indian government has named me* Government Architectural Advisor *for the dam. Mr. Nehru has become a friend. The building for the High Court of Justice, inaugurated three years ago, remains a topic of discussion even today. Financial problems hampered the realization of the pools. Now, after three years, it is done. The miracle is accomplished: the building shines in true splendor. The visitors exclaim their admiration. After seven years, here in the city, the gardens have attained a first-class beauty; and the people understand.*

The Philips Electronic Poem will be a great work with a scenario of considerable reach and content. It cannot be carried out with any other than the strange music of Varèse. The Philips instruments, Philips' masterful techniques will appear in a symphony: rhythm, color, image, music. In 1925, I made the Pavillon de L'Esprit Nouveau in Paris; in 1937, the Pavillon de Les Temps Nouveaux. Works now considered historic. Public opinion was violently challenged. The 1937 guest book, with its 1500 declarations, proves it.

Tell the Philips' directors that they must have confidence and be proud. The Electronic Poem *will be engraved in the memories of those who will come to Brussels, as an incredible souvenir. There can be no question, not even for a minute, that I will renounce Varèse. If that should happen, I will withdraw from the project. This is a very serious matter. My reputation is as much at stake as that of Philips. My life has been a succession of efforts and battles. The* UN *[building] in New York was built on my plans (stolen). The* UNESCO *[building] in Paris was built according to fundamental [ideas] which I provided (a committee of five of which I was member). I have written more than 40 books: they have sold out or are reprinted continuously, and some of them have been translated into 6 languages. Varèse is a great name in modern music. Look at the case of Picasso, he fought for 50 years, and today he has become the lighthouse of modern painting. Look at "The Rite of Spring" by Stravinsky, booed for hours. Look at our journal* L'Esprit Nouveau *1920–25, prophetic; its issues sell for high prices. Etc, etc.*

Philips has, too, with courage risked [in choosing] the names for [this] battle. Into this, we have poured (you with us) all our courage, all our faith. Brought to Philips the fruits of lives completely dedicated *to art. At the Brussels 1958 Expo, dedicated to human progress, will Philips retreat from the fight? What noise, what scandal [that would be] in the opinion of the public. What defeat! We are reaching the port, and you are capsizing [the boat].*[65]

Le Corbusier's firm stance seems to have quieted the waters. There would be no further question of another composer, although the directors never accepted Varèse or the nonmelodic character of the music. For now, the problem was to get the organized sound[66] on tape, and to prepare the controls for its distribution throughout the pavilion. Varèse was steadfast. "Do not worry; when I have decided something and I know that I am right, and know my rights, I will not [give in]. As far as concessions are concerned, I never make them."[67]

Presumably a short segment of the music was played for a meeting with the Philips management in January 1958. The directors were not thrilled with the music, although there were divergent opinions. Some members of the board were relieved that the demonstration was not to be considered as modern electronic music, but as a "production of sounds," while others characterized the demonstration as clumsy. Mr. Guépin had no objections to leaving both Le Corbusier and Varèse free to pursue their ideas, as long as a sufficient press campaign explained those ideas to the general public. Most of the board members hedged their bets and thought they should postpone judgment until the work was completed. Most resoundingly against the musical segment was Frits Philips who wanted to have noted in the minutes that he was in complete disagreement with all the other gentlemen: "He considers the sounds composed by Varèse as not representative of the ideals of Philips and the free Western world. If we are spending money to add to the confusion, he is not willing to support the plan."[68] In a second memo, they again raised the issue of a second musical score in case the Varèse work proved unacceptable.[69]

A "shadow" score, also called the *Poème électronique*, was clandestinely commissioned from the French composer Henri Tomasi [1901–1971] in case the Varèse work was deemed completely unacceptable [figure 5-8]. Tomasi received the Prix de Rome in 1927 and composed works inspired by Ravel and only of rare dissonance. Regarded as a brilliant orchestrator, he was known primarily for his music for theater. Tomasi's *Poème électronique* was scored for baritone, bass, soprano, mezzo-soprano, and contralto, and orchestra and featured flute, oboe, clarinet, and bassoon. According to Bart Lootsma, the piece was actually performed only once in spring 1958, with Le Corbusier and Xenakis present; all concerned, including the Philips representatives, found it unacceptable.[70]

Given the makeshift conditions of the test situation, the roughness of the production could be expected and addressed thereafter. The intransigent managerial stance against the "ordered sound" of Varèse, however, would remain throughout the months of the pavilion's brief existence, softened only to a small degree by the public and professional reaction to the pavilion. The following month, in March, a more comprehensive presentation was planned, al-

5-8
The shadow musical score, also called *Poème électronique*. 1958.
Henri Tomasi.
Cover page.
[Philips]

5-9
Determining the final positions of speakers. Unidentified, Le Corbusier, Kalff, Tak, Xenakis, left to right. [Philips]

though this was to center on the film segment shown to the Philips directors for the first time [figure 5-9].

After a visit to the construction site of the pavilion the day following the February meeting, Varèse wrote Vivier once again:

> Here, the work continues. On the sixth of this month, we—Philips, Le Corbusier, and Xenakis, for whom I have a great deal of affection and respect—spent the day in Brussels in the pavilion, splashing in the mud, up to our ankles in dirty snow, numbed by the cold. . . . [I]nside the pavilion we suffered a counterpoint of glacial drafts and winds—everything needed to make a visit memorable. . . . The result for Le Corbusier . . . and myself, a horrible flu from which I haven't emerged; nor have I emerged from my house either. The job goes on, however, and I think that around mid-March we will start experimenting to finalize the spatial distribution of my score. (Keep this to yourself, please. The Philips people would not want to leak their open secret.)[71]

As the tape produced by Varèse was created independently from the visual imagery scored by Le Corbusier and prepared by Jean Petit and Phillipe Agostini, Varèse's organized sound was created first on tape and then distributed spatially thereafter. Key figures in this work were Tak[72] and de Bruin [figures 5-10, 5-11].

The aural aspect of the Philips Pavilion was intended to exploit the full capabilities of the company's technical resources, and perhaps even to coerce the invention of new equipment. In Varèse's exploration of spatialized sound, stereophony played a crucial role. Just as we are equipped to perceive depth through the differential offset of the eyes (parallax), so are we able to discern direction by interpreting differences of clues to our two ears. The eyes, however, are limited to a very narrow cone of acute foveal vision, expanded to perhaps 270 degrees through peripheral vision. The ears, on the other hand, can sense sound from 360 degrees, thereby offering a far more developed and complex experience. The sound engineers worked under instructions from Varèse. Relying on stereophony as the key sonic ingredient, they distributed the *Poème électronique*'s aural clues throughout and around the space, thereby destroying any sense of fixed limits or dimensions. Tak described the intentions thus:

> The listeners were to have the illusion that various sound-sources were in motion around them, rising and falling, coming together and moving apart again, and moreover the space in which this took place was to seem at one instant to be narrow and "dry," and at another to seem like a cathedral.[73]

Tak admitted in the same article that the technical staff did not fully understand the composer's plan, but they were certain that "stereophony and

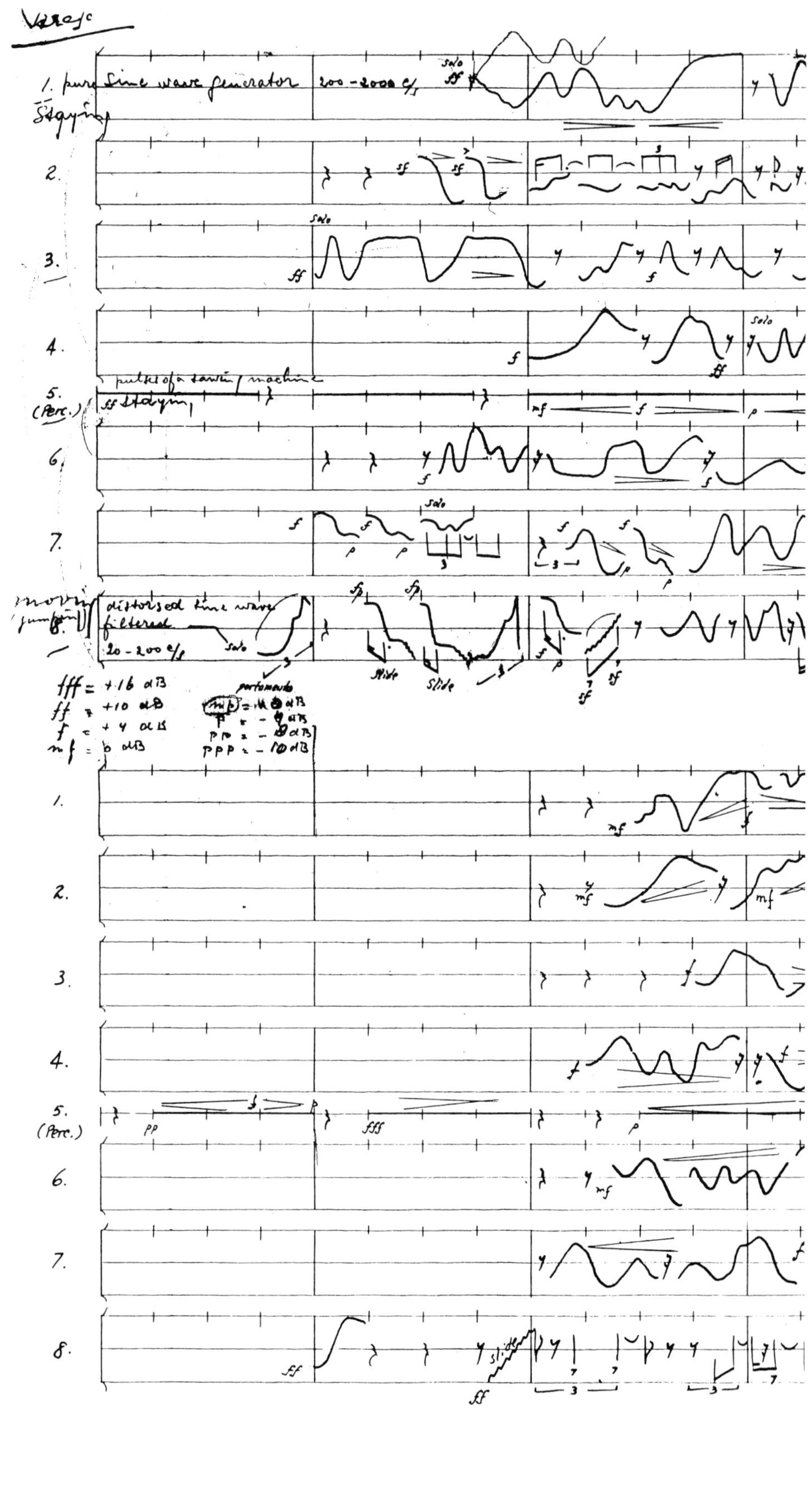

5-10

Sketch for the *Poème électronique*.

Edgard Varèse.

[Philips]

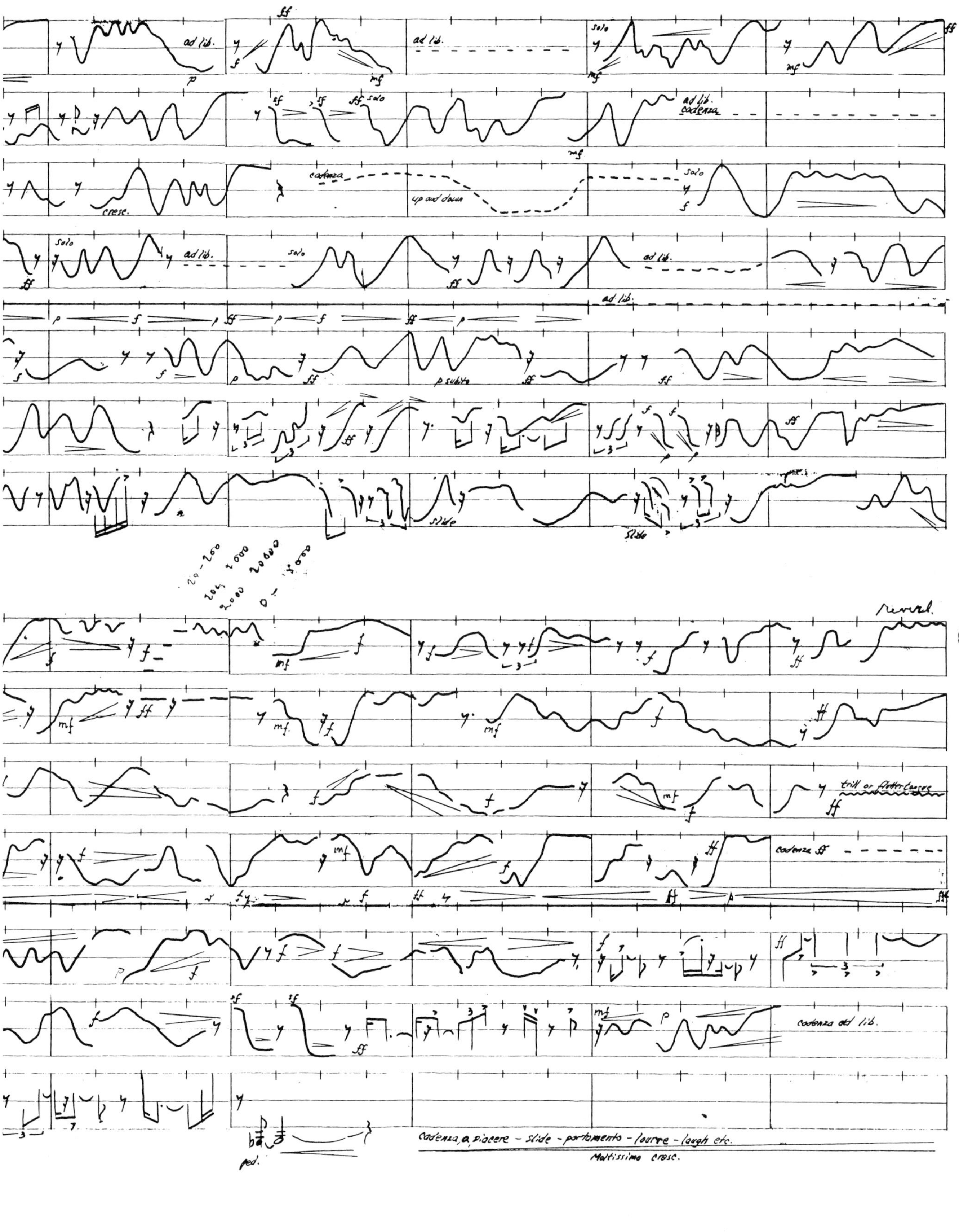

ad lib.
solo
ad lib. cadenza
cadenza
up and down
cresc.
p subito
slide
reverb.
trill or fluttertongue
cadenza ff
cadenza ad lib.
ped.
cadenza a piacere - slide - portamento - laugh etc.
Moltissimo cresc.

These look to be mixing instructions, perhaps for the technicians who were assisting Varèse, or perhaps to organize his own thoughts. They possess, without question, the closest similarity to a musical score (in the Western sense of visually organizing polyphonic strands in time) of any of the examples. There are eight audio channels along the left edge and two such scores per double page. If this follows what became a relatively standard procedure, the "measures" are actually groupings of four seconds, each internal second indicated by an incision at the top of the channel-line. Horizontally, the center-line of each channel represents mid-range of frequency, the outer lines the top and bottom limits of that range. The score is thus able to provide a guide as to how to mix a number of audio channels together in the proper synchronization to arrive at the final composite. For the most part, **pitch** (frequency) exists in the *Poème électronique* as a *continuum* of availabilities from low to high (a glissando is one example) rather than a selection from a limited scale of finite pitches. This said, it is easy to see why pitch is indicated by continuous, wavy contours above and below the center-line; a traditional staff, limited to 12 finite pitches, would be clearly inadequate to the complex and fluid nature of Varèse's sound-objects. **Time** is conceived more in terms of duration in seconds than traditional rhythmic notation; when the latter occurs, it is of strictly local significance, a means of describing the internal character of a specific, fleeting sound object. **Dynamics** (ff, crescendo-diminuendo, etc.) are indicated in the traditional manner. This is the least precise and most subjective parameter of music in traditional notation (precisely how loud is "loud"?). In electronic music, however, one is obligated to deal with dynamics numerically, for numbers are the *lingua franca* of the equipment one must use. In the left margin, between the two systems, can be seen Varèse's attempt to deal with this issue: he has assigned zero decibels to middle loudness (mf) and created a graduated dynamic scale both above and below that reference point.

fff = +16 dB
ff = +10 dB
f = +4 dB
mf = 0 dB
mp = no dB (originally −4 dB, apparently, then overwritten)
p = −4 dB (originally −6 dB[?], apparently, then overwritten)
pp = −8 dB (originally −10 dB, apparently, then overwritten)
ppp= −10 db (originally −16 dB, apparently, then overwritten)

This chart is both interesting and instructive, for it shows Varèse beginning in logical, systematic fashion, building a dynamic scale whose steps below the reference point would almost exactly mirror those above. Apparently, this did not yield the subjective result he anticipated—a lesson in psycho-acoustics before the advent of that field. To remedy the situation, he seems first to have equalized middle-softness (mp) with middle-loudness (mf), ridding himself of what every musician knows as the most problematic nomenclature distinction in traditional dynamic markings. Having done this, he seems to have moved the values below the reference point down by (in most instances) one notch, ppp ending as −10 dB, not −16 dB, the number required to match fff. He seems to have realized that losing one decibel below the reference point does not produce the psychological inverse of gaining one decibel above it. One final note about dynamics: they constitute one area where the technician can have final control and, indeed, Varèse may have been dependent on the technician following instructions as to how the volume level of each track should be modulated in the course of mixing. On the other hand, the dynamics indicated here may represent the actual composer's score, reminding him of the shape the dynamics must take within each track as he created them. **Timbre**, sound-quality, and specific sound-object identification also occur here, some of the tracks beginning with indications as to the type of sound employed:

Track 1: "pure sine wave generator, 200 to 2000 c/s (cycles per second)." This establishes the range limits which correspond roughly to the lower and upper limits of the track diagram.

Track 5: "pulses of a sawing machine" and "ff staying," which may mean *sempre ff* (always loud) or relate to track 8 (see below). Given these indications, it is not surprising that track 5 is labeled "(Perc.)" at the outset and that its track chart has only the center timeline, not the two additional lines needed to indicate the upper and lower pitch limits. Varèse has here converted a practice from standard orchestral scoring, where nonpitched percussion instruments are indicated on a single line rather than a five-line staff. Track 8: "distorted sine wave filtered" and "20-200 c/s." The beginning of the track is labeled "moving jumping," which suggests that the word "staying" in track 5 above may have been a character indication in opposition to "moving jumping."

Various tracks contain other indications: "solo, ad lib., slide, cadenza, reverb., trill or fluttertongue," and in the lower system, track 8, the most elaborate indication of all, "cadenza, a piacere - slide - portamento - lourré (an interesting misspelling of his native French for *louré*, a bowing technique using slurred repetitions) —and finally "laugh etc." [Richard Felciano]

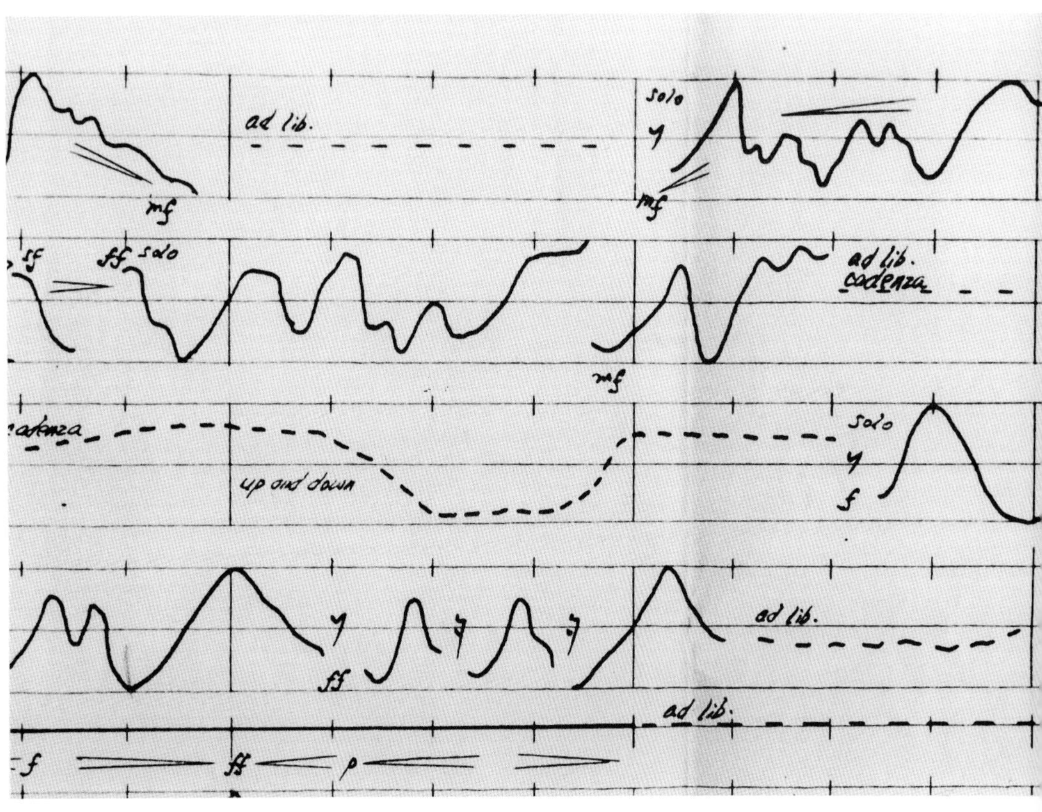

5-11
Sketch for the *Poème électronique*. Detail of Figure 5-10.
Edgard Varèse.
[Philips]

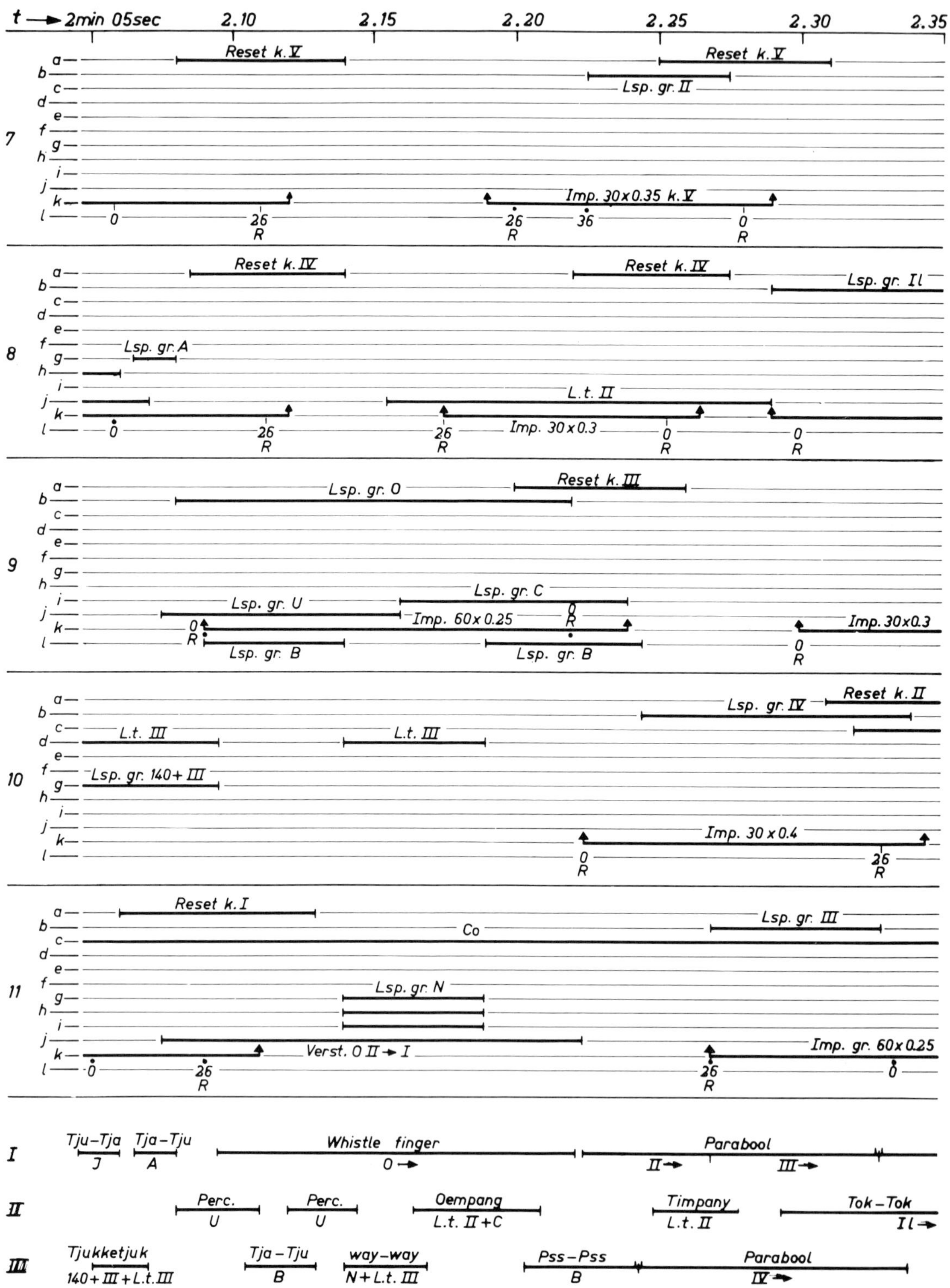
t → 2min 05sec
2.10
2.15
2.20
2.25
2.30
2.35
7
a b c d e f g h i j k l
Reset k.V
Reset k.V
Lsp. gr. II
Imp. 30x0.35 k.V
0
26 R
26 R
36
0 R
8
a b c d e f g h i j k l
Reset k.IV
Reset k.IV
Lsp. gr. Il
Lsp. gr. A
L.t. II
Imp. 30x0.3
0
26 R
26 R
0 R
0 R
9
a b c d e f g h i j k l
Lsp. gr. O
Reset k.III
Lsp. gr. C
Lsp. gr. U
Imp. 60x0.25
Imp. 30x0.3
0 R
0 R
0 R
Lsp. gr. B
Lsp. gr. B
10
a b c d e f g h i j k l
Lsp. gr. IV
Reset k.II
L.t. III
L.t. III
Lsp. gr. 140+III
Imp. 30x0.4
0 R
26 R
11
a b c d e f g h i j k l
Reset k.I
Co
Lsp. gr. III
Lsp. gr. N
Verst. 0 II → I
Imp. gr. 60x0.25
0
26 R
26 R
0
I
Tju-Tja J
Tja-Tju A
Whistle finger 0 →
Parabool II → III →
II
Perc. U
Perc. U
Oempang L.t. II+C
Timpany L.t. II
Tok-Tok Il →
III
Tjukketjuk 140+III+L.t.III
Tja-Tju B
way-way N+L.t.III
Pss-Pss B
Parabool IV →

artificial reverberation would turn out to be important elements of the 'Electronic Poem.'"[74] With this prediction, they set about developing equipment before the music was created. Tak also wrote that "Varèse concentrated primarily on the character of the tonal pattern, and for the most part left us to decide the 'intonation' (distribution of the sound over the loudspeakers, i.e., the spatial effect).[75] At this point, there seems to have been at least grudging respect between Varèse and Tak, who was left with ample responsibility for giving the work its perceivable form. Varèse wrote to Le Corbusier: "[Tak] will tell you that we have considered different possibilities so as to maintain the homogeneity of the whole while preserving the integrity of its parts."[76]

The recording was made on one track, reserving the second and third tracks for "reverberation and stereophonic effects."[77] A second tape, with fifteen tracks, was used to control the distribution of the sounds among the several hundred speakers positioned throughout the building [figure 5-12]. The tracks were 35mm wide, punched with the same perforations as cinematic film. Each of the tapes was scanned on its own playback machine (with up to fifteen playback heads); each unit was duplicated, yielding a total of four playback machines in the control room [figures 5-13, 5-14]. Twenty amplifiers, each with 120 watts, powered the production.[78]

The *Poème électronique*, including silence, extended exactly eight minutes. Having been composed using tapes and electronically generated sound, there was no score as such, although spectrometer charts suggest the variation of sonic color and intensity.[79] The piece opens with a gong, like one in the body of a church, and suggests a day of reckoning. The expectation is not corroborated, however, because the sounds that follow do not develop in any traditional progression, nor do they coordinate with the sequence of images projected on the warped surfaces of the pavilion walls. Rising sequences, blips, tap-tap-taps, many of which were set apart in space constituted the elements of the work. The human voice, solo or grouped, reinstilled a human presence into the work: "I wanted it to express tragedy—and inquisition," Varèse stated.[80] The gong returned, its repetition and cadence providing a tentative suggestion of parts within the total sweep of the piece. Having utilized a multitude of sound sources joined through electronic collage, Varèse realized this work to a large degree through trial and error, constantly adjusting each aural element toward the goal of an integrated work.

While the nature of the music was unknown during the design of the pavilion, the architects and engineers could venture a fairly accurate prediction of the type of sound—mixed concrete, vocal, and electronic—that Varèse would create. Early on, realizing that the speakers might have to be of considerable dimension to produce the low as well as high frequencies, Xenakis had proposed one scheme in which the speaker enclosures would project

5-12
Diagram of the automated tracks. The three musical channels are at the bottom. [*Philips Technical Review*]

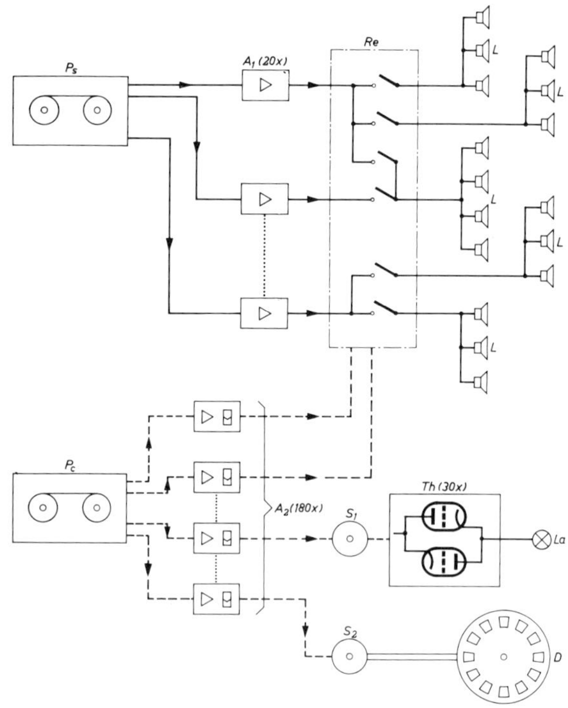

5-13
The playback and automation equipment for the visuals and sound.
[*Philips Technical Review*]

5-14
The automation equipment in the control booth.
[*Philips Technical Review*]

through the building's skin, with the speaker enclosures on the exterior of the building.[81] This plan would have left the interior as the desired smooth surface for the projections, but it was discarded as overcomplicating the external form of the structure, weatherproofing conditions, and expense.

Following the architect's indications, the Philips staff proposed to mount speakers in constellations on the walls; the excessive dimensions were unacceptable, however, as being too disruptive for the projections and colored light. Instead, Xenakis proposed dividing the speakers by frequencies into two groups. Those for high frequencies would be mounted directly on the wall in irregular clusters that countered the geometric curvatures of the hyperbolic paraboloid shells.[82] In certain places, the arrangement followed more or less linear patterns, since one of the sonic ideas was to create "sound routes" that would zip the sound spatially on, up, and over the walls [figure 5-15]. Being small, the high-frequency speakers were acceptable within the overall design, used to visually punctuate the walls and the projections.

The low-frequency speakers, on the other hand, were fixed directly to the floor and screened from view by concrete parapets on both sides of the space. This created two sinuous "orchestra pits" that accommodated not only the sound devices, but also the projectors and part of the light filtration system. An enclosed control room housed and silenced all the equipment for sound, projection, lighting, and automation so that the human presence was hidden during the presentation of the *Poème électronique*.[83]

Taking advantage of his talents as a composer as well as an architect, Xenakis was commissioned to compose the two minutes of music that would accompany the changes of audience between presentations; this had been decided by the time Le Corbusier first outlined the project in a letter to Varèse on 12 June 1956: "The Pavilion has an entrance and an exit (a sort of screen) to accustom the persons entering into its ambiance and departing into the daylight; two minutes for entering and leaving. These two minutes I have entrusted to Xenakis (so that he should have a part in all this) and so that he can loose the din of St. Polycarpus on all the devils."[84]

The younger composer's charge was left vague at first, although somewhat typically, Le Corbusier already had a firm idea of what he believed would be the proper solution. In a letter (or memo) written from Paris, 27 November 1957, Le Corbusier laid out what he believed to be the suitable program for the music:

> *I have thought very seriously about your two minutes of music. What is it about?*
>
> *1st/ A human voice announces, several times, "This way, the entrance; this way, the exit."*

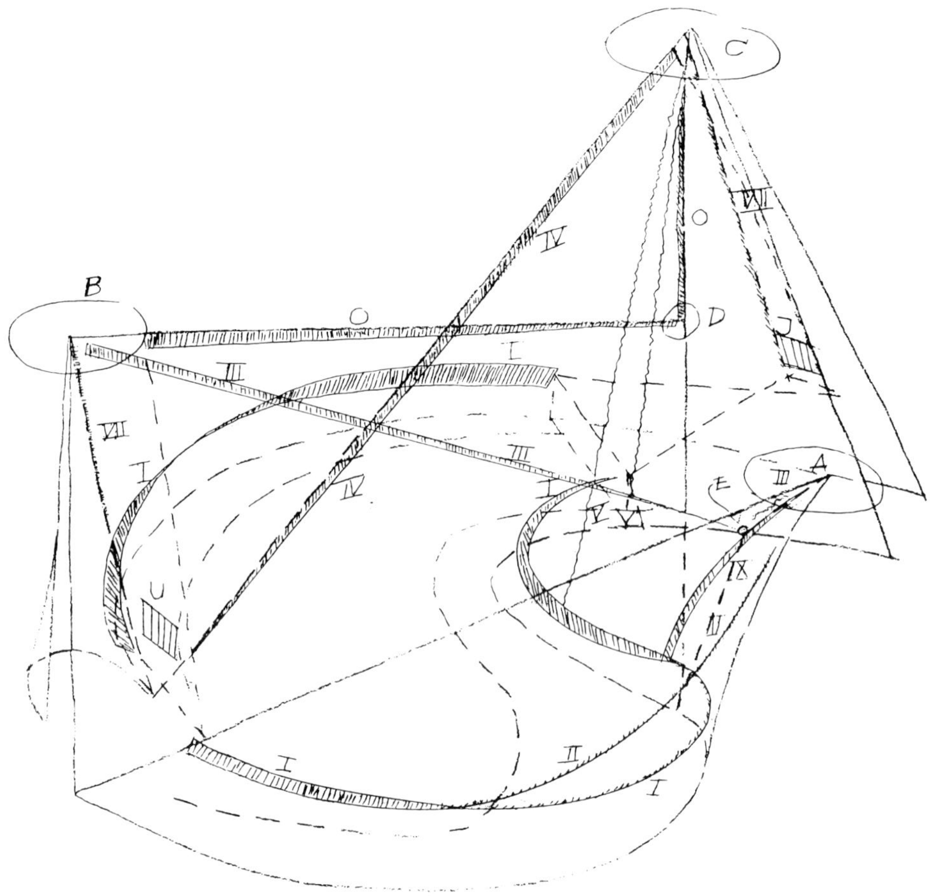

5-15

Diagram of the "Sound Routes." Iannis Xenakis. [*Philips Technical Review*]

Route	I	horizontal
	II	I ▸ A
	III	B ▸ E ▸ A
	IV	U ▸ C
	O	B ▸ D ▸ C
	VI	along the warped surface
	VII	B ▸ U
	VIII	C ▸ J
	IX	A ▾

PLAN OF THE INTERLUDE
120 sec. duration

End of the Poème électro-nique	5 to 10 sec.	10 to 15 sec.	90 to 100 sec.	5 sec.	Start of the Poème électro-nique
	of Absence= Silence + Reflections of still, diffused light which allows the audience to move in the darkness	A voice announces the presentation is over, ordering the audience to to leave the hall, in a quite sharp yet normal voice, in French or English	Clouds of intermittent sounds, varying in density and intensity, and moving within the space of the pavilion.	of a rhythmic signal understandable by an international audience, for example: the blows of a hammer which announce raising the curtain at the theater.	

5-16

Diagram of the interlude between presentations of the *Poème électronique*. 29 November 1957. Le Corbusier?

2nd/ And then it announces: "You are going to witness the presentation of the Electronic Poem, organized by the Philips company (Mr. Kalff, Artistic Director, and realized by the technicians in Eindhoven). The Electronic Poem is by Le Corbusier, the building is also by him [in association with] one of his collaborators, Jean [sic] Xenakis, the music being by Varèse, the montage of images by Agostini, etc . . . to show the new resources that electronic technology has placed at the disposition of the arts, sound, color, image, and rhythm."

All of this will take place during the hubbub of entering and exiting. "This way, the exit; this way, the entrance" will be the words repeated numerous times. Consequently, it is a sort of carnival hawking, in which it is possible to pack a lot of wit and content that can touch a crowd that by definition is inattentive.

In addition, it is necessary that there should be no confusion between these two minutes of introduction and the eight minutes of music by Varèse.

Consequently, it is necessary to use other techniques, some common sources of sounds set splendidly into music and with intellectual content.[85]

Two days later, Le Corbusier wrote to Kalff with the second sequence of the *minutage*. The letter included a diagram of the two-minute interlude [figure 5-16], although the character of the music is left unspecified.[86]

The realized interlude would not follow Le Corbusier's premature prescriptions. For one, Varèse did not accept the presence of Le Corbusier's voice within his work; he would not have his music serve as background music.[87] In addition, Xenakis had higher ambitions for his music than being used as a replacement for a carnival hawker. The matter was not easily resolved. Xenakis already chafed at his unenviable position of managing the project in Le Corbusier's frequent absence while receiving no credit, and there was disagreement in the conception of the score. Trained as an engineer and almost an autodidact as a composer, Xenakis, like Varèse, was moving toward compositions inherently based on sound and structure rather than melody or imagistic associations.

Privately, Xenakis was critical of Le Corbusier's use of photographic images to create the electronic poem. Instead, he believed, the entire production should address light and space and sound, a theme he would pursue in certain of his own projects. "Abstraction is taken in the sense of conscious manipulation of laws and pure notions," he later wrote, "and not of concrete objects. In fact the plays of forms and colors detached from their concrete context imply conceptual relations of a higher level."[88] He has told his biographer that Le Corbusier had asked him to compose some "chanson à la [Maurice] Chevalier," to which Xenakis replied that he should commission Chevalier directly:

He [Le Corbusier] was furious and accused me of vanity and pride. This

> discussion took place in front of the Philips representatives who were listening, astonished at my confidence before Le Corbusier. Later he asked me to make music with non-melodic sounds, noises and lots of rhythms. I reminded him of our conversation and he was annoyed again because he saw that I was right.[89]

Xenakis had hoped to spend three weeks in Eindhoven in order to use the Philips electronic facilities to realize his score. Le Corbusier, with righteous indignation, denied the request while reasserting both his control and his position, despite a basically paternalistic tone:

> *It is impossible [for you] to go Eindhoven for three weeks. Don't even think of it! We have an architectural office here; we are not an academy with students who may spend their time as they wish. We have strictly defined duties: these should be directed by the exceptional nature of your work and your earnings.*
>
> *I hope that you understand the situation. I will end my remarks with these words which are addressed to you as well as to your comrades in the office: the notion of what an architect is, is not clear in this office; the work is not efficient. It is nicely done, I'll agree, and I thank you for it. But when one claims the salary of the boss, one must really wear the hide of the boss.*
>
> *I am sending a copy of the first part of this letter to Mr. Kalff so as to any avoid confusion.*[90]

Forced to use the facilities of the Philips Paris laboratory, Xenakis produced a piece with a quite simple structure, later renamed as *Concrète P.H.* after the pavilion's hyperbolic paraboloid structure (from the French term) and the concrete nature of the sounds. Its source was the sound of burning charcoal, although the sounds were electronically reformed in the studio:

> It was based on the concrete sound of burning charcoal with its characteristic tingling, scintillating texture. The tape manipulation was simple, since the studio had limited resources; Xenakis confined himself to transposing and filtering the concrete sounds. He then created a "canon of electrons" by mixing one layer of differently transposed, distinct impulses to another layer in free and regular rhythms, the denser moments faster than the thinner ones.[91]

Xenakis had written Kalff in March 1958, outlining his plan to record the burning charcoal on three tapes of differing speeds, the lags in time and variation in sound then distributed over the full field of the pavilion's speakers.[92] The technical limitations of the studio, however, forced him to restrict his work to a single tape. Nonetheless, he remained hopeful about the resultant sonic effect.[93] Three weeks later, he contacted Kalff anew, describing the potential distribution of sound, but also noting that he had added thirty seconds of silence to the interlude. Two minutes was not sufficient for changing the audience, "even if you use tear gas."[94]

Xenakis was not pleased with the quality of the sound, which had been remastered in Eindhoven at a level below his specifications and his likings. He wrote to Kalff, expressing displeasure. But Kalff, although responding in sympathetic tones on 28 July 1958, seems to have regarded this as a minor matter compared to the other problems facing him: the lateness of the opening of the pavilion; technical problems with the sound and projection equipment, and no doubt the hot breath of the Philips management on his neck. The technicians had planned to have four months for recording and perfecting the sound distribution; in actuality, it had to be done—however imperfectly—in a matter of weeks. Kalff did acknowledge the problem with the Xenakis music, however, and thought that with the bugs finally resolved, they could re-record the piece. He invited Xenakis to Eindhoven to work on the revisions in August should he desire—the office would be on vacation, so there would be no trouble with the master this time. Tak could rework the tape and bring it up to acceptable standards.[95] The final versions, however, would be completed only in early August, quite late considering that the fair closed in October.

Within several months after the opening of the pavilion, Xenakis would break definitively with Le Corbusier. Xenakis's departure was due in part to the normal process of maturation, when a young architect realizes that he or she will always have a limited role in the creation of a building, for which another person will take all or most of the credit.[96] In this particular case, however, the departure was a turning point in a choice of career. Although he would create a multimedia structure at the Pompidou Center in Paris in 1978, and another building for a musical conservatory as late as 1982—both based on the use of hyperbolic paraboloids—Xenakis turned to music as his primary vocation.[97]

Due to the late refinement of the sound equipment, the pavilion was inaugurated 17 April, closed immediately, and re-opened the following month on 2 May—which enraged the composer. Varèse heard the performance of the piece in situ at the reopening and wrote Xenakis thereafter:

> Your piece . . . comes over as sound and spreads admirably—and mine, . . . comes over well too. Everything will be ready [for the public] next week, and I think we shall have all the sound and fury required.[98]

The Philips management, on the other hand, remained troubled by the music to the very end. In some ways, they seem to have been puzzled that the public was less outraged by the entire electronic poem, for it was something that most of the Eindhoven directors could not understand. The people did come, and the Philips pavilion, presented its *poème* to nearly two million people in the six months that the building was open. Clearly, the artists, architects, and technician had achieved their goal in terms of corporate public relations. As an artistic creation, however, the production was not beyond criticism.

The public was often mystified by the presentation they watched and heard within the pavilion. But perhaps in a curious way, the strange film of effigies and statues set and rotated in deep black space, all encompassed within the shaped walls of a strange building, provided the score within its own suitable context, far from the concert hall where the classical symphony still reigned. While few—if any—would claim to understand the Varèse score, or to find it beautiful, a handful of critics felt that they were in the presence of something profound: "Here, one no longer hears the sounds, one finds oneself in the heart of the sound source. One does not listen to sound, one lives it."[99]

Le Corbusier was always certain that the selection of Varèse had been the correct one, and he retained only the highest praise for the musician. In 1960, he told Fernand Ouellette, Varèse's biographer:

> *Varèse is a craftsman, truly. He knows that the job is [to put] sound after sound, relentlessly, without faltering, from one to the other with a relentless drive. He hasn't encountered the muses of music during his life! Their wings haven't brushed him. I am in the same situation: no brush of the wings but [instead] the work of the damned or a dog or a monk, in any case [working with] perseverance and minutiae.*[100]

Albert Jeanneret, seventy-two years old at the time, in fact, had been completely taken with the music, and preferred it attended alone, unfettered by his brother's visuals:

> De Bruin let me hear the music alone between two complete performances and I was filled with joy by that powerful and varied score. I like it much more than when linked to the visual elements, because the two together lessen and dissipate each other. The human voice at the end, takes on the feeling of a song of liberation, the diagrams for the Radiant City appear as programmatical answers to the images of destruction; the shells, the stone figures as elements of possible perfection. One feels a sense of progress in all this already encountered previously, whereas the music on its own is truly a spontaneous creation endowed with that infinity toward which all music tends. It draws its strength from an originality untouched by a propaganda whose methods have already been encountered elsewhere. I repeat: for the public, for society, for Philips, it is a success.[101]

At the close of the fair in October 1958, as the pavilion was about to come down, some limited efforts were made to maintain the pavilion for electronic works by other composers. Given the amount of technical equipment and the unusual configuration of the struture, the architecture was essentially a sounding room. It is such an obvious idea, that one wonders why it hadn't been thought of earlier. Some proposals were made to preserve the building, but it was equipped neither for weathering nor for the Belgian winters. The pavilion had been constructed without insulation and its multitude of wires were designed to endure for only a short time. Retaining the pavilion proved to be impossible.[102] The building was to pass, and with it the true setting for the Varèse *Poème électronique.*

This music would last, however, in recorded form, first as a tape and later as a record, and it is again available on compact disc.[103] By many, it is regarded as a landmark in the early history of electronic music, if ranked below other, more fully developed Varèse compositions such as *Déserts* and *Ionisation.* In other ways, however, the music no longer exists. Conceived and executed for a specific space, and perhaps more importantly, utilizing a specific sound system related to that space, no recorded performance today can equal the sounds experienced within the pavilion. A recording can on rare instances capture the excitement of a live performance as well as the sound made by the instrument or voice. But no recording can conjure the space as well as the sound—until, perhaps, virtual reality is perfected.

Like the building, the music is gone. There are spectrographs and there are recordings and reconstructed recordings, just as there are photographs of the building. But the experience of the music, like that of a performer deceased, can be only suggested and not re-created. There are some sketches, but there is no written score—one of the reasons that the work possesses an obscure fame. Like the building, the full experience of the music can be described as demolished.

In a 1959 lecture at Sarah Lawrence College, Varèse reflected on his work in Brussels. He described the play of images and lights, and the means by which the sound was distributed along the "sound routes." In the Philips Pavilion, Edgard Varèse had been able to realize a venture that had begun in the first decades of the century, and that he could realize only late in life, at the age of seventy-five. He closed his talk with the reflection: "For the first time I heard my music literally projected into space."[104]

AN ANALYSIS OF THE *POÈME ÉLECTRONIQUE*

RICHARD FELCIANO

Varèse seems first to have chosen—presumably for unity—a limited vocabulary of "sound objects," which nevertheless contain a substantial variety of shapes and gestures capable of projecting a variety of psychological effects. The *source* of the sound was probably not of much importance to him. The gong or bell that opens the piece, for example, may have been recorded in Holland, where there are many, but once recorded he seems interested in it only from the standpoint of abstract acoustical properties. The bell is heard as a pronounced attack followed by slow decay—in traditional terms, a diminuendo. Later, the process is reversed, with a slow crescendo ending in an abrupt cutoff. In other places, the two are combined, for example, as a sustained note gains slowly in loudness and then becomes slowly softer again. In this attempt to make much out of little as he built a cohesive form, Varèse can be seen as an artist of traditional skills in spite of the radical nature of his materials. His obvious desire for formal control separates him from most of the others who created *musique concrète*.

His approach, which also appears in some of his orchestral works, limits the number of ideas, then constantly, yet unpredictably, refers back to earlier musical events during the progress of the work. None of this appears in any predictable way, however. In the *Poème électronique*, these ideas include bells, sirens, percussive sounds including snare drum, sustained oscillator sounds, mellow staccato sounds, ascending and descending glissandi, motoric sounds, water-like sounds, aviary-like sounds, glass-like sounds, female voice, and male chorus.

The piece is eight minutes long and divides itself, to my ear, into roughly two minute segments. All the basic ideas except the voice appear to be exposed in the first two minute segment. Then follows what could be termed a sort of "development": many of the ideas return (glissandi at 2′; the bell at 2′35″, the sustained crescendo/diminuendo oscillator sounds at 3′20″), but they are now presented in thick textures made of higher and lower versions of the same sound, or by presenting diverse sound objects simultaneously. In contrast, at the beginning of the piece, they were presented one at a time in a straightforward expository manner. The fragmenting and compacting of data is a classic developmental technique, though in the past the data was melodic in nature, whereas here it is sound object.

The "climax"—more a structural downbeat—takes place around 4′20″ when a sudden increase in loudness accompanies a fast upward glissando and then ends with a sustained diminuendo. This might be intended to function as a psychological equivalent of a recapitulation, though surely not a return to the beginning. Speaking abstractly, the section that follows the 4′20″ mark can be seen as presenting new variants of previously heard material (glissandi, crescendos, voice moans) but in a less compacted manner than at 2′.

To render the climax (structural downbeat) at 4′20″ effective, a structural upbeat would have to be created in the previous passage; *this*, I think is the significance of the introduction of new material (the voice moan) at 3′45″ and 4′; the newness draws attention to the sound, and the moan is a *soft, downward* glissando, a "set-up" for being interrupted by a **loud,** *upward* sweep at the 4′20″ climax. In terms of the larger form, this means that the upbeat preparation at 3′45″ and the climax itself at 4′20″ straddle the 4′ halfway point of the piece.

To my mind, the other pronounced point is the Grand Pause, a musical term for an exceptionally long cessation of the musical proceedings: it occurs at 5′45″ and is quite noticeable. I believe it to be the upbeat or psychological preparation for the final section, which begins at 5′50″—again roughly a 2′ sectional division of the piece. That section begins with the *longest* restatement of the crescendo/diminuendo oscillator sound, suggesting a slowing of action and finality.

The midpoint (4′) voice moan upbeat followed by the glissando downbeat discussed earlier is echoed midway through the final section, although this time it is actually music sung by a soprano at 6′50″ and a male chorus at 7′05″. These are followed by glissandi, as at 4′20″, but they are at first shorter and both ascending and descending, then lead to the longest (again suggesting finality) upward glissando of the work: for me, the concluding declamatory gesture. This means that in the final 2′ section occur the longest crescendo/diminuendo figure and the longest upward glissando. There seems little likelihood that this was other than to instill a sense of arrival.

In over-simplified terms we have:

0′ Presentation of Ideas (Exposition)

2′ Compacting of those ideas (Development)

4′ Climax and thinning out of continuously varied material ("Recapitulation")

6′ Grand Pause and broadening (extending) of material for conclusion (Coda)

6. REACTION

In the American press, the Brussels exposition was seen as nothing less than a cold war competition between the United States and the Soviet bloc [figure 6-1]. There were to be an estimated 175,000 to 200,000 visitors at the fair each day. Each had come to see the roughly two hundred pavilions, and they were simultaneously wooed by the two major antagonists and the multitude of other nations and corporations. The battle was waged on cultural grounds, in the form of the pavilions, the presentations of performing arts, and the content of the exhibitions.[1] Round in shape, the United States pavilion by Edward Durrell Stone featured an open interior roofed with a suspended structure, tightened almost as the of the spokes of a bicycle wheel [figure 6-2]. The perforated concrete block that sheathed the vast interior announced the decorative style Stone would use in the coming decade. Those of more purely structural or modernist taste were disappointed in the building's exterior. "In my own case," Stone stated in his defense, "I feel the need for richness, exuberance and pure, unadulterated freshness."[2] The planners of the American displays, on the other hand, were more pragmatic and tried to "suggest that 'something happens to be living in America . . . which may be called progress, real freedom, and better living.'"[3] To the credit of their makers, the exhibits included both the voting booth intended to represent democracy and photographs of the social unrest that accompanied the civil rights movement. A fifteen-minute film, shown in 360-degree projection, wowed the audience with grand scenery, and eventually became a permanent fixture at Disneyland in Anaheim, California. In addition, an early computer, the IBM 305 Ramac, answered questions in "ten languages in ten seconds."[4]

The Soviet Union displayed a model of the recently launched Sputnik, which had become a symbol of the Socialist eclipse of American space technology and education in the sciences [figure 6-3]. The experience of the fair ran the gamut, however, from the highly intellectual to the broadly popular, and the visitor could easily ignore the political undertones that propelled the various national exhibitions. One could "stroll through the sculpture garden, view the large and somewhat controversial exhibition of American art, or sample the latest U.S. atoms-for-peace proposals, including one of the world's first nuclear clocks."[5] After all, it was a world's fair, and fairs should be fun.

The public and professional reaction to the Philips Pavilion was understandably mixed, ranging from baffled mystification to criticism of its form. The press seemed to understand that this project had been a serious enterprise, and yet both audience and critic alike could not help but laugh at the whining or blips of the Varèse score. A journalist for the Dutch newspaper *Eindhoven Dagblad*, who had attended the official opening, noted that a high-ranking director now had marks on his arm where his wife had pinched him hard during the presentation of the visual poem. The reporter was a bit ner-

6-1
Aerial view of the Exposition.
[*L'Architettura*]

vous, not that many visitors might have their arms pinched, but that they would laugh at the presentation—when in fact there was nothing at which to laugh. The images of concentration camps and nuclear explosions reflected Le Corbusier's concern for humankind and were hardly humorous. The form the presentation took, however, puzzled the reporter. While he was not put off by the pavilion itself, he sensed that it couldn't "contain" the music, some of it pleasant to listen to, much of it horrible. As a totality, the journalist termed the *Poème électronique* a "modern nightmare," ostensibly discussing the content of the visual and aural presentation rather than its challenging form.[6]

Quite understandably, the pavilion received its greatest coverage in the Belgian newspapers—most of it published just after the press preview and apparently drawn from what must have been a standard publicity release. In spite of the confusion the spectacle caused, the power of the *Poème électronique* touched the critics: "One gets a choking and gasping feeling and the eyes and ears undergo the sensation of a chaotic struggle. . . . The intensity of this poem is so baffling that the eight minutes seem to be an eternity."[7] The premonition was created by the building itself, the shape of which "in itself is like a fierce cry."[8] One journalist, although clearly moved by the experience, queried: "Is this art, where so many volts are required?"[9] Another writer ambivalently questioned whether this was a presentation to be "contemplated or endured"; his final judgment, however, was that "it will leave no one unmoved."[10] The novelty of the presentation and the mixed media by which it was realized outweighed any contemplation of the poem's message, although virtually every writer was palpably affected.

The building's form was strange, bizarre, even provocative. What did it do and what did it mean? Jean Girard, writing in *La Cité*, declared that:

> the answer will not be given until the spectacle has started. For in the Pavilion, void, naked, baffling, is a spectacle to be seen. It's there, moreover, that its greatest interest lies, its greatest originality. This spectacle is not in the least banal. It has never been seen before, hitherto unknown, sensational. What is certain is that, technically speaking, the spectacle presented at the Philips Pavilion is a fantastic realization. Here, undoubtedly, technology must stand aside to art. [It is] undeniably a work of genius.[11]

The correspondent for *La Lanterne*, in an article titled "Le Corbusier: The Man of the Day," recalled, like Le Corbusier, the discord and rioting that had accompanied the premiere of Stravinsky's noted composition, and concluded that the *Poème électronique* was "*The Rite of Spring*, but in 1958."[12]

One writer described the impression of the whole as "hallucinatory," noting that "you plunge into an extraordinary world."[13] Only the correspondent for *Echo du Central* sought the origins of the spectacle, or parallel explorations, in works of art.

6-2
United States Pavilion, Brussels. 1958. Edward Durrell Stone, architect. [*L'Architecture d'Aujourd'hui*]

6-3
The Pavilion of the U.S.S.R., called "The Refrigerator." Y. Abramov, A. Borestki, V. Doubov and A. Polanski, architects; Y. Ratksevitch and K. Vassilieva, engineers. [*L'Architecture d'Aujourd'hui*]

> It seems that Le Corbusier and his collaborators were inspired by the connections [André] Malraux had developed in his "Musée Imaginaire." But they have gone much further than the writer, since they have linked photographic reproductions of human documents and works of art, not only to light and color, not only to sounds, music, noises and cries, but also to speed, everything with accelerated motion. The use of this movement adds to the impression of the instability, and the insecurity that the modern world feels to the point of despair.[14]

Howard Taubman, the correspondent for the *New York Times*, characterized the entirety as "the strangest exhibit at the Brussels's World's Fair," presented in a "building that looks like an improbable and gigantic seashell frozen into silvered concrete," and added that the "effect it has on the thousands who visit it each day ranges from bewilderment or distaste to admiration."[15] As to the score: "The sounds that accompany these images are as bizarre as the building. . . . One hears rattles, whistles, thunders and murmurs." Almost surprised, Taubman offered that "at one point there is a sound that seems to emerge from a human throat." Compared to the controversy debating the meaning and aesthetic value of the Philips Pavilion, the differences between the U.S. and Soviet exhibits were minor.

The editor of Philips's own *Announcer* expressed justifiable pride in the company's presence at the fair. Yes, there had been problems with synchronizing the sound and illumination, and the consequent late opening of the pavilion had forced the fair's scanty coverage. There was triumph nonetheless:

> There has been some delay. But that hardly counts if measured against the grandeur and advanced conception of our many-sided effort. . . . It will not take Philips people very long to be . . . proud of what has been achieved with the "Poème électronique." As yet their numbers may still be small, but they include, at any rate, the editor of the Announcer.[16]

It took a bit longer to convince the Philips management.

While the newspaper reports described experience rather than meaning, the professional press tended to dwell on the technical rather than poetic aspects of the pavilion and its multimedia presentation—and the bewilderment and ambivalence of professional reportage equalled that of the lay press. Presumably due to the early publication date required by the April issues of the respective journals, many of the pavilions were visited before they were officially opened. In many instances, one must assume, the exhibitions were not yet complete. Most coverage of the Philips project was restricted to a factual description of the building and its involved geometry. There was virtually no probing inquiry into what it all meant. Note was made of Le Corbusier's authorship of the projected poem, but presumably none of the contributors to the English, French, or Italian architectural journals had, in fact, experienced the eight minutes of "modern nightmare" because the pavilion had closed for a month of adjustment and reconfiguration immediately after its formal inauguration. Philips was not alone in being incomplete at the opening of the fair, however: the pavilions for France, Italy, Brazil, Morocco, and Spain, among others, were not finished.[17]

The appropriateness of structural expression for architecture in general, and exhibition pavilions in particular, was the principal issue discussed by the architectural press. The production of new alloys and improved means of calculation and fabrication fostered buildings that muscularly expressed their structural systems. The theme of the entire Brussels exhibition, paralleling its times, was scientific progress and the relation of humankind to technology. The symbol of the fair, the Atomium, was based on the form of the steel molecule enlarged to gigantic proportions and intended, like the Eiffel Tower, to function as a sign and place marker. Unlike Eiffel's structure, however, the Atomium was more literary than structural, a fact that did not go unnoticed by critical observers: "[I]n the case of the Atomium . . . the structural side of it is, to say the least, a makeshift."[18] Not all the architectural writers were charmed by the exaggerated constructions found in Brussels. Several structures, including those for Brazil and the O.E.C.E. (Organisation européenne de coopération économique, the predecessor of the European Community),[19] used suspended roofs, and these immediately suggested that the structural system or the dimensions of the span were more important than the requirements of the internal spaces or the exhibitions within. While the French pavilion,[20] in the tradition of the Galerie des Machines at the 1889 Paris fair, used a structure of hyperbolic paraboloids to enclose a considerable volume free of interior columns, many of the buildings, Philips among them, restricted the potential of their structures to formal applications alone.

If the structure of the French pavilion was an impressive tour de force, as architecture the building was judged far less a success. In *L'Architettura*, Renato Pedio condemned the building as a structural exhibition that lacked any intrinsic quality. Towering 65 meters above its entrance level, its hyperbolic paraboloid roof "forms a metallic cage of this 'machine' representing the existing entirety of French society. . . . The internal reality does not correspond to the spectacular 'facade,' and it all ends in bad taste."[21] René Sarger, who designed the suspended roofs of both the French and the Marie Thumas pavilions,[22] contributed an article, "The Structural Worth of Structure," to *L'Architecture d'Aujourd'hui*'s coverage of the fair. Philips was singled out by the engineer for praise:

> Has not Le Corbusier, from the beginning, thought of a tent whose skewed interior surfaces disappear, curve, twist, and fade; the suggested dynamic

> volumes becoming part of the "Poème électronique" presented there? But doesn't the technical design correspond to the essential [architectural] concept?
>
> Here is a beautiful spatial investigation that seems to be hampered by the lag between architect and builder.[23]

Little did Sarger know of the role played by the builder in the realization of the pavilion.

In reviewing the full worth of the exposition architecture, Bruno Zevi found few true architectural developments among the structural noise that assumed the role of cultural symbols at the fair. Zevi's bias towards the "organic" architecture of designers such as Frank Lloyd Wright and Carlo Scarpa left little room for appreciating buildings that were interesting primarily as advancements in engineering. Lacking were both true spatial developments and architectural form whose worth was more than its structure. To Zevi the sculptured symbolism of Le Corbusier's chapel at Ronchamp was, without doubt, legitimate, but:

> At Brussels 1958 you can't find even one [spatial idea], and the enclosed exterior spaces around the Swiss, German, Finnish, and Italian pavilions do not sufficiently compensate for this deficiency. At the architectonic market of Brussels, they sell all sorts of merchandise: the rational in all sauces, including the hexagonal; the "gestural," the rustic, the folkloric, the brutalistic, the nuclear and the traditional. Missing is only the organic, because this does not express itself through extrinsic symbols, but grows from the inside out and does not humiliate architecture with intellectual metaphors or with functions that flatter.[24]

Craig Ellwood, a Southern California architect noted for his clean, modernist contributions to *Arts and Architecture*'s Case Study House program, offered an overview of the fair for *L'Architecture d'Aujourd'hui* and challenged the overly expressive forms and structural gymnastics that informed pavilion designs throughout the fair. Ellwood's own house designs often used a steel structure to order the plan and its spaces; but structure complemented, rather than dominated, the architecture. Ellwood, too, looked to Ronchamp to understand the architectural quagmire of the present fair, and was clearly troubled by the many bastard offspring of Le Corbusier's creation.

> The chapel of Ronchamp expressed a profound movement of revolt against machine construction. It marked a point. Already among us have appeared numerous pseudo-Ronchamps, determined adepts of the bearing wall and of the need for plastic investigation.[25]

The plasticity of Philips was not that of Ronchamp. Although based on complex geometry, the pavilion was executed by artisans. The architecture expressed the era of the machine without requiring its participation, and one could say that in spite of its nonrectilinear form, the Philips Pavilion expressed less the machine than its successor, electronic technology.

In contrast to the expressionism of the more structuralist pavilions were several designs that pursued the narrow path of refined elegance. Among these were the classic simplicity of the Norwegian pavilion by Sverre Fehn and the clean sophistication of the Transportation Pavilion by T. and F. Hoet, Segers, H. Montois, and R. Courtois.[26] With a simple plan and structure, this tranquil building employed cigar-shaped trusses representing accepted construction that gained a presence through design refinement and detail.[27]

The Dutch Pavilion—officially by Van den Broek, Bakema, Boks, and Rietveld, although the last had been forced from the project—combined a number of prismatic masses with a bubble-shaped auditorium spanned by lateral arched trusses [figure 6-4]. Its theme, "Man Against the Sea," was frequently cited as one of the few that addressed reality free of political propaganda. During the course of design and construction, it will be recalled, there was some mounting apprehension within the Dutch committee that the exuberance of the Philips Pavilion and the reputation of its architect would overshadow the simplicity of the Netherlands section. There was no need to have worried: the Dutch building and exhibitions were perhaps the only elements of the fair that received unanimous praise. Bruno Zevi, who reproached buildings throughout the fair, found in the Dutch pavilion a model: "There is a single exception to the rule of symbolism, and that is the Netherlands pavilion which illustrates something real—the fight against the sea. The most serious point is that symbolism cannot create interesting interior spaces. In order to create true human spaces, architecture (and the architects) must be freed from the function of adulation."[28] Why is the Dutch Pavilion better than the others? questioned Renato Pedio. "Perhaps because at its roots there is no great idea to project into a pure architectural form. The four prisms are not exceptional in conception, but their sunscreens are sufficient to enclose their space, and the sculpture in brick to animate it."[29] The British, too, were impressed by the quiet dignity of the Dutch work and its ability to balance construction and expression:

> Those pavilions which are the most satisfactory from the broad point of view, are, like the Dutch pavilion, those which combine into a single entity the problem of providing a satisfactory architectural shell, and which, at the same time, carry a massive load of popular imagery.
>
> The Netherlands pavilion is interesting because, here, the display has created out of its sequences a series of excellent buildings. The detailing is always careful and precise, and the use of rough materials with exposed

6-4
The Pavilion of the Netherlands. Brussels 1958. Van den Broek, Bakema, Boks, and Rietveld, architects. [*L'Architecture d'Aujourd'hui*]

6-5
The rear of the Philips Pavilion seen over its neighboring buildings. [*L'archittectura*]

steelwork and plate glass and slatted fascias gives the whole exhibit a lightness which is not shared by the other pavilions.[30]

Given that reportage of the fair was essentially photographic, the Philips Pavilion did well. Featured in dramatically composed images that captured the raked lines and skewed entrance and exit portals, the pavilion was widely published as one of the most striking works at the fair. The awkwardness of the building's form is apparent only in other views that illustrate the structures of several nations and that record the Le Corbusier work seemingly by accident [figure 6-5]. As an image, however, captured and edited by professional photographers, the Philips Pavilion achieved the remarkable success Kalff had predicted at the initiation of the project. As he had also predicted, the Philips corporation garnered immense publicity in newspapers and periodicals around the world [figure 6-6].

Architectural journals used few words in describing the *Poème électronique*, probably because of early publication deadlines. Perhaps the most poetic evocation of the spectacle appeared in the Swedish architectural journal *Byggmästeren*. Kristian Romare, an editor, tried to understand the creators, the design, the presentation, and the experience in an article titled "Le Corbusier's Electronic Creation":

> The final picture—a child accompanied by a piercing sound and an unbridled crescendo—offers us Le Corbusier's evangel without pathos: the human's unique lot is, in the end, for good over evil and to continue.
>
> For me, this "electronic poem," although it stands as an ephemeral exhibition, is a deeply fascinating realization of a dream which has tempted artists since the Bauhaus, since Kandinsky—and even Wagner: the dream of the total work of art. It collects elements of the art of our time. It touches upon, with analytical understanding, the secrets and great myths such as graphics by Picasso and his *Guernica*. It developed the cubist technique of simultaneity; the surrealist use of shock montage; and more recently, kinetic and space-sculpture investigations. Le Corbusier is an architect, painter and sculptor. Here the seventy year old enters as a completely radical artist.
>
> The chapel at Ronchamp has become his own liturgy in Brussels. He has not made a piece of art for "high-brows." One is struck by the strength of these nearly banal effects: the tunnel of horrors; Tarzan films; and the electronic game machine. Le Corbusier has used the world exhibition to reach people, or more precisely, to preach [to them]. He is never solely an artist satisfied with creating indigestible subtleties. This is a poem without distance between the observer and the artwork. We are literally in its stomach; it is as if [the pavilion] is literally digesting us, and exposing us, against our will, to acids that etch us indelibly.[31]

The doubt felt by many of the pavilion's architecturally sensitive visitors, the ambivalence between appreciating the building's animated and dynamic accomplishments while recognizing its formal shortcomings, was poignantly expressed by the Italian architect Ernesto Rogers.

> Is it beautiful? Is it ugly? It matters little if one manages to acknowledge that this disconcerting artist is able nevertheless—having now reached a respectable age—to show us unexplored horizons behind which are sensed the promised lands of new experiences. This Philips Pavilion, let's come right out and say it, doesn't reach the heights dreamed by its author: to provide a consistent depth to his ideas, Le Corbusier has had to make a number of compromises, forced by the demands of the situation. Where the result should have been a fluid sequence of convexities and concavities—as in the [example of the] ear lobe, there are disturbing elements for reinforcing. Nevertheless, if at Ronchamp every external surface seems to reverberate with ambient pre-existences ("acoustic relations"), here, too, there collects on the interior the sound of musical instruments continuously filling the volume and enlivening even the most distant corners: like a wind that tenses and supports a sail in stable equilibrium. It is not a fulfilled architecture, it is not a clear composition: it is only the indication of new architectonic dimensions. It is a poetic seedling that at another time—either in the work of the author or another who will know how to reap and transform the message—will probably manage to flower.[32]

Rogers's comment is arguably the most perceptive of the many critiques cast upon the Philips building. Most critics approached their subject with a template through which they tried to measure their personal values. Rogers's own values had themselves evolved. A year after the fair, he retorted an unfavorable criticism by Reyner Banham on the historicism that had emerged from the purity of Italian Rationalism. Banham pondered how this was possible. Rogers recounted recent history, both cultural and political, as a means to explain stylistic developments:

> Little by little the thematic became richer, and as a consequence the exigencies became subtler. Formal issues, therefore, became more difficult, because they tried to encompass an ever greater number of propositions—the widening of the architectural problematic and the immediate effect of critical thought; the historicist revision of all historical periods and especially those closest in time, which had been distorted through the normal opposition caused by the dialectical traffic between generations.[33]

While Le Corbusier might not have consented to the investigation of recent style as such, he probably would have agreed with Rogers in his broad cultural investigation. With Le Corbusier, however, this broader view of culture

6-6
Louis Kalff, Le Corbusier, and Edgard Varèse before the Philips Pavilion. [Philips]

informed to a greater degree the electronic poem than the making of its architectural container.

It has been said—and sometimes attributed to the actress Elizabeth Taylor—that there is no such thing as bad publicity. While the Philips Pavilion and its *Poème électronique* mystified many of its visitors, few were left untouched by the experience. Since its visitors included scores of reporters and journalists, that sense of mystery was broadcast throughout the world. If professional writers were at times critical of the building's form and structural ideas, they did not fail to acknowledge this joint achievement by Le Corbusier and Philips. The popular press contributed to the notoriety generated by the project. Whether successful as architecture, as music, or as film, the Philips Pavilion project was an unqualified success as a source of publicity. And that was, in fact, the very reason the project had been conceived.

To Le Corbusier, however, the Philips project represented his confrontation with electronic technology, which had both taken from and given to the human spirit:

> *A medium of action has appeared: electronics, outside the human scale, drawing from physics extraordinary resources of power, volume, delicacy or violence, slowness or speed,—a physical and mechanical event that places in the hands of man almost unlimited forces, and amazing sources of psycho-physiological action through the medium of*
>
> *light color rhythm sound image*[34]

These were, in fact, not only the characteristics of electronic technology, but also the elements of the electronic poem presented at Brussels. What the medium offered was unlimited versatility coupled with universality:

> Offering *to the modern world (in all latitudes, all longitudes, all climates) the* electronic plays *capable of stirring the emotions of men and women, capable of reaching the heart of the individual as well as 1,000 or 10,000 or 100,000 spectators and listeners.*[35]

Although the pavilion at Brussels was small, revealing the *Poème électronique* to only 500 visitors at a time, it was intended to be, however modestly, the beginning of a new art form.

The building was neither intended nor destined to endure. Early discussions at Philips broached the possibility of demounting their Brussels structure and re-erecting it in Eindhoven as a museum. The discussion did not proceed very far; the parties concerned felt that the cost and the strain it would place on the design and construction of the building would outweigh any long-term benefits of relocation. In January 1957, as the design and construction brief became more precise, Kalff formalized the structure's temporal existence:

> After September 1958 the contractor will demolish the building, gather up all the debris and as much as possible, will return the site to its original state during the time provided by the directors of the exhibition.[36]

Having come this far, however, Le Corbusier wanted the pavilion to remain as a sort of research laboratory for "electronic play."[37] This was, after all, a restless period in the arts that witnessed the chance encounters of John Cage's music and Merce Cunningham's dance, the assemblages of Robert Rauschenberg, the Situationist spectacle of Guy Debord, the origins of the Fluxus group, and happenings by Alan Kaprow and Claes Oldenburg. Le Corbusier found a willing conspirator in Stanislas Jasinski, president of the Belgian society of architects. Having been profoundly moved by the Philips building and its spectacle, Jasinski wrote to "the master" in late September, offering to undertake the project of preserving the structure.[38] Realizing that time was of the essence, Le Corbusier responded almost immediately, outlining in detail his ideas for a facility that could be placed at the disposition of dramaturges, poets, people of theater, artists, all using the potentials of technology to "touch audiences small and large." He recounted how 50,000 people were to be embraced in the ceremony for laying the corner stone in Chandigarh—all due to electronic sound enhancement. Le Corbusier hoped Jasinski could convince the Belgian government to develop such a center in Brussels on the site of the exposition: the modest Philips Pavilion would be set within the parks' existing trees.[39] Additional letters were exchanged, with Le Corbusier further detailing his ideas and his cautions. He warned to keep the board members local: "Avoid members at long distance who cannot participate practically in the meetings. It is useless to be pompous! It suffices to be effective (ardent, lucid, courageous)."[40]

The efforts were in vain. Charles Spaens, head of Philips in Belgium, informed Jasinski that after discussing the matter with the Philips directorship, they had decided the building could not be maintained.[41] For one, the structure was without heat or insulation and could not protect the sensitive electronic equipment from the hazards of Belgian winters or maintain a comfort level for visitors. And more critical, the prestressing wires and their connections were in no way designed to withstand the elements or the onslaught of oxidation. From the start, everything had been designed to be short-lived. Le Corbusier's contribution would be maintained only through photographs and memories.

The following February, Kalff wrote to the architect informing him that demolition was under way. It did not prove to be an easy task. Although the building had been prestressed piecemeal, there was no way to effectively reverse the procedure. There was considerable apprehension that wires—wires under great tensile forces—might snap, hurling panels in all directions with grave danger to the workers. To circumvent these threats, the pavilion would be imploded. "[Demolition is] a difficult and fairly dangerous job," Kalff wrote, "but one which proves one more time the remarkable resistance provided by the hyperbolic paraboloid surfaces." An emotional element accompanied the work: "It is a task which makes us a bit sad; nevertheless, it is necessary since the avant-garde creations at the exposition are only remarkable for a very short time. Already, to see again the few pavilions remaining on the site of the exposition gives us very little pleasure."[42]

At 2:00 P.M., 30 January 1959, the pavilion for the Philips corporation was reduced to rubble.[43]

6-7
Completing the site work around the pavilion. [Philips]

7. THE ELECTRONIC PAVILION RECONSIDERED

Almost forty years have elapsed since the coming and passing of the Philips Pavilion, and today the building is cited only as an egotistical aberration in the life's work of a modern master. As a realized building project, it receives only summary treatment in the total oeuvre of Le Corbusier. Agostini's accomplishments are usually tallied with his work with Robert Bresson, Max Ophüls, Marcel Pagnol, and other directors[1]; Jean Petit is known as an editor, publisher, and graphic designer. Iannis Xenakis became a composer of renown, perhaps impelled by the Philips commission. For Edgard Varèse the Brussels venture was one of the most important of life's moments. In closing the chronicle, we should again consider the Philips project, its accomplishments, its shortcomings, and the reasons for its subsequent neglect.

Without much exaggeration, one could say that the pavilion—conceived as the "bottle" for the presentation of the *Poème électronique*—has been, with rare exception, forgotten. Certainly, the destruction of the building has made it difficult to preserve any of its original power or to judge the scale of its accomplishment. No artifact tempts, tantalizes, or greets the student, the architecture tourist, or the historian. Only photographs remain, and even these are relatively few. Given its constrained site and its elusive hyperbolic paraboloid form, the pavilion was difficult to capture on film. So ambiguous is photography as a medium that even knowledgeable Le Corbusier scholars have mistaken the precast concrete building for a fabric tensile structure.[2] Second, for Le Corbusier the pavilion never was an important building, although the electronic poem was the preeminent example of his attempt for a synthesis of the arts. From the first meetings he appeared reluctant to design the exterior of the structure; his thoughts were directed almost entirely on the electronic spectacle to be presented within. Even in the *Oeuvre Complète*, the multivolume anthology by which the architect shaped and controlled his image and immortality, the Philips project figures in only a small way. Two pages alone are devoted to it in the 1953–1957 volume; the collective 1910–1960 volume presents visions of the film projections as well as the architecture, featuring the monsterlike image of a African tribal statue: this very same image had appeared as early as 1925 in his book *The Decorative Arts of Today*, and probably appealed as much for its drama as its revelation of human genius. This also describes the architect's attitude toward the creation of the *Poème* as opposed to the building in which it would be screened.

If his sketchbooks and correspondence can be used as an accurate gauge, Le Corbusier was far more involved in the possibilities that electronic sound, light, and automation technique could offer the projected spectacle than in the constrained realism of construction. His interest in color translated into the projected chromatic *ambiances*, their effects augmented by the striking imagery of the film/*écran* and the "organized sound" by Edgard

7-1
The chapel of Notre-Dame-du-Haut at Ronchamp, France. 1955. Le Corbusier. [Marc Treib]

Varèse. In any event, Le Corbusier was involved in the creation of an entire city in India, a project that dwarfed the Philips commission in size, importance, and immortality. One could argue, in the architect's behalf, that due to other commitments and overseas travel he probably couldn't have done much more on the design even if he had wanted to. A monastery and new capital necessarily took precedence over a short-lived pavilion. But the architect's effort feels incomplete.

For a mind that had once seen architecture, autonomous and pure, as the magnificent play of masses and volume in light, the denial of perceived interior space was a radical departure. The Philips project thus represented a schism in Corbusian architectural thinking by separating the world of physical perception from the world of contrived illusion. That schism had recurred in his thoughts since the 1930s. Recall his statements on the effects of the mural and the painter's ability to "blow open all the doors to the deep land of dreams."[3] To the painter's tools he now added the photographic and cinematic apparatus. He was impressed with the scientific mysteries revealed by the camera and the motion picture, "mysteries of the universe which previously have been beyond our perception."[4] In 1933, writing about film, Le Corbusier noted:

> *But documentary revelation was only one aspect of the camera's power:* But I want the lens to disclose the intensity of human consciousness to us through the intermediary of visual phenomena which are so subtle and so rapid in nature that we have no means ourselves to discover and record them; we are unable to observe them, we simply feel their radiance.[5]

Light animated the external form and interior spaces of Ronchamp, whose "curved volumes [were] governed by rectilinear generators [figure 7-1]."[6] In architecture as in the motion picture: "The key [was] light and light illuminates shapes, and shapes have emotional power."[7] At Ronchamp the extenuated towers brought light deep into the chapels; splayed window openings dispersed the daylight while modeling the heavily textured wall surfaces [figure 7-2]. The solidity of the bearing walls and the ephemerality of changing light joined in a religious symbiosis.

In the Philips Pavilion the curving interior planes disappeared in the darkness of the interlude or beneath the projected colors and recognizable imagery of the *Poème électronique*. In response, the exterior of the building was forced to condense the symbolism necessary to carry the entire project; form, if intuitively derived, was mathematically tempered [figure 7-3]. Geometry, the creation of the mind, governed the architecture that contained a poem to humankind. Body and geometry; humanism and technology; chaos and order; despair and ultimate hope, remain the underlying oppositional messages of the poem and its pavilion.

7-2
The play of light that streams through the splayed windows of the chapel interior. Notre-Dame-du-Haut at Ronchamp, France. 1955. Le Corbusier. [Marc Treib]

Perhaps it was the disputed authorship that troubled the design midstream, that stained the purity of Le Corbusier's involvement with the building's forming. The existing documentation supports Iannis Xenakis's claim that he, and not Le Corbusier, was the major—if not the sole or instigating—architect of the Philips Pavilion. Le Corbusier's sketchbooks include precious few doodles for the building, instead exhibiting notes testing color combinations, and lists of image sources, objects, and buildings offered in sequence. Even as late as 1957, drawings in his sketchbook—if these are, indeed, sketches of the pavilion—are quite vague in portraying the geometry of the complex intersections of hyperbolic paraboloids.

And yet, as Kalff claimed, the very idea of the "stomach" form, the use of mathematics, and the concept for a poem dependent on and showcasing electronic media, was Le Corbusier's own—although the seeds for the idea may have first germinated in the Philips company. The sole drawing for the pavilion design in Le Corbusier's hand and signed by him, however, is a patent tracing of a perspective painstakingly plotted by Xenakis and redundantly included as a part of the first client presentation in October 1956. From that point on, the business of design development, refinement, and adjustment instigated by the exigencies of construction reality, were assigned to Xenakis, who executed his tasks studiously, competently, and with dispatch. As Xenakis wrote in his fateful letter to Kalff:

> It is I who entirely conceived the form and mathematical expression of the Philips Pavilion. Le Corbusier never denied it and it was his great gesture, which I very much appreciate, to have accepted in its entirety a creation of one of his own collaborators. . . . It [to give Xenakis credit as co-author of the design] is the least gesture of justice and truth which Philips owes me for the intellectual and moral qualities which I placed at its disposal.[8]

It was Xenakis who coordinated the work with the engineer Vreedenburgh in Delft and the constructor Duyster in Belgium. And it was Xenakis who remained in Paris, at work on the pavilion, when Le Corbusier was in India. Given this tremendous investment of energy and talent, Xenakis's bitterness at his forced anonymity, and later his dismissal, is completely understandable:

> I have been with Le Corbusier for ten years and like me there are two or three colleagues, and in his last volumes of complete works there is not a single mention of our names; it is Le Corbusier who designs and we do not exist. He works only half an hour on plans and the rest of the time he is concerned with personal publicity, exhibitions, etc. . . . and in the afternoon he never comes to the studio. All the responsibility for the work falls on our shoulders, for both design and sometimes execution; he only

signs and takes the credit. This is the firm Le Corbusier. All this for a miserable salary of 100,000 francs. Here is the reality of Le Corbusier and Philips.[9]

Yet in spite of Xenakis's faithful labor—which Le Corbusier himself acknowledged—we can question whether the authorship of the pavilion can be so easily and completely assigned. It is common that those at work in architectural offices labor under the illusion that a building design is truly theirs. They work eight hours or more each day on a single project. The master appears for a few minutes —perhaps more, perhaps less—and then demonstrates the annoying habit of passing off the building to the client or press as if it were the product of his or her exertion alone. Today, the entire design team usually receives credit in the professional literature, and few professionals actually believe that any architect—with rare exceptions—designs and builds without assistance. To the general public, however, the building is the product of a person or a firm, and not a team. In that, the situation has not substantially changed since 1958, and Kalff was quite correct in reminding Xenakis that the design team also included the engineer, the builder, and even Kalff—as client and consultant—himself. Yet it better suited Philips's purpose to credit Le Corbusier alone with the architecture and Varèse with the score. The world loves to credit those who have been previously credited, those who have received the mantle "master." As Le Corbusier vehemently retorted to Xenakis's assertion, although the master had not physically drawn a line, was the pavilion any less his vision, any less his creation?

It is interesting to note that shortly after the completion of the building, Xenakis was to leave architecture and devote himself almost exclusively to musical composition. The qualifier "almost" is used, however, because he did undertake two or three architectural projects in subsequent years. But his "Diotope" mixed-media work, realized in 1978 at the Centre Pompidou in Paris, was housed in a building that was itself a reflection of the Philips Pavilion, although two decades had elapsed [figure 7-4]. Clearly, Xenakis lacked the architectural genius of Le Corbusier although his musical creativity is unquestioned.

The precast system devised by Vreedenburgh and Duyster was a brilliant solution to a difficult problem and it succeeded in freeing the architects from a sticky situation. Like Ove Arup's insightful solution to the construction of the concrete shells for Jørn Utzon's 1963 Sydney Opera House, the engineer devised an eloquent way out of a troublesome predicament.[10] In both projects, however, one could ask whether the proposed architectural scheme should have been revised or even discarded when the construction difficulties so tested the expertise and cleverness of the design team. At what point should

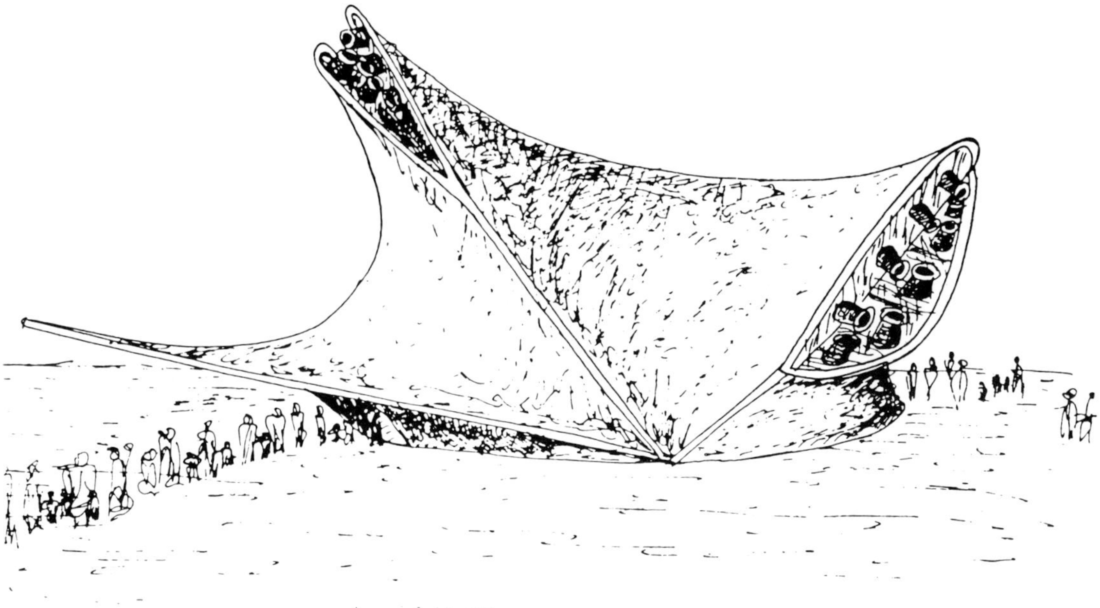

7-3
Philips at the Fair.
[HBG]

7-4
The Diatope, created for the opening of the Centre Pompidou in Paris. 1978. Iannis Xenakis.
[Courtesy Iannis Xenakis]

an architectural preconception be considered null and void, and a new one proposed in its place? Recall that the pavilion's design began as a flexible structure suspended within a steel frame. But given the time constraints operating at Brussels, this proved to be unviable. The problems of tardiness, however, were contingent to a great degree on Le Corbusier's chronic lateness and lack of attention. In the end, the construction "logic" of precast panels and prestressing wire, lashing the forms together as contingent shells, was a complicated way of executing what could have been a simple idea: the use of structural concrete shells. The ribs that integrated the shell fragments were ungainly at best, and they undermined the purity of the structure.

The use of hyperbolic paraboloids in itself dates the project to the late 1950s when the forms were all the rage. The building was a product of its times, even if the abstract idea behind the use of ruled surfaces was deeper than the superficial application of warped surfaces that appeared on a number of the exposition's pavilions [figure 7-5]. At the fair, most of the pavilions using the hyperbolic paraboloid and other structure forms did so for muscular display. The United States Pavilion used a radial structure to produce a long, clear span, and was thus primarily structural despite its glitzy exterior. Philips, in contrast, used a complex technique to encase a very small space. It was a formal spatial exercise rather than a structural investigation. But like Ronchamp and certain other Le Corbusier projects of the 1950s, the aspiration was for an "ineffable space" that joined intuition with rationality, and revealed "the action of the objective on the subjective."[11] And like Ronchamp the design of the pavilion demonstrated the affinity between music and nature, which Le Corbusier had described as an "acoustical phenomenon brought into the realm of forms. . . . It is an intimacy that should enter into each thing and bring out the effulgence of indescribable space. . . .The lyricism and the poetic phenomenon are loosed by . . . the clarity of the relations, everything being based on the faultless mathematics of the combinations."[12] On the hillside in the Vosges the form of the chapel resonates with the surrounding landscape, its interior feels the result of the process of these "visual acoustics." In Brussels, the architectural cacophony surrounding the site suggested nothing, and the form of the pavilion resonated instead with the electronic media presented within: mathematics and geometry in the service of humanism.

It is ultimately ironic then that Mr. Merkelbach's plea to Le Corbusier not to upstage the Netherlands pavilion with his Philips creation was completely unfounded. Of all the buildings at the fair, the Dutch work received the most consistent critical acclaim and would become one lasting model of exhibition architecture. The United States pavilion was cited for its structure and its multifaceted presentation of contemporary American life. The rectangular glass box of the Soviet pavilion, dubbed "The Refrigerator"

7-5
Pavilion "Marie Thumas." Brussels. 1958. Baucher, Blondel, Filiponne, architects; René Sarger, suspended roof consultant. [*L'Architecture d'Aujourd'hui*]

in the popular press, received attention for its mass, the density of its exhibits and ideology, and the extended hours it was kept open.[13] Structural accomplishment was the hallmark of the French pavilion, although critics found it overblown. The Dutch work, however, was appreciated for its elegance, its simplicity, its tactility and real materials, the contrast of its cupola form to the prismatic masses of the principal spaces—and the fact that its theme was real and not propagandistic.

Finally, it must also be suggested that the Philips Pavilion may have drifted into oblivion because, like a minor opus by a major composer, it just wasn't a great building. The thin concrete shells lacked both the material massiveness and sense of collective life embodied in either the *Unité d'Habitation* at Marseilles or the later monastery of La Tourette. It lacked the rational poetry of Le Corbusier's Utopian schemes like the Voisin Plan for Paris of 1925 or the sinuous multimile housing viaducts of the Obus scheme for Algiers from the 1930s. Philips also lacked the purely expressive poetry of the 1955 chapel at Ronchamp, with which it was frequently compared by critics reviewing the fair and its architecture.

Ronchamp grew from a dream of renewal; Le Corbusier drew upon the ruins of the earlier building and reconfigured rubble into an expression of faith. As a pilgrimage chapel set on a hillside Notre-Dame-du-Haut at Ronchamp is a sculptured object that can be seen from a multitude of distances and viewing angles; it is seen in progression from far to near [figure 7-6]. The solidity of its roughly stuccoed walls, with raking shadows cast by architectonic elements such as manneristically elongated scuppers, enclose a gently modeled space enlivened by light. An irregular yet harmonious patterning of incised windows punctuates the thick masonry walls; the radiant beams they admit catch dust and pilgrims alike in their light. In stark contrast to the pinpoint light rays that pass through the windows, two half-cylindrical shafts rise beyond the level of the roof, catch the sun, and transmute the masonry mass into columns of soft light. In all, Ronchamp feels like the work of a sculptor who has created architecture by modeling enclosure around a space. It feels to be the work of the heart.

Philips, in contrast, feels to be the work of the mind. The building's form has been generated by mathematics and structural requirements developed from the stomach-shaped plan. In places where the intersections of the resultant hyperbolic paraboloid shells were clumsy, they remained that way. At one of the peaks, exposed metal trusses, presumably needed because the slope was too low and the spreading forces too great, held the shells together in tension. Throughout the design of the building, the aesthetic resulted from intellectual process. Antoni Gaudí, as noted earlier, used models to develop the structure of his major works. But he never left them as wire models sub-

7-6
The chapel of Notre-Dame-du-Haut at Ronchamp, France. 1955. Le Corbusier. [Marc Treib, 1994]

sequently executed in stone. The wire models traced forces and stress; they determined the skeleton, not the body, of the church. It was the architect's task thereafter to elevate the structural diagram to the status of architecture.

In the end, architectural quality may have been sacrificed to a geometric ideal, much as Kandinsky had claimed that painting had suffered under cubism:

> The "emancipation" of today must advance on the lines of the inner need. It is hampered at present by external form, and as that is thrown aside, there arises as the aim of composition—construction. The search for constructive form has produced Cubism, in which natural form is often forcibly subjected to geometrical construction, a process which tends to hamper the abstract by the concrete and spoil the concrete by the abstract.[14]

Yet the novelty of its space and the power of its impact engaged both the visitor and the professional critic. It may have been a building flawed in its resolution, but its presence seems to have transcended its shortcomings.

The most frequently published views of the pavilion focus on the sharply skewed triangles of its entrance and exit [figure 7-7]. Featuring the leaning verticals and yawning voids of the entrances, the photos exaggerate the expressionist qualities of the building, as if they derived from the raked floors of Dr. Caligari's cabinet [figure 7-8]. By foregrounding these apertures, the sense of perspective is magnified, creating increased contrast between foreground elements and the rear parts of the structure. But in general views of the exposition site that catch the Philips Pavilion unawares, the ungainly portions of the form are only too apparent. The details that filled out the required facilities—the ribs and the intersection of certain shell segments, the entry and exit doors; the blocky mass and windows of the control booth, for example—are almost invariably crude and ill resolved. Here we must distinguish between resolution and roughness. La Tourette, for example, is left rough as an integral aspect of its character, but its pieces are completely resolved. The shells of the pavilion, on the other hand, aspired to a smooth coherence that integrated each element within a single unity. The roughness was the product of incongruities rather than intention. In spite of its glittering (painted) aluminum surface, the pavilion was far less than perfect in its realization. Here one misses the sure hand and inventive solutions, and the complete involvement, of the experienced master, Le Corbusier.

But then one can argue that this was not a building in the typical sense. It was defined as temporary from its inception; it would exist for some 180 days. Of course there needed to be *some* exterior, but the interior was of far greater consequence. Forget the awkward intersections that only Gaudí

might have solved; forget the inconsistencies of normal doors set in very abnormal walls; forget the structural problems that arose when the warping of the shell was too acute. These are minor when compared to the accomplishments of the interior as sculptured space, as a complex series of surfaces upon which to screen an electronic poem. Moholy-Nagy's description of the new sculpture could aptly apply to the Philips work:

> Openings and boundaries, perforations and moving surfaces, carry the periphery to the center, and push the center outward. A constant fluctuation, sideways and upward, radiating, all-sided, announces that man has taken possession, so far as his human capacities and conceptions allow, of imponderable, invisible, and yet omnipresent space.[15]

And within the stomach of the pavilion "a boundless depth open[ed] up, efface[d] the walls, [drove] away contingent presences, accomplish[ed] the miracle of ineffable space."[16]

The Philips Pavilion was an electronic corollary to German baroque creations like Dominikus Zimmerman's church at Wies (c. 1750), where architectural form modulated celestial light to destroy the finite boundaries of perceived space. But it was the darkness, rather than the light, that effaced all sense of perceived limits within; reality was superseded by the virtual images of cinema and the sensual hovering of ambient colored light. For Le Corbusier, who defined architecture as the magnificent play of masses in volumes in light, creating a space of darkness—a space in which reality became the virtual play of flickering images and colored light—was a radical departure in form if not from his established path toward a synthesis of the arts.

The building, like the spectacle, was intended to be a phenomenon that no member of the audience had ever experienced before. The audience was titillated, tantalized, impressed, and ultimately mystified by the building they had entered and what they had seen within it. They would remember the Philips Pavilion and its spectacle as a total work of art, and not the details of its architectural construction, its cinematic realization, and the production of its lighting effects. If one measures success in this way—and that was, after all, the measure that the Philips corporation intended to apply—then the building must be judged a success.

The eight minutes of film that provided the quasi-narrative structure for the *Poème électronique* also provided its most vivid memories. The ideas and the objects used to illustrate them were Le Corbusier's own; his sketchbooks reveal detailed if random thoughts on the composition of the poem, but no simple statement of an idea that underlies the entire work. In his early formulation of the spectacle, he defined it in terms of syntax rather than signifi-

7-7
The straining verticals recall the dynamics of expressionist architecture. [HBG]

7-8
Still from *The Cabinet of Dr. Caligari*. 1919. Film by Robert Wiene, with sets by Hermann Warm, Walter Reimann, and Walter Röhrig.
[Museum of Modern Art]

cance. His notes include sources for images from museums and from books; his detailed instructions allowed Jean Petit to work as Xenakis worked in the office: realizing yet subtly informing the content of the work. But what was produced by Petit and Xenakis was only the raw material.

Philippe Agostini's contribution to the film was beyond that of a cinematographer. Working on the project late in his career, he had surpassed mere filming and was termed a *cinéaste*, or filmmaker. For the *Poème électronique*, he produced a *mise-en-scène*, similar to the way a graphic designer working on a book executes the *mise-en-page*, or in English, the layout. Agostini had established his reputation during the 1940s with cinematic frames of striking composition and jarring montage. In the closing minutes of Robert Bresson's *Les Anges du Péché*, Agostini has the camera traverse space in jumps, with dramatically lit subjects themselves tableaux. The film did not progress linearly and realistically, and here director and filmmaker reinforced one another. In the *Poème électronique*, Agostini structured the images on the screen and manipulated them before the viewer. In the opening sequence, he took two images—the bull and its human adversary—and essentially animated their conflict by carefully opposing and framing the still images.[17] The film could have been only a boring rehearsal of the "poetic objects" and photos selected by the architect. Instead, Agostini posed flopped images in symmetrical balance; then, for example, shifted the axis of symmetry from vertical to horizontal. By selectively illuminating parts of a given photo—and almost all of the objects came to him already compressed into two-dimensional photographs—he dissected its elements to create a film-within-a-photo. At times, a visual vibration occurred between the positive and negative versions of the same image shown successively; at other times an object floating in black space would be set against a white ground, although due to the effect it would have on the color ambiances, this was kept to a minimum.

Agostini's contribution to the film was ultimately to establish relationships between the objects and photos of which it was comprised, working toward overcoming the inherent disparities between the constituent pieces. The *minutage* offered no smooth and obvious narrative, no simple timeline from "primitive" to "civilized." Instead, the alternations in time and geography, in joy and horror, from inanimate to organic, enriched the unfurling of visual sensations in the film. In its fragmentation and its assembly of opposing images, it evoked Eisenstein's ideas of film montage as the collision of discreet frames, gaining their power and their meaning through juxtaposition. If these oppositions were at times mysterious, approaching the surreal, they took hold of the audience for their eight minutes suspended in space.

With the exception of the "Ray-o-gram"–like shapes of abstracted machinery that appear in some sequences, the images did not test the limits of

filmmaking. Some thirty-five years earlier, Fernand Léger had created his ten-minute film, *Ballet mécanique*.[18] The artist/filmmaker wrote that there was "no scenario. The interactions of rhythmic images that is all. . . . Figures, fragments of figures, mechanical fragments, metallic [fragments], manufactured objects, enlargement with minimum perspective"—these were the elements of which the film was made.[19] Christopher Green explained that Léger used:

> the photographic image, mobile or static, as the basic unit of a series of measured rhythmic sequences. . . . Léger uses the repetition of images, several of the images which appear at the beginning of the film—the series of wine bottles, the straw-boater, the isolated smiling lips, the pendulum—reappear again and again; but most of all he uses combinations of contrasted or analogous images—pulsing metallic discs lead into the sudden contrasting image of a cockatoo reflected three times in mirrors, the sequence on the *collier de perles* leads into the sudden analogous image of a horse's harness. There are pure shapes—flat circles, squares, and triangles—but, as Léger insisted in *L'Esprit Nouveau*, "This film is objective realist and in no sense abstract"; for him, what mattered was the quality of the images in themselves.[20]

It was purely and simply a film constructed of frames assembled for their rhythmic potential. Green also explained what Léger ultimately learned from film: theater and dance could also be "visually dynamic in a compelling way, but the special capacity of film to create dynamic rhythm not only by the mobility of the images filmed in movement, but also by the effect of cutting from image to image, gave it even greater potential as an art of movement."[21] Le Corbusier may have been drawn to the film because it represented the possibility of a medium doubly rhythmic. In the case of the *Poème électronique*, however, the base images of which it was comprised were almost exclusively still, undercutting the kinetic potential of the cinematic medium.

Early on, Xenakis had questioned the "realism" of the photographs that would comprise the film part of the spectacle, and thought of abstraction as a higher vehicle for reaching the audience. Le Corbusier avoided complete abstraction, rooting his painting in the world of commonplace or natural objects and the human figure. The designer Edward Gordon Craig, under the influence of Adolphe Appia, had sought to banish realism from the theater stage. He looked back to the Greeks, whose stage scenery was architecture: "In those days they built their theatres for their dramas, not their dramas for their theatres."[22] Replicating "reality" within the theater denies the power of the imagination, and is thus a distraction.[23] Some critics regard photography and cinematography as forms of heightened imitation, but this attitude denies the transformational power of the two media. One does not take a picture; one

makes a photograph. The framed view, montage, and other essentially mechanical techniques embody the artistic vision through a distancing or intensification of film from "reality"—as Eisenstein so persuasively argued in his writings and in his films. The *Poème électronique* derived from flat, still images, but in Agostini's hands it was transformed into "living" cinema.

Le Corbusier's brother Albert, who had received a private showing of the *Poème électronique*, wrote to his brother and told him that he preferred the music without the images. To some degree this is understandable given that Albert was a musician and not an architect/painter. And given that the music and the film and colors were created independently and only juxtaposed at the last moment, the images in no way contributed to the "reading" of the music; in fact, they might have been a distraction.

But there is a more tangible explanation. The sound routes and dispersed loudspeakers that disposed Varèse's organized sound/music in space were inherently related to three or even four dimensions, and to the architecture of the pavilion itself. The cinematic presentation (recall that it was termed *écran* or screen), on the other hand, was created flat—indeed, from flat originals. Its flatness was distorted by the curving surfaces of the pavilion, but it remained a basically bisymmetrical planar display on the walls. Albert Jeanneret understood that the architecture of Varèse's music had never been experienced before; the cinema, on the other hand, was given only a warped plane rather than created as a projection for the original geometry of the space. In 1921, Jeanneret published a review of Erik Satie's *Parade* in *L'Esprit Nouveau*, in which he compared the success of Satie to the limits of Stravinsky:

> A musical composition is, in the flow of time, a definite fragment of time, animated by the elements of rhythm, melody, and harmony.
>
> But in the case of ballet the musical fact is bent by a particular need: the adaptation to the stage.
>
> The music adapts itself to the ballet, and is not suppressed by it. That is what makes "Parade" superior to "The Rite of Spring."
>
> The ballet was created for the eye, the dancer expressing with technique, the music maintaining the evolution, suppressing it, directing it, suggesting to him its rhythms, expressing the character of the dancer.[24]

Similarly, one could ask whether the architecture should adjust to the cinema, or the cinema to the architecture? For Gordon Craig, there had been no question: the scene must enhance the dramatic idea. Albert Jeanneret probably judged correctly when he found Varèse's score more innovative than the film; not as a film, perhaps, but as cinema.

Given the Philips corporate resistance to the selection of Edgard Varèse as the pavilion's composer, it is somewhat ironic that it is the score alone that remains most highly regarded. When writing for traditional instruments, a composer is at the whim and mercy of the performing group. Without instrumentalists to perform the work, the composer will never fully experience the piece, except if the work is written for a single instrument and the composer himself can perform it. In electronic composition, however, the sound is produced in the presence of the composer, who records it, thus operating as composer and performer simultaneously. Each time the piece is "played" it will be almost the same.

To John Cage, Varèse's achievements had both positive and negative aspects. That he attempted to control the process of composition, Varèse reflected the old order: "Rather than dealing with sounds as sounds, he deals with them as Varèse." On the other hand, Varèse "more clearly and actively than anyone else of his generation . . . established the present nature of music. . . . While others were still discriminating 'musical' tones from noises, Varèse moved into the field of sound itself, not splitting it in two by introducing the perception of its mental prejudice."[25] This was his true achievement. Varèse's ability to synthesize opposing tendencies in contemporary life also attracted the admiration of the composer Karlheinz Stockhausen. The *Poème électronique* represented a "modern formulation of compositional relationships whose true significance can only today be recognized: namely the sequential presentation and superimposition—even though sometimes abrupt and unmediated—of events of a heterogeneous nature (for instance, extremely realistic events, events resembling musical hoardings, and freely invented events)."[26] In uniting the diverse fragments of life and its music, Varèse represented not only modern musical ideas but contemporary existence.

The *Poème électronique* is unique in that it was composed for a specific space, and the space was designed acoustically to transfer sound patterns: to truly distribute sound in space. Technically speaking, the *Poème électronique* as organized sound/music was destroyed with the building because the aural experience can never be re-created without the exact configuration of the space that supported it. And yet recordings remain; Varèse himself played the tape to small groups after the exposition ended. Therefore, we must assume that the extant parts of the work are sufficient to be heard and judged.

Texts on modern music usually include Varèse as some sort of musicians' musician, a renegade of the highest moral stance, somehow removed from the normal flow of twentieth-century music. When the Philips directorship learned that Varèse termed his work "organized sound" they were much relieved; what they had heard in the one-minute demonstration in February 1958 was not what they would call music. The terminology was loaded, and

Kalff diffused some of the problems by redirecting the attention of the august board. Recall that Frits Philips, the chairman of the board, was against the music and referred to it as being against "free Western society": an objective statement to substantiate a subjective reaction. Henry Miller didn't see it that way. To him Varèse was a visionary, a leader:

> Back of every creation, supporting it like an arch, is faith. Enthusiasm is nothing: it comes and goes. But if one *believes*, then miracles occur. Faith has nothing to do with profits; if anything, it has to do with prophets. Men who know and believe can forsee the future. They don't want to put something over— they want to put something *under* us.[27]

Musicians and music scholars have been far kinder to the *Poème électronique* than Frits Philips. Paul Griffiths called it "one of the few great works of music on tape." Continued interest has been expressed in its mixture of concrete and synthesized sounds, mixed with the human voice and traditional percussion such as the gong. Virgil Thomson, one of the United States' major twentieth-century composers and music critics, wrote:

> *Le Poème électronique* is for the larger part of its time miraculously delicate, a chamber piece of ravishing quietness, and though played on that most blatant of all sound-sources, an ensemble of loudspeakers (thankfully each one tiny), a work of great delicacy. One wishes Varèse could have lived to make more like it, for all the opposition of the Philips Company, which Le Corbusier fought off for him, and for all the complexities of almost microscopic wiring that it took two months for the engineers to bring into tune and timing.[28]

Thomson raises a thorny question: Can Varèse be credited with the work entirely, given the amount of collaboration provided by Bruin and Tak in Eindhoven and Brussels? Or does one regard their contributions as one would a virtuoso violinist or conductor who might consult on the limits of what can be performed on the instrument?

In many respects, the electronic equipment necessary for realizing Varèse's score or musical idea reflected innovative means of generating sounds according to an organized plan. Marc Bredel enthusiastically regarded the *Poème électronique* in his biography of the composer: "It is a world premiere, a music box for the twentieth century! The work of two septuagenarians, the one world renowned, the other not yet assigned to posterity; a fantasy for the former, an undreamed of apotheosis for the latter; and the meeting of two sculptors in space, one working with 'solids', the other with the ephemeral."[29]

While the army of speakers (reported as being anywhere from 300 to 450, with 325 being a reasonable assumption), the vast installation of wiring necessary to link the elements of the reproduction system, and the curving concrete surfaces of the pavilion's wall no longer exist, the recording remains as a trace of the aural production of the *Poème électronique*. With the building gone and only photographs as witness, the four-dimensional spatial experience of the Philips Pavilion has been compressed to two. Perhaps the three-dimensional experience of the Varèse score, now realized only on stereo recordings, still retains more of its original form than does the architecture. It suffers less in its incompleteness; and it retains a greater part of its "scale."

With the project completed and judged a critical triumph by most of the parties who mattered to them, the three principle figures could sit back for a moment and reflect on the scope and nature of the accomplishment. In a gracious moment, Le Corbusier wrote to Kalff with some words of thanks, apparently forgetting or choosing to forget the crisis of December 1957:

> *For three years, you have been a perfect chief-of-the-works* ["Chef de file"] *and note that it is an essential role in human affairs. But, in general, when these men exist, we forget to appreciate their presence; and when there are no such men, we do not notice their absence because no work is realized!*[30]

He did not forget to mention Varèse's role in the creating the sensation that was the poem:

> *I think that all the same you are content to have taken Varèse on your team. My brother, a musician, told me that Mr. de Bruin gave him a personal performance of the music alone. My brother was taken by this performance and the music of Varèse. I am certain that without Varèse, we would have been purely and simply lost, including by the listeners and viewers of the "Poem."*[31]

For his part, Kalff was the cordial diplomat; but at this point, he had the right to feel content:

> Like you, we are convinced that Varèse gave us an excellent work and it would have been hard to find better at this time. The efforts made up to this point by the artists to express themselves through sound have been all less convincing than this creation by Varèse.[32]

Le Corbusier wrote to Varèse in New York, since he had not been able to meet the composer before he returned to the United States; both of them seem to have been quite busy. The architect wanted to express, however, his "great satisfaction with your brilliant collaboration on the Poème électronique for Philips. You will have to agree that I did not bother you, but I must say that you yourself have been the structure for this Poème with your magnificent music."[33] Indeed, of the entire creative team, it was Varèse who most benefitted from the Philips project. Having lived such a significant portion of his professional life in the shade, if not the shadows of public disapproval, in Brussels he was brought fully into the public spotlight. The *Poème électronique* was a pivotal moment in his career and in his life. "Your expression of 'great satisfaction' and Mr. Albert Jeanneret's generous approval . . . deeply touched me," Varèse replied. "I am proud and pleased to have been your collaborator and trust that I won't forget the expression of solidarity by one who doesn't let his pal down."[34]

In the essay "Nothing Is Transmissible but Thought," which introduced the final volume of the *Oeuvre Complète*, Le Corbusier interjected: "Only those who play are serious types!" Indeed, in much correspondence he referred to the Philips project as an "electronic game." Only those who play are serious: Le Corbusier was serious even in his seeming detachment from the pavilion and its electronic spectacle.

> *When a client of mine stuffs my head with such and such little requirements, I accept, yes, I accept up to a point where I say impossible! For then the thing gets outside the rules of my game, of the game in question, the game of this given house, of this combination whose rules have emerged at the moment of creation, have developed, affirming themselves, become commanding. Everything within the rules—that's my motto! Nothing exists outside the rules! If not, then I no longer have any reason for existing.*[35]

The rules of this electronic game for Philips joined tribal sculpture with rockets, geometry with intuition, light with darkness, horror with hope. It was all a game for which Le Corbusier established the rules.

In Milan Kundera's *Immortality*, the protagonist Paul soliloquizes on the development of European culture:

> High culture is nothing but a child of that European perversion called history, the obsession we have with going forward, with considering the sequence of generations a relay race in which everyone surpasses his predecessor, only to be surpassed by his successor. Without this relay race called history there would be no European art and what characterizes it: a longing for originality, a longing for change.[36]

Through the efforts of Le Corbusier, Edgard Varèse, Iannis Xenakis, Philippe Agostini, Jean Petit, Louis Kalff, H. C. Duyster, Willem Tak, S. L. de Bruin, and C. G. Vreedenburgh, Philips entered the race as it rounded the turn into the electronic age. Employing architecture as its principle medium—coupled with sound, film, light, and automation systems only recently available—the Philips Pavilion grabbed the baton, and grasped it firmly for precisely 480 seconds.

7-9
Commemorative metal model of the Philips Pavilion. 1958.
[HBG]

PHILIPS POEME ELECTRONIQUE
BRUXELLES 1958

INTRODUCTION

1 Fernand Léger, "Color in the World," *Europe*, 1938, reprinted in Léger, *Functions of Painting* (New York: Viking Press, 1973), p. 122.

2 Le Corbusier, "Synthesis of the Major Arts—Project 'Porte Maillot 50,'" in *Le Corbusier: 1946–52* (Zurich: W. Boesiger/Editions Girsberger, 1953), p. 67.

3 Le Corbusier, "Ineffable Space" (L'espace indicible), *Architecture d'Aujourd'hui* (January 1946), reprinted in Joan Ockman, ed., *Architecture Culture 1943–1968* (New York: Rizzoli International, 1993), p. 66.

4 Although I have tried to establish the first names of those involved in the pavilion's design, in some instances it has proved impossible. All Dutch publications and documents use only the last names or first initials.

CHAPTER 1: COMMISSION

1 Unless indicated to the contrary, communication between parties is in the form of a letter. D'Aboville to Le Corbusier, 16 January 1956, Fondation Le Corbusier (hereafter FLC).

2 Kalff, internal memo to Philips directors, 25 February 1956, Philips Archives (hereafter Philips).

3 Minutes of a directors' meeting, undated, probably February 1956, Philips.

4 Kalff to Le Corbusier, 25 June 1956, FLC.

5 Ibid.

6 A. J. Guépin, internal memo, 9 August 1957, Philips.

7 Ibid.

8 Kalff to Le Corbusier, 25 June 1956, FLC.

9 Kalff to Le Corbusier, 11 September 1956, FLC.

10 Le Corbusier to Kalff, 17 September 1956, FLC.

11 Memo from Kalff to Pugliesi-Conti, 8 October 1956, PA.

12 Kalff to Le Corbusier, 13 October 1956, FLC.

13 When the Philips group first heard the fee proposed by Le Corbusier, "they hit the ceiling." They apparently had in mind a sum one-tenth of that; Le Corbusier to Varèse, 24 October 1956, FLC. The architect stood firm in his demand, and in the end Philips acquiesced.

14 Le Corbusier to Varèse, 12 June 1956, FLC; cited in Nouritza Matossian, *Xenakis* (London: Kahn and Averill, 1986), pp. 110–11.

15 Varèse to Le Corbusier, 16 June 1956, FLC.

CHAPTER 2. DESIGN CHRONICLE

1 Le Corbusier, in Jean Petit, *Le Poème Électronique*, (Paris: Editions Vives, 1958) p.23. In a later recollection, Xenakis recalled that Le Corbusier told him: "The reason behind my design is not to add yet another building to my career, but to create with all of you a first 'electric game,' electronic, synchronic, in which light, line, color, volume, movement, and idea comprise a whole, surprising and accessible of course to the general public." Xenakis, "Le Pavillon Philips: A l'Aube d'une Architecture," in *Musique Architecture* (Paris: Casterman, 1976) pp. 128–29.

2 Kalff to d'Aboville, 9 January 1956, Getty Center Special Collections.

3 This and subsequent information on the firm, its factories, and its postwar recovery are drawn from Guus Bekooy, *Philips: A Century of Enterprise, 1891–1991* (Eindhoven: Philips International, 1991).

4 "Philips . . . received orders for cinema installations with which films could be projected with a three-dimensional effect. One year later, 20th Century Fox Film Corp. was producing films in Cinemascope, which required special projectors for the extra-wide picture. In 1953 Philips concluded a contract with American Optical Company for the development of a projector for the 70mm Todd-AO films with six-track stereo sound. By 1954 Philips had the projector ready. . . . In many countries, Philips provided the complete technology for sound and light shows for historical buildings and tourist attractions." Thus, the company possessed the means to realize elaborate audio-visual presentations. The automation of a presentation like the *Poème électronique*, however, would tax the

limits of their technology. Ibid., p.66.

5 Kalff, in Petit, *Le Poème Électronique*, (extract published for distribution at the fair), n.p.

6 Various files, Philips.

7 L. C. Kalff, "Het Licht op de 'Exposition Internationale' te Parijs 1937," *Philips Technisch Tijdschrift* (December 1937), pp. 361–66.

8 There was the additional incentive of a tax break given by the government to Belgian companies participating in the fair. Internal memo, 23 November 1955, Getty Center.

9 Internal memo from G. Van Kooten, 17 January 1956. Getty Center. It appears that by this date—before any discussions with Le Corbusier had begun—Kalff's idea to replace consumer products with an aesthetic and electronic spectacle had already convinced the upper management at Philips.

10 Internal memo from G. van Kooten, 17 January 1956, Getty Center. I thank Bart Lootsma for bringing this explanation to my attention in conversation. Rotterdam, 16 November 1992.

11 Eric Walter White, *Benjamin Britten* (London: Boosey & Hawkes, 1954), pp. 58, 70–71.

12 *Young Person's Guide to the Orchestra* is discussed in White, *Benjamin Britten*, pp. 56–58.

13 Ibid., pp. 32–33. The work was only partially completed.

14 Phoebe Douglas, "Benjamin Britten: Another Purcell," *Town & Country*, December 1947, cited in White, *Benjamin Britten*, p. 79.

15 Ossip Zadkine was born in Vitebsk, Russia, in 1890, but settled in France in 1909. Over the years his works were shown not only in France and Italy, but more successfully in Belgium and Holland. He received commissions for the 1937 International Exposition in Paris and won the Grand Prix for sculpture at the Venice Biennale in 1950. In any given period, however, his work was rarely near the forefront of artistic developments, and it would not be unfair to characterize his sculpture as making more advanced ideas accessible to the general public. He died in Paris in 1967. For a brief biography and description of Zadkine's work, see *Zadkine Museum: General Guide* (Paris: Paris Musées, 1991).

16 Rietveld to Kalff, 10 June 1956, Getty Center Special Collections.

17 Kalff to Bakema, 16 October 1956, Philips. Le Corbusier's proclamation is doubly ironic given that few, if any, buildings have no "exterior" and given the extreme shapes assumed by the Philips Pavilion as constructed.

18 During the remainder of the year, the work was directed by Le Corbusier's partner—and cousin—Pierre Jeanneret. The scope of the work ranged from the conception, siting, and planning of the city as a whole (definitive design, 1952) to the architectural designs for the Palace of the Governor (1954), the High Court Building (1956), the Secretariat (1958), and the National Assembly (1962), among other projects.

19 Iannis Xenakis was born in Braïla, Romania on 29 May 1922; he was schooled in Greece and attended the engineering program of the Polytechnic in Athens, completing his studies in 1946. The war years were tumultuous—the Polytechnic was closed—and the young Xenakis played an active part in the resistance, coming very close to losing his life in the explosion of a mortar shell. Remaining underground during the British occupation, he fled first to Italy and then in 1947 to France, where he luckily found a position working as an engineer in association with Le Corbusier's office. His interests in music continued to develop and, aided by Olivier Messiaen's interest in his work, and later Herman Scherchen's, he diligently pursued his studies whenever possible. In his early architectural work, Xenakis employed progressional systems closely related to Le Corbusier's Modulor system. See Matossian, *Xenakis*, especially pp. 11–43.

The first *unité*, at Marseilles, was hailed as a landmark building at the time. The apartment block, with its shops and services, was treated as a structural frame into which apartments of different configurations were "inserted." A corridor, termed a "street" by the architect, was to provide an active center for the building several stories in the air. Lacking sophisticated resources for casting concrete, Le Corbusier boldly exposed the wooden shuttering used for the formwork, an effect he called *béton brut*. This rough surface was widely admired and imitated internationally, and gave birth to a stylistic trend called the "New Brutalism" by the British architectural historian Reyner Banham. See Banham, *The New Brutalism* (New York: Reinhold, 1966).

20 Le Corbusier, *Modulor 2*, trans. Peter de Francia and Ann Bostock (London: Faber and Faber, 1958) p. 326.

21 Le Corbusier's Modulor proportional system displays parallels with both the classical Golden Section and the Fibonacci series of numbers. Whether one actually *designed* with the Modulor scales, or applied their numbers to any design *after* intuitively establishing pleasing spatial proportions or facade modulations, is a point raised by several critics. For

Xenakis's description of the La Tourette project, see Matossian, *Xenakis*, pp. 65–73.

22 Le Corbusier, *Modulor 2*, p. 327.

23 Ibid., p. 330.

24 Xenakis ultimately left the office and achieved a distinguished career as a composer. One of his few executed buildings was the *Diatope*, originally created in 1978 for the Centre Pompidou in Paris, which developed the ideas of multimedia first pursued in the Philips Pavilion. Unlike the Brussels spectacle, however, the Xenakis work avoided all figurative imagery. See Matossian, *Xenakis*, pp. 222–27.

25 Kalff retired in January 1960 and died 16 September 1976. Philips employee records.

26 He wrote a historical overview of exposition illumination in his detailed account of the Paris fair, which was one of the first times lighting had played such a major role and rendered the architecture diurnally brilliant.

27 Le Corbusier, in Petit, *Le Poème Électronique*, p. 25.

28 In "The Philips Pavilion at the 1958 Brussels World Fair," *Philips Technical Review* (English edition), reprint of vol. 20, nos. 1 and 2/3 (1958), p. 1.

29 For the biography of Edgard Varèse, see Fernand Ouellette, *Edgard Varèse*, trans. Derek Coltman (New York: Orion Press, 1968); and Odile Vivier, *Varèse* (Paris: Editions du Seuil, 1973).

30 Le Corbusier, in Petit, *Le Poème Électronique*, pp. 23–25.

31 Le Corbusier, *The Modulor* (Cambridge, Mass.: Harvard University Press, 1954; reprint 1980), p. 71.

32 Le Corbusier, *L'Architecture d'Aujourd'hui*, special issue (1948), p. 43.

33 Le Corbusier, *Towards a New Architecture* (New York: Dover Publications, 1986), p. 1. Reprint of the 1931 translation by Frederick Etchells.

34 Le Corbusier, *Precisions: On the Present State of Architecture and City Planning* (Cambridge, Mass.: MIT Press, 1930; reprint 1991), p. 133.

35 Le Corbusier to Director of Federal Technical Institute, Zurich, 15 October 1956, FLC. This personal contact may have stemmed from meetings connected with the construction of Le Corbusier's Swiss Pavilion at the Cité universitaire in Paris in 1930–31.

36 Van der Waerden to Le Corbusier, 17 October 1956, FLC.

37 *Mathematischer Modelle für den höheren Mathematischen Unterncht* (Leipzig: Martin Schilling, 1911).

38 Christopher Green, "The Architect as Artist," in Michael Raeburn and Victoria Wilson, eds., *Le Corbusier: Architect of the Century* (London: Arts Council of Great Britain, 1987), p. 112.

39 A. Ozenfant and C.-E. Jeanneret, "Sur la Plastique," *L'Esprit Nouveau* 1 (October 1920), pp. 42–43. Ozenfant would expand this thesis in his *Foundations of Modern Art*, English ed. (New York: Dover Publications, 1930; reprint 1952).

40 Le Corbusier and Amédée Ozenfant, "Formation de l'optique moderne," *L'Esprit Nouveau* 21 (1924).

41 Le Corbusier and Amédée Ozenfant, *La Peinture moderne* (1925); cited in James Thrall Soby, "Le Corbusier the Painter," in Stamo Papadaki, *Le Corbusier: Architect, Painter, Writer* (New York: MacMillan, 1948), p. 118. For a recent discussion of Le Corbusier as a painter, see Viveca Bosson and Beate Sydhoff, *Le Corbusier: Konstnären och arkitekten* (Stockholm: Kulturhuset, 1993).

42 Le Corbusier, *Le Poème de l'angle droit* (Paris: Fondation Le Corbusier / Editions Connivences, 1989). English trans., n.p.

43 Of course, Le Corbusier was familiar with the engineering accomplishments of Eugène Freyssinet (1879–1962), an early innovator in reinforced concrete construction. Two images of his hangars—compared favorably in dimension to the nave of Notre Dame—appeared on the closing pages of *Towards a New Architecture*.

44 Xenakis, "Le Pavillon Philips: A l'Aube d'une Architecture," p. 129.

45 Ibid., p. 32.

46 "Second decision: Because the audience is standing and looking forward, two concave, almost vertical walls almost vertical are needed, permitting the spectators to see above the heads of their neighbors." Le Corbusier, in Petit, *Le Poème Électronique*, p. 24.

47 Le Corbusier used this technology—which had been demonstrated at the 1925 Exposition Internationale des Arts Décoratifs et Industriels Modernes in Paris—in a number of projects from the late 1920s. The process proved to be inapplicable for constructing the housing at Pessac, but it was used at Ronchamp and for the interior finishing of the Philips Pavilion. Brian Brace Taylor wrote: "The Ingersoll-Rand company, which marketed the cement gun in France, had publicized the utility of the method for house construction in regions susceptible to earthquakes such

as southern California. The sprayed cement process was also widely used in mining operations to coat the walls of tunnels." See "Le Corbusier at Pessac: Professional and Client Responsibilities," in Russell Walden, ed., *The Open Hand: Essays on Le Corbusier* (Cambridge, Mass.: MIT Press, 1977), p. 183, n. 2.

48 In the FLC numbering system, the drawing is 28585.

49 The drawing is marked "Drawn by Xenakis." FLC 28583.

50 "[H]e was so impressed that he immediately tried to show me that my ideas were in the same direction as his sculptural ideas." Interview with Xenakis, 1977; cited in Matossian, *Xenakis*, p. 113.

51 Ibid., p. 134.

52 The Italian reviewer for *L'Architettura*, for example, commented upon the disjunction between architectural form and the contents of the exhibitions—and the overly expressive structural systems used without real purpose in many of the exposition's pavilions. Renato Pedio, "La crisi del linguaggio moderno nell'Esposizione Universale, Bruxelles 1958," *L'Architettura* (October 1958), p. 391. See chapter 6 below for further professional reaction to the fair.

53 It should be noted, however, that only the exterior of the building is presented in the 1952–57 volume of the *Oeuvre Complète*. In *Creation is a Patient Search* (1960), however, images of the projected electronic poem accompany those of the structure. Certainly, a limited number, but the most widely published, of photos of the pavilion are taken from its most photogenic station points.

54 Xenakis, "Le Pavillon Philips: A l'Aube d'une Architecture," pp. 127–28.

55 Ibid., pp. 124–25.

56 Le Corbusier, *Architecture d'Aujourd'hui*, special issue, 1948, p. 53.

57 Kalff? [unsigned] to Le Corbusier (via Xenakis), 13 October 1956, FLC.

58 Kalff to Le Corbusier, 31 October 1956, FLC.

59 Ibid. Perhaps ironically, the president of the Eiffel office wrote to Le Corbusier on 31 October expressing the company's interest in working on a project with a specialized steel construction.

60 Xenakis to Kalff, 9 November 1956, FLC.

61 At the beginning of the design phase, Xenakis and Le Corbusier were collaborating with the engineering firm founded by Gustave Eiffel (1832–1923), known for the tower in Paris named in his honor, the structure for Frédéric-Auguste Bartholdi's "Statue of Liberty," and, tainted by scandal, the first design of the locks for the Panama Canal.

62 Xenakis to Kalff, 13 November 1956, FLC.

63 Kalff to Xenakis, 17 November 1956, FLC.

64 Eiffel to Le Corbusier / Xenakis, 17 November 1956, FLC.

65 Ibid.

66 Xenakis to Le Corbusier, 26 November 1956, FLC.

67 Ibid.

68 Kalff to Le Corbusier, 29 November 1956, FLC.

69 The final structure comprised "[a] floor of reinforced concrete with ring beams, resting on 85 wooden piles, 9m long." L. C. Kalff, "Philips Pavilion," no date, no source; probably a press release published by the company in 1958. Philips.

70 Xenakis to Kalff, 3 December 1956, FLC.

71 Le Corbusier to Kalff, 31 December 1956, FLC.

72 Eiffel office (signature unintelligible) to Le Corbusier, 27 December 1956, FLC.

73 Le Corbusier to Lorin at Eiffel, 23 January 1957, FLC.

74 Merkelbach to Le Corbusier, 15 January 1957, FLC.

75 Le Corbusier to Kalff, 29 January 1957, FLC

76 Kalff to Le Corbusier, 4 February 1957, FLC.

77 Xenakis, minutes of meeting, with sketches, 21 February 1957, FLC.

78 Ibid. In the end, the film was made completely in black and white; color was added by lighting and projections.

79 Ibid.

80 Kalff to Le Corbusier, 4 February 1957, FLC.

81 The HBM was the principal contractor for the Dutch Pavilion at Brussels, but the contract was let with the stipulation that they should not be engaged on any other work at the fair. Because Strabed was an independent Belgian entity—in spite of its affiliation to the Dutch parent group—there was no legal conflict of interest. Anthony Van Kampen, *Bouwen op Mensen: Vijfenzeventig jaar Hollandsche Beton Groep*. (Rijswijk: Hollandsche Beton Groep, 1977), p. 244.

82 Kalff to Le Corbusier, 4 February 1957, FLC.

83 Le Corbusier to Le Lionnais, 18 February 1957, FLC.

CHAPTER 3. REFINEMENT AND CONSTRUCTION

1 This and other biographical information is taken from F. W. Keppler, "Duyster 40 Years," *Specie* (Mortar, the HBG house organ), vol. 6, no. 5 (5 October 1970). Duyster was born on 9 September 1907 and completed his doctorate in Ghent after he had worked for almost two decades with the HBM firm. His career with Strabed in Brussels culminated with the brilliant realization of the Philips project. After his successful tenure in Belgium, he moved back to Holland in January 1959 and became a director of the HBM in Den Haag. He retired in 1972, after over forty years with the firm. He was known as a sensitive as well as a canny engineer. The Evoluon, the Philips science museum in Eindhoven, reflects his concept and engineering expertise. He died 26 March 1987.

In December 1981 Strabed was absorbed by Copon Brussels, which in turn later became a part of CEI, a branch of the HBG. HSB management memo, 18 July 1983. Hollandsche Beton Groep Archives.

2 H. C. Duyster, "La conception, l'exécution et le comportement d'une coque mince en béton précontraint," unpublished paper presented at the International Colloquium on Construction Processes of Shell Structures, no date, probably 1958 or 1959. Hollandsche Beton Groep Archives.

3 H. C. Duyster, "Construction of the Pavilion in Prestressed Concrete," *Philips Technical Review*, reprint of vol. 20, nos. 1 and 2/3 (1958), p. 27.

4 "Forces lying in the median surface of a plate or shell are known as membrane forces." C.G.J. Vreedenburgh, "The Hyperbolic-Paraboloidal Shell and Its Mechanical Properties," *Philips Technical Review*, reprint of vol. 20, nos. 1 and 2/3 (1958), p. 11.

5 Ibid., pp. 11, 13.

6 "It is curious to note that Gaudí always designed taking into consideration the practical methods by which his plans could be carried out. He chose warped surfaces because model makers and masons can construct them easily, as both the hyperboloids and hyperbolic paraboloids are ruled surfaces which can be easily reinforced by straight steel rods, thus providing a considerable economy." James Johnson Sweeney and Josep-Luis Sert, *Gaudí* (New York: Praeger, 1960), p. 78. These properties also informed the design of the structure for the Philips Pavilion.

7 The laboratory, funded by the government and contract research, was founded in 1934; the subgroup specializing in buildings and materials remains one of the best known. Ivo J. Blanken, archivist at the Philips Company Archives, stresses that "in the Netherlands . . . a TNO report stands for independent, high quality research. Everything that comes out of the TNO organization will hardly be doubted." Letter to the author, 21 December 1993. I am grateful to Mr. Blanken for this and much other information.

8 A. L. Bouma and F. K. Ligtenberg, "Model Tests for Proving the Construction of the Pavilion," *Philips Technical Review*, reprint of vol. 20, nos. 1 and 2/3 (1958) pp. 17–18.

9 Ibid., p. 19.

10 Ibid., p. 22.

11 Ibid., p. 24.

12 Ibid., p. 23. The 1958 Strabed film on constructing the pavilion illustrates stress tests using translucent panels set within the plywood diamonds, examined under polarized light. Bouma and Ligtenberg make no mention of this method in their definitive article on the structural tests, however.

13 Ibid.

14 Ibid., p. 26.

15 On 2 April, Le Corbusier wrote to Kalff addressing the question of openings in the shell for ventilation, to which he was resolutely opposed. He also noted that the next phase of the design had been concluded. Le Corbusier to Kalff, 2 April 1957, FLC. The following day he submitted his bill for the second part of the fee: 3,500,000 francs. Le Corbusier to Kalff, 3 April 1957, FLC.

16 Xenakis in conversation with author in Paris, 15 June 1993, and by implication: memo from Xenakis to Le Corbusier, 2 May 1957, FLC.

17 Xenakis to Tak, 18 April 1956, FLC.

18 Tak also consulted on the acoustical design of the main assembly hall, proposing the acoustical panels that Le Corbusier referred to as the "clouds of Mr. Tak."

19 The number does not comply with the drawing reproduced as Figure 3-19. This discord could result because one of the segments is essentially a screen wall and does not actively contribute to the structure. It appears twice, once above the entrance, once above the exit. "Fundamental principles of the project and its engineering details." Typescript, no author, no date, Philips.

20 Xenakis to Le Corbusier, 2 May 1957, FLC.

21 Duyster, "Construction of the Pavilion in Prestressed Concrete," *Philips Technical Review*, p. 29.

22 Ibid., p. 30, 32, 33.

23 Ibid., Figure 6 caption, p. 31. "It may be pointed out that the tail blocks were in all cases *inside* the building. In the part of the building shown here, the wall bends outward. It may also be pointed out that as the

substructure can slide only horizontally and any rotational movement is completely prevented, an eccentrically-acting prestress is made to act centrally in the shell: the bending moment will be taken up by the foundation, leaving none in the shell."

24 Ibid., p. 29.

25 Ibid., p. 34.

26 Ibid.

27 Eighty-five wooden piles provided a stable foundation for the slab and the pavilion's concrete superstructure.

28 Kalff to Le Corbusier, 5 May 1957, FLC.

29 Ibid.

30 Telegram from Kalff to Le Corbusier, 11 June 1957, FLC.

31 Van Kampen, *Bouwen op Mensen*, p. 245. According to Van Kampen, Matthijs, like Le Corbusier himself, agreed to supervise the work only if he could do it "in his own way." The Strabed team seems to have accepted the challenges posed by the building as a test of their skill and their pride.

32 Kalff to Varèse, 13 June 1957, Getty Center Special Collections.

33 Louis Kalff, meeting notes or summary; no date. February 1956?

34 Despite his residence in the United States, Gabo was responsible for a major outdoor sculpture adjacent to Marcel Breuer's Bijenkorf department store in Rotterdam from 1955 (*Construction for the Bijenkorf*, 1954–55). Gabo's work was known by Philips management.

35 Kalff also proposed that the sculpture be displayed in Eindhoven after the fair ended. Internal memo from Kalff, 18 September 1957, Getty Center Special Collections.

36 Le Corbusier to Kalff, 11 June 1957, FLC.

37 Ibid.

38 Xenakis to Kalff, 3 October 1957, FLC.

39 Kalff to Xenakis, 8 October 1957, FLC.

40 Le Corbusier to Kalff, 12 October 1957, FLC.

41 Le Corbusier to Xenakis, 13 August 1959, Xenakis Archives; cited in Matossian, *Xenakis*, p. 132.

42 Matossian, *Xenakis*, p. 144. "Project Architect" or "Chief Draftsman" would probably be a more reasonable translation; in all probability the "Architect-in-Chief" could only be Le Corbusier himself. However, Le Corbusier did append the listing of "those who helped" at the end of *Creation is a Patient Search*: "Xenakis, Massonier, and Tobito [the three architects dismissed as a group], who left the studio in 1959, asked to be called architectes chefs-d'études." p. 309.

The log of drawings from the Le Corbusier office records that Xenakis was employed from 11 December 1947 until October 1959. In addition to his participation on the design of the Philips Pavilion, Xenakis prepared drawings for the High Court and Assembly Buildings at Chandigarh, the sports complex at Firminy, the monastery of La Tourette, the stadium at Baghdad, and the Unité at Nantes-Rezé.

43 Ibid., p. 144. Interview with Matossian and Xenakis, 1977.

44 On 25 June 1957, Xenakis unburdened his frustration to his mentor, the conductor Herman Scherchen:

"You ask me why my name is not mentioned.

1. Because Philips approached Le Corbusier, architect of world-renown, and not me.

2. Because I am an employee of Le Corbusier and have no firm: "Xenakis - Architect."

3. Because Le Corbusier is a miser, an egotist, an opportunist capable of trampling upon the corpses of his friends.

Not only did I do the project, which Le Corbusier would be incapable of imagining for he has remained twenty years behind the times, because he knows nothing of mathematics and is a conservative (see Chandigarh and Ronchamp) but also because I followed the project to its realization to such a point that while I was in Munich for the concert, Le Corbusier phoned Philips to say that he could do nothing without me and he had to wait for me in order to continue."

Cited in Matossian, *Xenakis*, pp. 119–20.

45 Le Corbusier to Kalff, 19 November 1957, FLC. "-Rama" seems to have been in the air in 1958: the pavilion for the [then] Belgian Congo was called "Congorama." It attracted 200,000 visitors. Robert Bordaz et al., *Le livre des expositions universelles 1851–1989* (Paris: Editions des arts décoratifs, 1983), p. 174. Philips ran a competition to find a name for the pavilion: they received over 700 hundred entries.

46 Kalff to Le Corbusier, 12 December 1957, FLC.

47 Kalff to Le Corbusier, 18 February 1958, FLC.

48 Although I have discovered no true color images of the geometric sign, the original hues were probably closer to primaries than the pastel tints of the restoration.

CHAPTER 4.
IMAGES, COLOR, AND LIGHT

1 Isadore Ducasse, known as Count de Lautréamont (1846–70) was a precursor of the surrealists and the author of *Chants de Maldoror*, published in 1869—the source of the definition.

2 The translation of this term remains problematic. Literally taken, the *écran* refers to the screen upon which the images are projected. On the other hand, the *scoring* of the imagery in the *minutage* (Le Corbusier's scenario) is called *écran*, although these are, in fact, the images that comprise the film. Thus, I have chosen to translate *écran* as "film" or "film images" throughout, although this is far from literal. Similarly, *ambiance* could be translated as "chromatic illumination," and *tri-trous* (which refers to the technique) as "projected images."

3 Moholy-Nagy, *The New Vision*, 1928, 3d rev. ed. 1946, p. 51.

4 Ibid.

5 Le Corbusier, "Ineffable Space," 1946, reprinted in Ockman, ed., *Architecture Culture 1943–1968*, p. 66.

6 Jean Petit, *Le Poème Electronique*.

7 In a letter from Le Corbusier to Kalff, 31 December 1956, the architect relates his recent conversation with Agostini "who has recently framed the documentary film on Le Corbusier soon to be completed." Lucien Hervé, who photographed Le Corbusier's buildings beginning with the Marseilles block in the early 1950s, told me that he had seen a film by Agostini at the inauguration of the Marseilles project (conversation with the author, 7 December 1993 in Paris). I have as yet been unable to find confirmation of this film project at the Fondation Le Corbusier, however.

8 Agostini was born in 1910 and in the last printed dictionary of French cinema (1991) was still living, although I have not been able to confirm this. He began working as assistant cameraman in 1933, and within ten years had become one of the leading cinematographers in France, working with directors such as Jules Dassin as well as Bresson and Ophüls. He produced films for television as well as for the commercial screen, but his work as a director never reached the level of his work as a cinematographer. Jean-Loup Passek, *Dictionnaire du Cinéma* (Paris: Larousse, 1991), p. 6.

9 Le Corbusier to Agostini, 15 June 1957, FLC.

10 Le Corbusier to Agostini, 27 June 1957, FLC.

11 Agostini to Le Corbusier, 30 June 1957, FLC.

12 Le Corbusier to Agostini, 1 July 1957, FLC.

13 Le Corbusier to Agostini, 26 July 1957, FLC.

14 Le Corbusier's wife, Yvonne, died on 5 October 1957. They had married on 18 December 1930.

15 Le Corbusier to Kalff, 21 October 1957. In the end, no narration was included in the score, the presentation's intention and structure were instead explained in the opening text accompanying the credits. See also Chapter 5.

16 Le Corbusier to Kalff, 26 October 1957, FLC.

The letter continues:

"and that everything is expressed in two diagrams:

a) a round of 8 minutes comprising the 7 sequences of the Poème expressing the symphonic composition: polychromy, rhythm, projection.

b) eight rounds, each of 60 seconds, totalling 480 seconds and containing the 7 sequences with their assigned time, whether successive or simultaneous: images; polychromy.

c) and these 7 onionskin sheets of manuscript provide the details of each sequence: illumination, film, projections.

d) 7 folders containing about 150 photographic documents destined for the work to be produced jointly by Agostini and myself.

e) the description of the Philips equipment for realizing the polychromy and the images and the indication of their positioning.

Our intention (Agostini's and mine) is to not schedule a meeting until we have realized one of the 7 sequences that we have selected so that we can talk with proof. Agostini will discuss this with me in the next few days.

On the other hand, you should know that Agostini is involved with another project until 15 December and thereafter he will finalize the 7 sequences. He seems to be extremely interested in the project.

I have already warned you that during the sixth and seventh sequences, there will be a few seconds [of narration] *spoken* by me, over the music of Varèse or, on the contrary, biting into the music of Varèse."

17 Le Corbusier, *Sketchbook*, 5 November 1957. The entry was headed "Varèse, Tak, Lighting [person]." Notes on the placement of the spoken portion of the soundtrack and the overall development in seconds bears the accompanying note "to confirm with Varèse."

18 Le Corbusier to Kalff, 25 November 1957, FLC.

19 Le Corbusier to Kalff, 29 November 1957, FLC.

20 Le Corbusier, in Petit, *Le Poème Electronique*, p. 25.

21 Le Corbusier to Petit, 2 December 1957, FLC.

22 See letter from Le Corbusier to Kalff, 24 December 1957, cited in its totality in Chapter 5.

23 A. Viguier of Alga-Cinema to Kalff, 23 December 1957, Philips.

24 Petit to Le Corbusier, 20 December 1957, FLC.

25 Kalff, internal memo, 18 November 1957, Philips.

26 Agostini to Kalff, 5 January 1958, Philips.

27 Le Corbusier to Agostini, 6 January 1958, FLC.

28 Le Corbusier to Kalff, 6 January 1958, FLC.

29 "I will take care of the story of the mannequin myself. It must be made by the tinsmith quickly with a wire chosen for its small scale. Philips will make the two enlargements [of the maquette] (fountain and ceiling)." Memo from Le Corbusier to Xenakis, 11 February 1958, FLC.

30 The page is undated but was probably written in mid- to late March 1957.

31 Kalff to Le Corbusier, 21 January 1958, FLC.

32 Kalff to Le Corbusier, 13 February 1958, FLC.

33 Agostini, who wanted to re-edit the film after the screening, also needed more time. Alga-Cinema required up to three more weeks to complete their work. Kalff, internal memo, 9 January 1958, Philips.

34 Program for the 6 February 1958 visit and test of film and musical score in the Philips "Garage," Eindhoven. Philips.

35 Internal memo; no date, probably mid-February 1958, Philips.

36 Vivier, *Varèse*, p. 165.

37 Kalff to Le Corbusier, 18 February 1958, FLC.

38 Le Corbusier to Kalff, 5 March 1958, FLC.

39 Kalff to Xenakis, 9 April 1958, FLC.

40 Le Corbusier to Agostini, 16 April 1958, FLC.

41 According to Bart Lootsma, the problems with synchronization, caused to a considerable degree by a missing piece of equipment, required that each of the automatic tracks required simultaneous starting by hand. Lootsma, "Poème Electronique, Le Corbusier, Xenakis, Varèse," in Vowinckel and Kessler, eds., *Le Corbusier*, p. 140. He also suggests that because Xenakis had destroyed several costly microphones in the course of recording the burning charcoal, that Philips was probably not too pleased with him! Conversation with author, Paris, July 1994.

42 That point was not lost on the thousands of visitors, and many of the reviews and coverage of the pavilion stressed that the spectacle was almost completely automated. Philips itself publicized the pavilion's automatic controls in the brochure handed to visitors as they waited in line: "The electronic poem will be repeated many thousands of times. That is why the equipment has been automatized to such an extent that the fallible human factor has been virtually eliminated. This synthesis between humanity and inventiveness has resulted from a co-operation between artists and technicians which has lasted for months, and with which Philips aspires to take a prominent share in this great manifestation of Modern Man." In spite of their claims, problems with light intensity, sound distribution, and synchronization persisted. Ibid.

43 Le Corbusier to Kalff, 12 July 1958, FLC.

44 Le Corbusier to Kalff, 22 July 1958, FLC.

45 Kalff to Le Corbusier, 28 July 1958, FLC.

46 Ibid.

47 In 1955, the Museum of Modern Art in New York mounted a monumental and well-publicized exhibition entitled "The Family of Man." Curated by Edward Steichen, the museum's curator of photography, the exhibition presented images of humankind—from childhood to advanced age—taken from around the world. The themes were brotherhood and shared human experience, messages diffused throughout the world in the book which issued from the exhibition. (The book has recently been reprinted and has rarely been out of print since its initial publication.) A number of the images in the *Poème électronique* share the general tone of *The Family of Man*, although nothing in the written documentation or in Le Corbusier's library suggests the New York exhibition as a source. More likely, any commonalities can be explained by a shared reaction to the horrors of the war and the general tone of life in the 1950s.

The exhibition was shown in Paris at the Musée d'Art Moderne from 20 January to 26 February 1956 under the title "La Grande famille des hommes." It received extensive press coverage and was treated as a major social and artistic event.

48 André Malraux, *The Voices of Silence* (Garden City, N.Y.: Doubleday and Company, 1953), p. 124.

49 Like Walter Benjamin, Malraux was also concerned with the loss of "aura" as central to the effects of photographic and printed reproduction. Walter Benjamin, "The Work of Art in the Age of Mechanical Reproduction" (1936), in Hannah Arendt, ed., *Illuminations* (New York: Schocken Books, 1969). According to Malraux, when the museum removed the object from its parent context, it fundamentally changed its purpose and reduced the full scope of its reality: rather than an *embodiment* of Christ within the church, the statue becomes a *representation* of Christ in the art museum or in reproduction. Malraux, *Voices of Silence*, p. 13.

50 Malraux, *Voices of Silence*, p. 24.

51 Ibid., p. 21. "Alongside the museum a new field of art experience, vaster than any so far known (and standing in the same relation to the art museum as does the reading of a play to its performance, or hearing a phonograph record to the concert audition), is now, thanks to reproduction, being opened up. And this domain—which is growing more and more intellectualized as our stocktaking and its diffusion proceeds and methods of reproduction come nearer to fidelity—is for the first time the common heritage of all mankind." Ibid., p. 46.

52 "It is not surprising that an art so much obsessed with theatrical effect, all gestures and emotions, should end up in the motion picture." Ibid., p. 122–23.

53 Ibid., p. 123.

54 Ibid., p. 633.

55 Le Corbusier, *Sketchbooks*. Published as *Le Corbusier: Sketchbooks*, vol. 3, 1954–57. Notes by Françoise de Frandieu. Each of the sketches is numbered and is indicated by square brackets. In the translation, "//" is used to indicate a line break. I have used the English translations from the Architectural History Foundation / MIT Press edition.

56 The full entry for 28 September reads:

"some funny people // Laurel and Hardy Fernandel Walt Disney. Hitler

some handsome people // athlete Don Quixote

The Silent World (fishes) C[ommander] Cousteau

Natural // 1 eye of a man of a women of a rhinoceros

Artificial // a set of gears a rope a chain" [704]

57 The entry read:

"7 *army* modern

battlefields concentration camps // 8 Aviation Wright [brothers], Santos-Dumont Voisin etc + the super planes of today projected or real // 9 microbiology [micro] mineral-ogy // 10 cinémathèque Charlie Chaplin + Fernandel Laurel and Hardy opera singer. // 11 3 Establishments of Man the routes *URB* Rio [de Janiero] 1929 UN 23 A Algiers 1930" [804]

58 Another scenario, which belonged to Louis Kalff, is found in the Getty Center Special Collections. Numerous changes for the program of ambiances are noted.

59 The introduction and credits also appeared in English.

60 Le Corbusier's personal library included copies of Albert Camus's *The Plague* and *The Rebel*.

61 Elie Faure, "Charlot," *L'Esprit Nouveau* 6 (March 1921), p. 660.

62 I must acknowledge Bart Lootsma for this interpretation.

63 Le Corbusier, *Le Poème de L'Angle Droit* (1955), no pagination.

64 Charles and Ray Eames' work on films that were ostensibly filmed slides began in 1955 with *The House: After Five Years of Living* and *Two Baroque Churches*.

65 "Square of the distance dictates *intensity*." Le Corbusier, Sketchbook entry [717].

66 Fernand Léger, "A Propos du Cinéma," *Plans* 1, (January 1931), pp. 80–81. The article is jointly dedicated to S. M. Eisenstein and Von Strohen.

67 Arthur Honegger, "Du Cinéma Sonore à la Musique Réelle," *Plans* 1 (January 1931), p. 78.

68 René Clair, "Étape du Cinéma," *Plans* 5 (May 1931), p. 70.

69 Pierre Devaux, "Le Cinéma a 100 Ans," *Rond-Point* 4 (1945), pp. 6–11.

70 Jean-Louis Cohen, *Le Corbusier et la Mystique de l'URSS* (Brussels: Pierre Mardaga Éditeur, 1987), p. 72.

71 Le Corbusier, dedication to *L'Art Décoratif d'Aujourd'hui* (1925), cited in Cohen, *Le Corbusier et la Mystique de l'URSS*, 72.

72 Cited in Cohen, *Le Corbusier et la Mystique de l'URSS*, 72.

73 "Eisenstein was charged by the government to make a propaganda film to teach the peasants how to use modern machinery and how to establish an economy for a model farm. Eisenstein looked for an architect in Moscow to realize his ideas in reinforced concrete." Petit, *Le Corbusier: Lui-Même* (Geneva: Editions Rousseau, 1970), p. 69.

74 Sergei Eisenstein, "The Cinematographic Principle and the Ideogram," in *Film Form*, trans. Jay Leyda (New York: Harcourt, Brace & World, 1949), p. 41. Lessons in framing composition appear in Lo Duca, *Technique du Cinéma* (Paris: Presses

Universitaires de France, 1943), a copy of which remains in Le Corbusier's personal library.

75 Sergei Eisenstein in Peter Wollen, *Signs and Meaning in the Cinema* (Bloomington, Ind.: Indiana University Press, 3rd ed., 1972) p. 32. No original source given.

76 Ibid., p. 40.

77 Images of Pudovkin's film *La vie est belle* appeared in the glossy pages of *Plans* 3 (March 1931), n.p.

78 Fernand Léger, "Contemporary Achievements in Painting," 1914, in Edward F. Fry, *Cubism* (New York: McGraw-Hill, n.d. [ca. 1967]), p. 64.

79 Eisenstein in Wollen, *Signs and Meaning in the Cinema*, p. 39, no citation of original source.

80 Ibid., pp. 49, 51.

81 Eistenstein, ibid., p. 48. No original source given.

82 Ibid., p. 65. No original source given.

83 Years later, Eisenstein reflected once again on the nature of montage. He had broadened his views to include questions of content within the image and its relation to the greater entity of the film. His basic idea did not change, however: "The basic fact was true, and remains true to this day, that the juxtaposition of two separate shots by splicing them together resembles not so much a simple sum of one shot plus another shot—as it does a *creation*. It resembles a creation—rather than a sum of its parts—from the circumstance that in every such juxtaposition *the result is qualitatively* distinguishable from each component element viewed separately." In *The Film Sense*, trans. Jay Leyda (London: Faber and Faber, 1986), p. 17.

84 Hugo Münsterberger, *The Film: A Psychological Study* (New York: Dover Publications, 1916; reprint, 1970), p. 30.

85 Ibid., p. 38. Italics in original.

86 Ibid., p. 39.

87 Jeanneret, "La Rythmique," *L'Esprit Nouveau* 2 (November 1920), p. 184.

88 Albert Jeanneret, "La Rythmique," *L'Esprit Nouveau* 3 (December 1920), pp. 332–33. For further discussion of Appia's work in theater and its relation to the ideas of Dalcroze, see Chapter 7.

89 Ibid., p. 334.

90 Ibid., pp. 334–35.

91 Ibid., pp. 335–36.

92 Cited in Albert Jeanneret, "L'Intelligence dans l'oeuvre musicale," *L'Esprit Nouveau* 2 (April 1921), p. 839.

93 Arthur Honegger, "Du Cinéma Sonore à la Musique Réelle," pp. 74–75.

94 Unsigned jacket notes, "Les Musiciens des Années 20, Volume 2." Praga, no date.

95 Varèse, "The Liberation of Sound," in *Contemporary Composers on Contemporary Music*, p. 202.

96 Albert Jeanneret, "Revue de L'Année: Musique," *L'Esprit Nouveau* 11–12 (November 1921, p. 1294.

97 Ibid., p. 1295.

98 L. C. Kalff, "The Light Effects," *Philips Technical Review*, reprint of vol. 20, nos. 1 and 2/3, 1958, p. 38.

99 Ibid., p. 39.

100 Kalff, internal memo, 15 November 1957, Philips Archives.

101 Le Corbusier sketchbook entry, dated 12 October 1957, Baghdad. The note was retyped, dated 5 November, and used at a meeting of the collaborators on 24 November 1957, FLC.

102 Kalff, "The Light Effects," *Philips Technical Review* p. 38. The low-intensity of the lighting caused its own problems, however.

103 For the list of equipment for this and below, see ibid., pp. 39–40.

104 The list of lighting elements is adapted from ibid., pp. 38–42. See figure 4–35 for the positioning of each element.

105 Fernand Léger, "The Wall, the Architect, the Painter," unpublished essay, 1933, printed in Léger, *Functions of Painting*, p. 96.

106 Ibid., p. 99.

107 Le Corbusier, "The Quarrel with Realism, the Destiny of Painting," in Martin et al., eds., *Circle*, (New York: Praeger, 1971), p. 73.

108 Le Corbusier, *L'Architecture d'Aujourd'hui*, special issue (April 1948), p. 53.

109 Kalff, "The Light Effects," *Philips Technical Review*, p. 38.

110 Fernand Léger, *Functions of Painting*, p. 150.

111 Charles Henry, "La Lumière, La Couleur et La Forme," *L'Esprit Nouveau* 6 (March 1921), p. 607.

112 Luisa Martina Colli, "La couleur qui cache, la couleur qui signale: l'ordonnance et la crainte dans la poétique corbuséenne des couleurs," in Claude Prelorenzo, ed., *Le Corbusier et la couleur* (Paris: Fondation Le Corbusier, 1992), p. 26.

113 Ozenfant and Le Corbusier, "Idées personnelles," *L'Esprit Nouveau* 27 (1925); cited in Colli, ibid.

114 Le Corbusier, "Polychromie = joie," in *Des Canons, des munitions. Merci! Des logis . . . s.v.p*, (ca. 1938), reprinted in *Le Corbusier: le passé à*

réaction poétique (Paris: Caisse nationale des Monuments historiques et des Sites, 1988), p. 74.

115 Luisa Martina Colli, "Couleur," in Jacques Lucan, editor, *Le Corbusier: une encyclopédie* (Paris: Centre Georges Pompidou, 1987), pp. 104.

116 Ibid., pp. 107–8. He prepared a second series, using more vibrant hues, in 1957.

117 Richard C. Beacham, Introduction to Part Three, *Adolphe Appia: Essays, Scenarios and Designs*, p. 119.

118 *Le Corbusier: le passsé à réaction poétique*, p. 209.

119 Beacham, *Adolphe Appia*, p. 121.

120 Cited in Beacham, *Adolphe Appia*, p. 121. "Although Dalcroze was [of] rather conservative taste aesthetically," Beacham explains, "he displayed a certain Calvinistic skepticism towards the theatre in particular." After Appia had joined him at his institute in Hellerau, Dalcroze's attitudes toward theater underwent a gradual development. Appia's interest in lighting effects produced a system of flexible illumination that was termed a "light organ," allowing him what he termed "luminous sound."

121 Beacham, *Adolphe Appia*, p. 124.

122 Adolphe Appia, "Eurhythmics and the Theatre," *Der Rhythmus, Ein Jahrbuch* 1 (1911), reprinted in Beacham, *Adolphe Appia*, p. 136.

123 Ibid., p. 150.

124 Ibid.

125 Ibid., p. 138.

126 Ibid., p. 139.

127 Appia's call was answered more immediately in the work of the British theater designer Edward Gordon Craig.

128 Fernand Léger, "The Spectacle: Light, Color, Moving Image, Object-Spectacle," (1924), reprinted in Léger, *Functions of Painting*, pp. 40, 44.

129 Wassily Kandinsky, *Concerning the Spiritual in Art*, trans. M.T.H. Sadler (New York: Dover Publications, 1977; reprint of 1914 edition), p. 39.

130 Ibid., p. 41.

131 Ibid., p. 38.

132 Ibid., p. 48.

133 Eisenstein took issue with pure abstraction in service of only itself. In examining an early Kandinsky color scenario from *Der blaue Reiter*, Eisenstein reaches the conclusion: "The contents of this work cannot be satisfactorily conveyed, due to the total absence of content—as well as theme." *The Film Sense*, p. 93.

134 Virginia Woolf, *Orlando*, 1928 (London: Virago Press, 1993), p. 169.

CHAPTER 5.
ORGANIZED SOUND

1 On the invention and development of the record jacket, see Steven Heller, "For the Record," *Print* (March/April 1992), pp. 64–73, 114.

2 Kalff, internal memo to Philips directors, 25 February 1956, Philips. Ouellette suggests that Walton, Copland, or Landowski were also considered, although existing documentation does not support this assertion. Ouellette, *Varèse*, p. 197.

3 Letter from Kalff to Le Corbusier (contract), 13 October 1956, FLC.

4 In spite of this claim, a letter from Le Corbusier to Varèse (in Paris) of 6 January 1955, FLC, proposes getting together for dinner. The meeting may never have taken place, however.

5 The text reads: "Marseilles construction-site. Soundtrack // by Varèse New York // music / the Theme would be called *Unité d'habitation.*" Le Corbusier, *Sketchbooks*, vol. 2, 1950–54, entry #487.

6 Le Corbusier to Varèse, 27 January 1954, Getty Center Special Collections.

7 Iannis Xenakis mentioned that Le Corbusier appreciated only two composers of modern music: Olivier Messiaen and Edgard Varèse. Conversation with the author, 15 June 1993, Paris.

8 Le Corbusier requested that Varèse receive three million francs, exclusive of his travel expenses. Le Corbusier to Ouellette, 4 June 1960. Cited in Ouellette, *Varèse*, p. 196.

9 Varèse to Le Corbusier, 16 June 1956; Varèse to Le Corbusier, 30 July 1956, FLC.

10 Le Corbusier to Varèse, 10 October 1956, FLC.

11 Le Corbusier to Varèse, 24 October 1956, FLC.

12 Varèse to Le Corbusier, 30 October 1956, FLC.

13 Kalff to Le Corbusier, 29 November 1956, FLC.

14 Xenakis to Kalff, 3 December 1956, FLC.

15 Varèse to Le Corbusier, 24 February 1957, FLC.

16 Ibid.

17 Le Corbusier to Varèse, 1 March 1957, FLC.

18 Xenakis to Varèse, 11 June 1957, FLC.

19 Le Corbusier to Varèse, 22 June 1957, FLC. At this time, the film was to end optimistically, with images of Laurel and Hardy and the smile of a baby.

20 Varese to Kalff, 11 July 1957, FLC, citing Le Corbusier's letter to Varèse, 16 June 1957. FLC.

21 Le Corbusier to Varèse, 29 July 1957, FLC.

22 Kalff to Varèse, 24 July 1957, FLC.

23 Ibid.

24 Le Corbusier to Varèse, 29 July 1957, FLC.

25 Le Corbusier to Ouellette, 4 June 1960. Cited in Ouellette, *Varèse*, p. 197. The letter continues: "I'd had the idea of asking for only one thing. . . . Right in the very middle of the *Poème électronique*, I thought of having an abrupt and total silence, and white light at the same moment—something to twist the audience's guts inside them. Result: one day as the car was taking us back into Brussels from Eindhoven, Varèse confessed to me: 'My dear Corbu, I haven't been able to fit in that silence of yours; that's exactly the moment in my thing where there's the most noise.' My reply: 'Much good may it do you. It doesn't matter! There were supposed to be a few words spoken at that moment in the *Poème électronique*, now they won't be, and perhaps it's just as well for the unity of the work.'

26 John Cage, *A Year from Monday* (Middletown, Conn.: Wesleyan University Press, 1969), p. 32.

27 John Cage, *Silence* (Cambridge, Mass.: MIT Press, 1966), p. 71.

28 Varese, "Music as an Art-Science," in Eliot Schwartz and Barry Childs, editors, *Contemporary Composers on Contemporary Music* (New York: Holt, Rinehart and Winston, 1967), p. 198.

29 Ibid., pp. 198–99.

30 Ibid., p. 200.

31 Ibid. In actuality, the composer would be given even further control over the perception of the work; it would be more like an author's reading aloud his or her own book.

32 "When I was about twenty . . . I went to the *Marché aux Puces*, where you can find just about anything, in search of a siren, and picked up two small ones. With these and using children's whistles, I made my first experiments in what later I called *spatial music*." Varese, "Spatial Music," in *Contemporary Composers on Contemporary Music*, pp. 204–5.

33 Ibid., p. 205.

34 Ibid., p. 206.

35 Ibid.

36 Varèse in Vivier, *Varèse*, p. 166, no source.

37 Varèse, "Music as an Art-Science," in *Contemporary Composers on Contemporary Music*, p. 200.

38 Varèse, "Rhythm, Form and Content," in *Contemporary Composers on Contemporary Music*, p. 201.

39 Ibid., p. 202.

40 H. Badings and J. W. de Bruyn, "Electronic Music," p. 193.

41 Bernard does not claim that Varèse directly adopted aspects of cubist method or ideas; he establishes parallels. Jonathan W. Bernard, *The Music of Edgard Varèse* (New Haven: Yale University Press, 1987), pp. 7–12.

42 Varèse, in an interview with Fred Grunfeld, WQXR, 13 December 1953. Cited in Frederic Waldman, "Edgard Varèse: An Appreciation," *Juilliard Review* 1/3 (1954).

43 Ibid.

44 H. Badings and J. W. de Bruyn, "Electronic Music," p. 193.

45 Badings and de Bruyn, "Electronic Music," pp. 191–201. Unsubstantiated rumor has it that Philips' interest in its Acoustic Group, which produced *Cain and Abel*, was focused less on the group's artistic achievements than on the possibility of its developing a silent switch for tape recording. In addition, having heard the ballet's electronic score, Philips management may have come to the conclusion that the company's one-time sponsorship of this sort of enterprise was enough. The company remains active in its sponsorship of classical music in Eindhoven, however.

46 Bart Lootsma suggests that Badings was somewhat politically compromised during the Nazi occupation of Holland, precluding his broad exposure at the Brussels exposition. Conversation with author, July 1994, Paris.

47 Composer Richard Felciano suggests: "It mays also be that working with the proximity and distance clues which the brain uses to determine near/far in perceiving antiphonal

sound led Giovanni Gabrielli to become the first composer to write down dynamic (loudness) indications for his music. The composition which broke this ground was called *Sonata piano e forte* [Soft and loud sonata]." Letter to author, 8 February 1994.

48 "A sometimes unsatisfactory solution occurs when Gabrielli's 8-part and 12-part canzonas are played in large halls, with wide separations between the groups. In the final sections of these works, where the harmonic changes are fairly rapid and the counterpoint is intricate but harmonically related, a total and exact rhythmic coordination is required. This is destroyed by a time lag, which cannot be accurately rectified by having the more distant groups attempt to 'anticipate' the conductor's beats." Harry Partch, "Space as an Essential Aspect of Musical Composition," in *Contemporary Composers on Contemporary Music*, p. 235.

49 Richard Felciano to author, 8 February 1994.

50 Ouellette, *Varèse*, p. 49.

51 This is a French parallel to, or version of, the Wagnerian leitmotifs that comprised the Ring cycle and other operas. Thematic fragments are used to symbolize personages or emotions, and are woven together as a conglomerate in which the sum far exceeds the limits of its parts.

52 For many of the years Varèse lived in the United States, he earned his living by conducting rather than by composing. He entered the New York musical scene with his 1 April 1917 conducting of the Berlioz *Requiem* at the Hippodrome, an auditorium seating nearly six thousand. The orchestral and choral resources nearly equalled those at the first performance in Paris. The newspapers reacted favorably to the performance: "Mr. Edgar Varèse who has made his American debut as a conductor seemed to possess the inspiration of genius." Ouellette, *Varèse*, p. 49.

53 Jacques Barzun has reconstructed the composer's perceptions:

"Now Berlioz had heard Mozart's *Requiem* at the Madeleine and one of Cherubini's at the Invalides. Comparing their effect in those large churches with that in the acoustic hall of the Conservatoire, he could see that the volume of sound must be brought into scale with the place, and that sharp contrasts must be established between successive numbers to sustain attention during the rites. For his purposes Berlioz found he needed 190 instruments and 210 voices, with additional timpani and brass choirs to be used in the *Tuba mirum*." *Berlioz and His Century* (New York: Meridian Books), p. 164.

54 From the liner notes by David Cairns, *Berlioz Requiem*, Sir Colin Davis, London Symphony Orchestra and Chorus, Philips CD 416 283–2 (1969), p. 64.

55 Hector Berlioz, *The Life of Hector Berlioz*, trans. David Cairns (London: Callanz, 1969), p. 231.

56 Ouellette, *Varèse*, p. 99.

57 Matossian, *Xenakis*, p. 121.

58 Varèse, "New Instruments and New Music," in *Contemporary Composers on Contemporary Music*, p. 197.

59 Petit, *Le Poème Electronique*, pp. 201–203.

60 Bart Lootsma, "Le Poème Electronique: Her verlangen naar sythese," *Wonen-TABK* (2 January 1984), p. 28. The story comes from the engineer H. C. Duyster, whom Lootsma interviewed on 3 October 1983. Stories reveal that Tak also had an intuitive side and that when consulting on acoustics, his analyses of spaces might begin simply by clapping his hands to test for reverberation.

61 In an undated letter Varèse to Odile Vivier, in Vivier, *Varèse* (Paris: Editions du Seuil, 1973), p. 162. The phrase "more than slow" may be a pun: *La plus que lente* is the title of a Claude Debussy piano piece written in 1910 in response to the popular *La Valse Lente*.

62 Varèse to Vivier, December 1957. Vivier, *Varèse*, p. 161.

63 Kalff to Le Corbusier, 12 December 1957, FLC.

64 Xenakis to Kalff, 17 December 1957, FLC.

65 Le Corbusier to Kalff, 24 December 1957, FLC.

66 Varèse's preferred term. See Varèse, "Spatial Music," in *Contemporary Composers on Contemporary Music*, pp. 204–7.

67 Varèse to Xenakis, December 1957?, in Vivier, *Varèse*, p. 162.

68 Internal memo, no date, probably late January or early February 1958, Philips Archives.

69 Ibid.

70 Lootsma, "Poème Electronique: Le Corbusier, Xenakis, Varèse," p. 137. I have found no records of the performance in the Philips or Le Corbusier archives, however.

71 The aftermath of that illness would trouble the composer for the rest of his life. Varèse to Vivier, probably February 1958, in Vivier, *Varèse*, p. 165.

72 That Le Corbusier was impressed with Tak's work is noted in a letter from Le Corbusier to Varèse, 29 July 1957, FLC.

73 Tak, "The Sound Effects," *Philips Technical Review*, reprint of vol. 20, nos. 1 and 2/3, p. 43.

74 Ibid.

75 Ibid. He also commented on the absence of sufficient language needed to describe the musical effects and sounds that Varèse wanted. Lacking a notation system, "Varèse frequently indicated his wishes by such expressions as 'more nasal,' 'less biting,' 'more rasping,' and it was our job to meet his wishes and frequency-shifting circuits. To define the necessary operations we had to resort repeatedly to onomatopoeic words, such as 'wow wow,' 'poowhip,' 'tick tock,' 'whoop,' and 'choochah.'"

76 Varèse to Le Corbusier, 24 February 1958, FLC.

77 Ibid., p. 44.

78 S. L. de Bruin, "The Electronic Control System," *Philips Technical Review*, reprint of vol. 20, nos. 1 and 2/3, p. 45.

79 The spectrograph traces of the recorded sounds appear like the tracks of the seismograph. To Robert Cogan, the varying density of the markings indicate two opposing, alternating sonority types with the *Poème électronique*. He suggests that Part I of the piece, the first 180 seconds, favors percussion and machine-like sounds, while harmonic or quasi-harmonic spectra dominate seconds 181–480. Notes to the CD *Electro Acoustic Music: Classics*, Neuma Records (1990), 450–74.

80 Cited by Karlheinz Stockhausen in a lecture, which is quoted in Karl H. Wörner, *Stockhausen: Life and Work*, trans. Bill Hopkins (Berkeley: University of California Press, 1976), p. 139.

81 Conversation with the author, 15 June 1993, Paris.

82 This is described in a letter from Xenakis to Tak, 18 April 1957, FLC.

83 The staff included four radio-technicians, who would execute and maintain the sound equipment, and two projectionists; in addition, the company estimated that four hosts would be required—possibly chosen from Philips pensioners; and two cleaners would work three hours daily. The idea of having a café on site was dismissed as impractical. Meeting notes, 6 September 1957, Philips Bureau de Presse et d'Information, Philips.

84 Le Corbusier to Varese, 12 June 1956, FLC. Cited in Matossian, *Xenakis*, p. 111.

85 Le Corbusier to Xenakis, 27 November 1957, FLC.

86 Le Corbusier to Kalff, 29 November 1957, FLC.

87 The explanation and credits for the presentation would be given in text at the opening moments of the film. Kalff, internal memo, 18 November 1957, Philips.

88 Xenakis, "Notes sur un 'Geste Electronique,'" in *Musique Architecture*, pp. 143–45.

89 Xenakis to Scherchen, 25 June 1957, Xenakis Archives. Cited in Matossian, *Xenakis*, p. 120.

90 Le Corbusier to Xenakis, 27 November 1957, FLC.

91 Matossian, *Xenakis*, p. 122.

92 Xenakis to Kalff, 11 March 1958, Getty Center Special Collections.

93 Xenakis to Kalff, 17 March 1958, Getty Center Special Collections.

94 Xenakis to Kalff, 7 April 1958, Getty Center Special Collections.

95 Kalff to Xenakis, 28 July 1958, FLC.

96 In conversation, Xenakis agreed that this was a frequent occurrence in architectural practice, but that office anonymity was not categorical. He cited, for example, Walter Gropius's involvement with team projects and his crediting each member of the work group. Conversation with the author, 15 June 1993, Paris.

97 See the comprehensive biography by Matossian, *Xenakis*, for the complete story.

98 Varèse to Xenakis, 2? May 1958, cited in Ouellette, *Varèse*, p. 199.

99 Unnamed, uncited quotation in Ouellette, *Varèse*, p. 201.

100 Le Corbusier to Ouellette, 4 June 1960, FLC.

101 Albert Jeanneret to Le Corbusier, 4 October 1958, cited in Ouellette, *Varèse*, p. 202. Like Albert Jeanneret, Xenakis was critical of the realism of the visuals. He says that tried to influence Le Corbusier to work in a more abstracted manner, using the various electronic media more purely. Conversation with author 15 June 1993.

102 Internal memo, 20 October 1958, Philips Archives.

103 In addition to the Neuma CD noted above, there is a second disk by the Asko Ensemble available in Europe.

104 Varèse, "Spatial Music," in *Contemporary Composers on Contemporary Music*, p. 207.

CHAPTER 6.
REACTION

1 "It will be the place where the cold war, fought with the weapons of art and drama, music and dance, architecture, books and films, will reach a climax." Howard Taubman, "Cold War on the Cultural Front," *New York Times*, 13 April 1958, p. 12.

2 *Time*, 31 March 1958, p. 56.

3 U.S. fair officials, quoted in *Newsweek*, 14 April 1958, p. 54.

4 *Time*, 28 April 1958, p. 29.

5 *Newsweek*, 14 April 1958, p. 53.

6 Unnamed author, *Eindhoven Dagblad*, 28 April 1958, Philips.

7 Jaak Veltman, "An Electronic Slap in the Face," *De Standaard*. From a summary of the Belgian press, translated by Philips, Philips.

8 Ibid.

9 Ibid.

10 "An Adventure signed Philips and Le Corbusier," *La Libre Belgique*. Summary, in English, of the Belgian press reviews. No date. Philips.

11 Jean Girard, "A Fantastic Realization," *La Cité*. From a summary, in English, of Belgian and French press commentary on the pavilion. Philips.

12 "Le Corbusier: The Man of the Day," *La Lanterne*, no date. Included in English summary of Belgian and French newspaper coverage. Probably April 1958, Philips.

13 "Le Pavillon Philips et le 'Poème Electronique' de Le Corbusier," *La Dernière Heure* (23 April 1958), Philips.

14 "L'inauguration du pavillon Philips," *Echo de Central*, no date (probably April 1958), no author. Philips.

15 Howard Taubman, "Fairgoers Hear Electronic Poem," *New York Times*, no date (probably late May 1958). Philips.

16 D. W. Bilderbeek, "Reason for Delayed Pride," *Announcer* (March 1958), p. 62.

17 *Time*, 28 April 1958, p. 29.

18 The architects for the Atomium were A. and J. Polak with André Waterkeyn, engineer. "Brussels Universal Exhibition 1958," *Architectural Design*, August 1958, p. 312.

19 The O.E.C.E. Pavilion was designed by architect Karl Schwanzer, with Robert Krapeenbauer, engineer.

20 The design team for the French Pavilion included Guillaume Gillet, architect; René Sarger and Jean Prouvé, consulting engineers. Anciens ETS Eiffel, builder.

21 Renato Pedio, "La crisi del linguaggio moderno nell'Esposizione Universale, Bruxelles 1958," *L'Architettura* (October 1958), p. 386.

22 Pavilion "Marie Thumas": Baucher, Blondel, Filiponne, architects; René Sarger, suspended roof consultant.

23 René Sarger, "Valeur Plastique des Structures," *L'Architecture d'Aujourd'hui* 78, vol. 29 (June 1958), p. 7.

24 Bruno Zevi, "The Closing at Brussels: Symbolism: Maggots of Architectonic Expression," *L'Architettura* vol. 4, no. 6 (October 1958), p. 365.

25 Craig Ellwood, "L'architecture et la machine," *L'Architecture d'Aujourd'hui* 78, vol. 29, (June 1958), p. 8.

26 Transportation Pavilion credits also: J. Goosens, Bara, and R. Moens de Hase, associate architects; A. Lipski, engineer.

27 It received the Reynolds Award for Aluminum for buildings erected in 1958.

28 Zevi, "Symbolism," p. 364.

29 Pedio, "La crisi del linguaggio moderno nell'Esposizione Universale, Bruxelles 1958," p. 391.

30 *Architectural Design*, August 1958, pp. 312–13.

31 Kristian Romare, "Le Corbusiers Elektroniska Skapelse," *Byggmästeren*, no. 8, (1958), p. 175.

32 Ernesto N. Rogers, "All'Expo '58 il futuro (dell'architettura) non è cominciato," *Casabella* 221 (September 1958), p. 4.

33 Ernesto Rogers, "The Evolution of Architecture: Reply to the Custodian of Frigidaires," *Casabella-Continuità* (June 1959), reprinted in Ockman, ed., *Architecture Culture 1943–1968*, p. 306.

34 Le Corbusier, "Electronic Poem," no date, no publisher; probably a press release issued by Philips, Philips Archives.

35 Ibid.

36 Kalff, construction brief, 29 January 1957, FLC.

37 On 24 July 1958 Radio Sweden wrote to Kalff about the possibility of re-enacting the *Poème électronique* in a Stockholm concert hall. Kalff agreed, but warned them that it was a difficult and technically demanding proposition. 2 August 1958. Nothing seems to have come of the request. FLC.

38 Jasinski to Le Corbusier, 22 September 1958, FLC.

39 Le Corbusier to Jasinski, 25 September 1958, FLC.

40 Le Corbusier to Jasinski, 3 October 1958, FLC.

41 Spaens to Jasinski, 9 October 1958, FLC.

42 Kalff to Le Corbusier, 2 February 1959, FLC.

43 Bart Lootsma, "Poème Electronique: Le Corbusier, Xenakis, Varèse," in Vowinckel and Kesseler, eds., *Le Corbusier: Synthèse des Arts, Aspekte des Spätwerks, 1945–65*, p. 141.

CHAPTER 7. THE ELECTRONIC PAVILION RECONSIDERED

1 Agostini's efforts on the *Poème Electronique* are not recorded in any of his articles or in his filmography. See *Film Dope 1*, 1 (December 1972), no pagination.

2 "Le Corbusier did not restrict his researches to concrete and rough brick, however much one may associate these materials with his late years. In the Phillips [*sic*] Pavilion for the Brussels World's Fair of 1958 he employed steel tubes and tension cables to create a dramatically pointed tent with curving surfaces." William J. R. Curtis, *Le Corbusier: Ideas and Forms* (Oxford: Phaidon Press, 1986), p. 214. Charles Jencks also implies that it was a tent in his *Le Corbusier and the Tragic View of Architecture* (Cambridge, Mass.: Harvard University Press, 1974): "This shift in emphasis from form to content or medium to message became stronger in Le Corbusier's next exhibition tent constructed for Philips at the Brussels World [*sic*] Fair in 1958" (p. 163). The confusion could derive not only from the ambiguity inherent in small-scale photographs, but also from Le Corbusier's Pavillon des Temps Nouveaux at the 1937 Paris fair, which was a tension structure supported by anchored cables. See "Pavillon des Temps Nouveaux," *L'Architecture d'Aujourd'hui* (September 1937), pp. 46–49.

3 Le Corbusier, "The Quarrel with Realism, the Destiny of Painting," in Martin et al., eds., *Circle*, p. 73.

4 Le Corbusier, "Esprit de vérité," *Mouvement* I (June 1933), in *French Film Theory and Criticism*, trans. Richard Abel, p. 113.

5 Ibid. Italics in the original.

6 Le Corbusier, *The Chapel at Ronchamp*, trans. Jacqueline Cullen (London: Architectural Press, 1957), no pagination.

7 Ibid., sketchbook note.

8 Xenakis to Kalff, 3 October 1957, FLC. Cited in Matossian, *Xenakis*, p. 118.

9 Xenakis to Scherchen, 25 June 1957. Xenakis Archives. Cited in Matossian, *Xenakis*, p. 120.

10 In his original sketches for the opera house, Utzon had lyrically proposed a series of fluid concrete shells. As the design was developed in greater detail, the realization of these forms became more and more problematic. Structural performance was one question; construction rationalization—like with the Philips Pavilion—was another. The cost of formwork for shells of continually varying slope and curvature would have been prohibitive. Arup's ingenious solution—although said to have been based on an idea by Utzon—was to select spherical triangles each taken from different sections of the same sphere. Thus, the wooden formwork could be based on a consistent radius.

11 Le Corbusier, catalog to his 1938 exhibit at the Galerie Balaÿ et Carré. Cited in Green, "The Architect as Artist," p. 111.

12 Le Corbusier in Jean Petit, *Texts and Sketches for Ronchamp* (Ronchamp: Association Oeuvre de Notre-Dame-du-Haut, 1965, 1990), no pagination.

13 The architects for the Russian pavilion were Y. Abramov, A. Borestki, V. Doubov, and A. Polanski; the engineers were Y. Ratskevitch and K. Vassilieva.

14 Kandinsky, *Concerning the Spiritual in Art*, p. 52.

15 László Moholy-Nagy, *The New Vision* (New York: Wittenborn and Company, third revised ed., 1946), p. 64.

16 Le Corbusier, "Ineffable Space," p. 66.

17 The sequence illustrates the tension between man and animal, and by extension, man and nature. The bull as a theme in Le Corbusier's paintings probably derives from the paintings and prints of Pablo Picasso.

18 The film was created over the period of about a year, from 1923–24, in collaboration with the American cameraman Dudley Murphy. Christopher Green, *Léger and the Avant-Garde* (New Haven: Yale University Press, 1976), p. 281.

19 From Léger's text about the film in *L'Esprit Nouveau* 28, cited in Green, *Léger and the Avant-Garde*, p. 281.

20 Ibid., p. 282.

21 Ibid., p. 284.

22 Edward Gordon Craig, *Towards a New Theatre* (London: J. M. Dent and Sons, 1913), p. 6.

23 "The Imagination and the Emotions are not for purposes of mimicry—they are for creation. When the Imagination and the Emotions can, through art, create the age, how should they stoop to mimic it? Art is useless after events: it must precede them." Ibid., p. 85. Craig also wrote that "Photography and Phonographic realism injure the minds of the people. They project upon them a grotesque and inaccurate representation of the outward and visible life—with the divine essence—the spirit—the beauty of life left out." Ibid., p. 89.

24 Albert Jeanneret, "Parade," *L'Esprit Nouveau* 4 (January 1921), pp. 449–50.

25 Cage, "Edgard Varèse," in *Silence*, p. 84.

26 Karlheinz Stockhausen in Wörner, *Stockhausen*, 1976, p. 139.

27 Henry Miller, "With Edgard Varèse in the Gobi Desert," in *The Air-Conditioned Nightmare*, vol. I, pp. 163–78 (New York: New Directions Books, 1945), p. 169.

28 Virgil Thomson, *American Music Since 1910* (New York, Holt, Rinehart and Winston, 1971), p. 47.

29 Marc Bredel, *Edgar Vàrese* (Paris: Editions Mazarine, 1984), p. 172.

30 Le Corbusier to Kalff, 31 October 1958, FLC.

31 Ibid.

32 Kalff to Le Corbusier, 14 November 1958, FLC.

33 Le Corbusier to Varèse, 17 November 1958, FLC.

34 Varèse to Le Corbusier, 6 December 1958.

35 Le Corbusier, "Nothing Is Transmissible but Thought," in Willy Boesiger, editor, *Le Corbusier: Last Works* (New York: Praeger Publishers, 1970), p. 174.

36 Milan Kundera, *Immortality*, trans. Peter Kussi, (New York: Harper Perennial, 1991), p. 122.

ARCHIVES

Fondation Le Corbusier, Paris [FLC]
Philips Pavilion: Files J2-19
Edgard Varèse: File R3 06

Getty Center for the History of Art and the Humanities, Special Collections, Santa Monica, California [Getty Center]

Philips International B.V., Eindhoven [Philips]

Hollandsche Beton Groep N.V., Rijswijk [HBG]

BIBLIOGRAPHY

Abel, Richard. *French Film Theory and Criticism: A History/Anthology, 1907–1939.* Volume II: 1929–1939. Princeton, N.J.: Princeton University Press, 1988.

Angerer, Fred. *Surface Structures in Building.* New York: Reinhold, 1961.

L'Architecture d'Aujourd'hui, special issue on Le Corbusier, 1948.

Badings, H., and J. W. de Bruyn. "Electronic Music." *Philips Technical Review*, vol. 19 (1957/58), pp. 191–201.

Banham, Reyner. *The New Brutalism: Ethic or Aesthetic?* New York: Reinhold, 1966.

Barzun, Jacques. *Berlioz and His Century.* New York: Meridian Books, 1956.

Beacham, Richard C., editor. *Adolphe Appia: Essays, Scenarios and Designs.* Translated by Walter R. Volbach. Ann Arbor, Mich.: UMI Press, 1989.

Bekooy, Guus. *Philips: A Century of Enterprise, 1891–1991.* Eindhoven: Philips International, 1991.

Berlioz, Hector. *The Life of Hector Berlioz.* Translated by David Cairns. London: Gollanz, 1969.

Bernard, Jonathan W. *The Music of Edgard Varèse.* New Haven: Yale University Press, 1987.

Bilderbeek, D. W. "Reason for Delayed Pride." *Announcer* (March 1958), p. 62.

Bordaz, Robert et al. *Le livre des expositions universelles 1851–1989.* Paris: Editions des arts décoratifs, 1983.

Bosson, Viveca, and Beate Sydhoff. *Le Corbusier: Konstnären och arkitekten.* Stockholm: Kulturhuset, 1993.

Bouma, A. L., and F. K. Ligtenberg. "Model Tests for Proving the Construction of the

Pavilion." *Philips Technical Review*, reprint of vol. 20, nos. 1 and 2/3 (1958), pp. 17–26.

Bredel, Marc. *Edgar Varèse*. Paris: Editions Mazarine, 1984.

"Brussels Universal Exposition 1958." *Architectural Design*, vol. XXVII/8 (August 1958), pp. 312–36.

Cage, John. *Silence*. Cambridge, Mass.: MIT Press, 1966.

Cage, John. *A Year from Monday*. Middleton, Conn.: Wesleyan University Press, 1969.

"Le Cinéma en l'an 2000." *Rond-Point*, n.d., n.p.

Clair, René. "Étape du cinéma." *Plans* 5 (May 1931), pp. 69–72.

Cohen, Jean-Louis. *Le Corbusier et la mystique de l'URSS*. Brussels: Pierre Mardaga Éditeur, 1987.

Colli, Luisa Martina. "La couleur qui cache, la couleur qui signale: l'ordonnance et la crainte dans la poétique corbuséene des couleurs." In Claude Prelorenzo, editor, *Le Corbusier et la couleur*, pp. 19–34. Paris: Fondation Le Corbusier, 1992.

——. "Vers une polychromie architecturale." In Jacques Lucan, editor, *Le Corbusier: une encyclopédie*, pp. 104–10. Paris: Centre Georges Pompidou, 1987.

"Comment Ils Font Leurs Films: Phillippe Agostini." *Photo-Cinéma* (January 1956), pp. 18–19.

Craig, Edward Gordon. *Towards a New Theatre*. London: J. M. Dent and Sons, 1913.

Curtis, William J. R. *Le Corbusier: Ideas and Forms*. Oxford: Phaidon Press, 1986.

De Bruin, S. L. "The Electronic Control System." *Philips Technical Review*, reprint of vol. 20, nos. 1 and 2/3 (1958), pp. 45–49.

Delluc, Louis. "Photogénie." *L'Esprit Nouveau* 5 (February 1921), p. 589.

Devaux, Pierre. "Le cinéma a 100 ans ou quelques scènes de la vie future." *Rond-Point* 4 (1945), pp. 6–11.

Duca, Lo. *Technique du cinéma*. Paris: Presses universitaires de France, 1943.

Duyster, H. C. "La conception, l'exécution et le comportement d'une coque mince en béton précontraint." Unpublished paper presented at the International Colloquium on Construction Processes of Shell Structures, [1958?].

——. "Construction of the Pavilion in Prestressed Concrete." *Philips Technical Review*, reprint of vol. 20, nos. 1 and 2/3 (1958), pp. 27–36.

——. "De Bouw van het Philips Paviljoen." *Binddraad*, Hollandsche Beton Maatschappij, Den Haag (15 June 1958), pp. 1–5.

Eisenstein, Sergei. *Film Form*. Translated by Jay Leyda. New York: Harcourt, Brace & World, 1949.

——. *The Film Sense*. Translated by Jay Leyda. 1943. London: Faber and Faber, 1986.

Ellwood, Craig. "L'architecture et la machine." *L'Architecture d'Aujourd'hui* 78, vol. 29 (June 1958), p. 8.

"Entretien avec Philippe Agostini." *Cinématographe* (February 1983), pp. 30–31.

Fabre, Gladys C. et al. *Léger and the Modern Spirit*. Houston: Museum of Fine Arts, 1982.

Faure, Elie. "Charlot." *L'Esprit Nouveau* 6 (March 1921), pp. 657–66.

Friebe, Wolfgang. *Buildings of the World's Exhibitions*. Translated by Jenny Vowles and Paule Roper. Lepizig: Edition Leipzig, 1985.

Fry, Edward F. *Cubism*. New York: McGraw-Hill, 1966.

Green, Christopher. "The Architect as Artist." In Michael Raeburn and Victoria Wilson, editors, *Le Corbusier: Architect of the Century*, pp. 110–31. London: Arts Council of Great Britian, 1987.

——. *Léger and the Avant-Garde*. New Haven: Yale University Press, 1976.

Griffiths, Paul. *Modern Music*. New York: Thames and Hudson, 1985.

Hanoun, Marcel. "Entretien avec Philippe Agostini." *Photo Cinéma* (January 1956), pp. 18–19.

Henry, Charles. "La Lumière, La Couleur et La Forme." *L'Esprit Nouveau* 6 (March 1921), p. 607.

Hoerée, Arthur. "Une oeuvre nouvelle de Igor Strawinsky." *Plans* 4 (April 1931), pp. 95–96.

Honegger, Arthur. "Du cinéma sonore à la musique réelle." *Plans* 1 (January 1931), pp. 74–79.

——. "Pour prendre congé." *Plans* 7 (July 1931), pp. 43–46.

Hoover, Jeffrey. "Phenomenology and the New Music." *ComposerUSA* (Spring 1994), pp. 4, 13.

Jahnke, Eugen and Fritz Emde. *Tables of Functions with Formulae and Curves*, 3d edition. Leipzig: B. G. Teubner, 1938.

Jeanneret, Albert. "L'Intelligence dans L'Oeuvre Musicale." *L'Esprit Nouveau* 7 (April 1921), pp. 839–42.

——. "Parade." *L'Esprit Nouveau* 4 (January 1921), pp. 449–52.

——. "Revue de L'Année: Musique." *L'Esprit Nouveau* 11–12 (November 1921), pp. 1294–96.

——. "La Rythmique." Part I, *L'Esprit Nouveau* 2 (November 1920), pp. 183–89; Part II, *L'Esprit Nouveau* 3 (December 1920), pp. 331–38.

Jencks, Charles. *Le Corbusier and the Tragic Vision of Architecture*. Cambridge, Mass.: Harvard University Press, 1974.

Joedicke, Jürgen. *A History of Modern Architecture*. New York: Praeger, 1960.

Kalff, L. C. "Het Licht op de 'Exposition Internationale' te Parijs 1937." *Philips Technisch Tijdschrift* (December 1937), pp. 361–66.

——. "The Light Effects." *Philips Technical Review*, reprint of vol. 20, nos. 1 and 2/3 (1958), pp. 37–42.

Kandinsky, Wassily. *Concerning the Spiritual in Art*. Translated by M.T.H. Sadler. 1914. Reprint, New York: Dover, 1977.

Keppler, F. W. "Duyster 40 Jaren." *Specie* (*Mortar*, the Hollandsche Beton Groep house organ), vol. 6, no. 5 (5 October 1970).

Kloos, Maarten. "Iannis Xenakis: muziek, architectuur, ruimte." *Wonen/TABK* (2 January 1984), pp. 18–19.

Kundera, Milan. *Immortality*. Translated by Peter Kussi. New York: Harper Perennial, 1991.

Lamour, Philippe. "City Lights de Charlie Chaplin." *Plans* 5 (May 1931).

Le Corbusier. *L'Art décoratif d'aujourd'hui*. 1925. Reprint, Paris: Les Editions Arthaud, 1980.

——. *Des canons, des munitions. Merci! Des logis . . . s.v.p.* Boulogne: Editions de l'Architecture d'Aujourd'hui, [1938?].

——. *The Chapel at Ronchamp*. Translated by Jacqueline Cullen. London: Architectural Press, 1957.

——. *Creation is a Patient Search*. Translated by James Palmes. New York: Praeger, 1960.

——. "Esprit de vérité." *Mouvement* I (June 1933). In Richard Abel, *French Film Theory and Criticism: A History/Anthology, 1907–1939*. Volume II: 1929–1939. Princeton, N.J.: Princeton University Press, 1988.

——. "Ineffable Space." *L'Architecture d'Aujourd'hui* (January 1946), pp. 9–10. Reprinted in Joan Ockman, editor, *Architecture Culture: 1943–1968*, p. 66. New York: Rizzoli International, 1993.

——. *Last Works*. Edited by Willy Boesiger. New York: Praeger, 1970.

——. *The Modulor*. 1950. Translated by Peter de Francia and Ann Bostock. Cambridge, Mass.: Harvard University Press, 1954.

——. *Modulor 2*. 1955. Translated by Peter de Francia and Ann Bostock. London: Faber and Faber, 1958.

——. *Oeuvre Complète: 1946–52*. 2d edition. Zurich: W. Boesiger/Editions Girsberger, 1961.

——. *Oeuvre Complète: 1952–57*. 2d edition. Zurich: W. Boesiger/Editions Girsberger, 1958.

——. *Le Poème de l'angle droit*, 1955. Paris: Fondation Le Corbusier/Editions Connivences, 1989.

——. *Precisions: On the Present State of Architecture and City Planning*, 1930. Translated by Edith Schreiber Aujame. Cambridge, Mass.: MIT Press, 1991.

——. "The Quarrel with Realism, the Destiny of Painting." In J. L. Martin, Ben Nicholson, N. Gabo, editors, *Circle*, pp. 67–74. 1937. Reprint, New York: Praeger, 1971.

——. *Le Corbusier Sketchbooks*. Volume 2, 1950–54; Volume 3, 1954–57; Volume 4, 1957–64. Notes by Françoise de Frandieu. New York: Architectural History Foundation/MIT Press, 1982.

——. *Towards a New Architecture*, 1923. New York: Dover, 1986; reprint of the 1931 translation by Frederick Etchells.

Léger, Fernand. "À propos du cinéma." *Plans* 1 (January 1931), pp. 80–84.

——. "Color in Architecture." In Stamo Papadaki, *Le Corbusier: Architect, Painter, Writer*, pp. 78–80. New York: Macmillan, 1948.

——. *Functions of Painting*. New York: Viking Press, 1973.

Lootsma, Bart. "Een ode van Philips aan de vooruitgang." *Wonen/TABK* (2 January 1984), pp. 10–17.

——. "Le Poème Electronique: Her verlangen naar sythese." *Wonen/TABK* (2 January 1984), pp. 20–31.

——. "Poème Electronique: Le Corbusier, Xenakis, Varèse." In Andreas Vowinckel and Thomas Kesseler, editors, *Le Corbusier: Synthèse des Arts, Aspekte des Spätwerks, 1945–65*, pp. 112–47. Berlin: Ernst & Sohn, 1986.

Lucan, Jacques, editor. *Le Corbusier: une encyclopédie*. Paris: Centre Georges Pompidou, 1987.

Luigi, Gilbert. "Les dernières oeuvres de Le Corbusier et sa postérité." In Michel Ragon, editor, *Le temps de Le Corbusier*, pp. 187–212. Paris: Editions Hermé, 1987.

Malraux, André. *Le musée imaginaire de la sculpture mondiale*. Paris: Gallimard, 1954.

——. *The Voices of Silence*. Garden City, N.Y.: Doubleday, 1953.

Mathematischer Modelle für den höheren Mathematischen Unterricht. Leipzig: Martin Schilling, 1911.

Martin, Marcel. "Rencontre, à propos de l'évolution esthétique de l'image, avec Philippe Agostini." *Cinéma Pratique* 147, vol. 22 (August/September 1976), pp. 146–47.

Matossian, Nouritza. *Xenakis*, London: Kahn & Averill, 1986.

Miller, Henry. "With Edgard Varèse in the Gobi Desert." In *The Air-Conditioned Nightmare*, Volume I, pp. 163–78. New York: New Directions Books, 1945.

Milne, Tom. "Angels and Ministers." *Sight and Sound*, vol. 56, no. 4 (Autumn 1987), pp. 286–87.

Moholy-Nagy, László. *The New Vision*. 1928. 3d edition, New York: Wittenborn and Company, 1946.

Münsterberger, Hugo. *The Film: A Psychological Study*. 1916. Reprint, New York: Dover, 1970.

Newsweek, April 14, 1958, pp. 53–58.

"No Apology" [about Varèse], *Time*, May 12, 1961, p. 34.

Ockman, Joan, editor. *Architecture Culture: 1943–1968*. New York: Rizzoli International, 1993.

"L'Oeuvre de Robert Bresson." *Télé-Ciné* 25 (1951), pp. 1–8.

Ouellette, Fernand. *Edgard Varèse*. Translated by Derek Coltman. New York: Orion Press, 1966.

Ozenfant, Amédée. *Foundations of Modern Art*. 1930. Reprint, New York: Dover, 1952.

Ozenfant, Amédée, and Charles-Edouard Jeanneret. "Sur la Plastique." *L'Esprit Nouveau* 1 (October 1920), pp. 42–43.

"Il padiglione della Philips." *Casabella* 211 (September 1958), pp. 14–15.

"Padiglione della Philips all'Expozitione di Bruxelles." *L'Architettura* 29, vol. 3 (March 1958), pp. 781–83.

Papadaki, Stamo. *Le Corbusier: Architect, Painter, Writer*. New York: Macmillan, 1948.

Partch, Harry. "Space as an Essential Aspect of Musical Composition." 1949. Reprinted in Eliot Schwartz and Barney Childs, editors, *Contemporary Composers on Contemporary Music*, pp. 209–42. New York: Holt, Rinehart and Winston, 1967.

Passek, Jean-Loup. *Dictionnaire du Cinéma*. Paris: Larousse, 1991.

"Pavillon Philips." *L'Architecture d'Aujourd'hui* 78, vol. 29 (June 1958), pp. 14–15.

"Pavillon des Temps Nouveaux." *L'Architecture d'Aujourd'hui* (September 1937), pp. 46–49.

Pedio, Renato. "La crisi del linguaggio moderno nell'Esposizione Universale, Bruxelles 1958." *L'Architettura* 36, vol. 4, no. 6 (October 1958), pp. 384–93.

Petit, Jean. *Le Corbusier*. Lausanne: Editions Rencontre, 1968.

——. *Le Corbusier: Lui-Même*. Geneva: Editions Rousseau, 1970.

——. *Le Poème Électronique*. Paris: Editions Vives, 1958.

——, editor. *Texts and Sketches for Ronchamp*. 1965. Reprint, Ronchamp: Association Oeuvre de Notre-Dame-du-Haut, 1990.

"Philippe Agostini." *Film Dope* 1 (December 1972), n.p.

"Philippe Agostini." *Cinématographe* (June 1981), pp. 34–39.

Prédal, René. "La dimension plastique de l'oeuvre de Bresson." *Jeune Cinéma* 201 (May-June 1990), pp. 4–11.

Prelorenzo, Claude, editor. *Le Corbusier et la couleur*. Paris: Fondation Le Corbusier, 1992.

Ragon, Michel, editor. *Le temps de Le Corbusier*. Paris: Editions Hermé, 1987.

Rogers, Ernesto. "The Evolution of Architecture: Reply to the Custodian of the Frigidaires." *Casabella-Continuità* (June 1959).

Reprinted in Joan Ockman, editor, *Architecture Culture: 1943–1968*, pp. 301–7. New York: Rizzoli International, 1993.

——. "All'Expo il futuro (dell'archittectura) non è cominciato." *Casabella* 221 (September 1958), pp. 3–5.

Romare, Kristian. "Le Corbusiers Elektroniska Skapelse." *Byggmästeren*, no. 8 (1958), p. 173–76.

Roud, Richard. "The Early Work of Robert Bresson." *Film Culture* 20 (1959), pp. 44–52.

Saddy, Pierre, and Claude Malécot, editors, *Le Corbusier: le passé à réaction poétique*. Paris: Caisse nationale des Monuments historiques et des Sites, 1988.

Sarger, René. "Valeur Plastique des Structures." *L'Architecture d'Aujourd'hui* 78, vol. 29 (June 1958), pp. 6–7.

Soby, James Thrall. "Le Corbusier the Painter." In Stamo Papadaki. *Le Corbusier: Architect, Painter, Writer*, pp. 115–32. New York: Macmillan, 1948.

Streichen, Edward. *The Family of Man*. New York: Doubleday, 1955.

Sweeney, James Johnson, and Josep Luis Sert. *Gaudí*. New York: Praeger, 1960.

Tak, W. "The Sound Effects." *Philips Technical Review*, reprint of vol. 20, nos. 1 and 2/3 (1958), pp. 43–44.

Taubman, Howard. "Cold War on the Cultural Front." *New York Times*, April 13, 1958, pp. 12–37.

Taylor, Brian Brace. "Le Corbusier at Pessac: Professional and Client Responsibilities." In Russell Walden, editor, *The Open Hand: Essays on Le Corbusier*. Cambridge, Mass.: MIT Press, 1977.

Thomson, Virgil. *American Music Since 1910*. New York: Holt, Rinehart and Winston, 1971.

——. "Varèse, Xenakis, Carter." In *A Virgil Thomson Reader*, pp. 487–97. New York: E. P. Dutton, 1984.

Time, April 28, 1958, p. 29.

Time, March 31, 1958, pp. 56–64.

Van Helmond, Marcel et al. *Muziekstad Eindhoven*. Hapert/Eindhoven: Kempen Pers B.V., 1992.

Van Kampen, Anthony. *Bouwen op Mensen: Vijfenzeventig jaar Hollandsche Beton Groep*. Rijswijk: Hollandsche Beton Groep, 1977.

"Varèse et Xenakis." In "Varèse, Vingt Ans Après . . . " *La Revue Musicale* (December 1985), pp. 383–85.

Varèse, Edgard. "The Electronic Medium." 1962. Reprinted in Eliot Schwartz and Barney Childs, editors, *Contemporary Composers on Contemporary Music*, pp. 207–8. New York: Holt, Rinehart and Winston, 1967.

——. "Music as an Art-Science." 1939. Reprinted in Eliot Schwartz and Barney Childs, editors, *Contemporary Composers on Contemporary Music*, pp. 198–201. New York: Holt, Rinehart and Winston, 1967.

——. "New Instruments and New Music." 1936. Reprinted in Eliot Schwartz and Barney Childs, editors, *Contemporary Composers on Contemporary Music*, pp. 196–98. New York: Holt, Rinehart and Winston, 1967.

——. "Rhythm, Form and Content." 1959. Reprinted in Eliot Schwartz and Barney Childs, editors, *Contemporary Composers on Contemporary Music*, pp. 201–4. New York: Holt, Rinehart and Winston, 1967.

——. "Spatial Music." 1959. Reprinted in Eliot Schwartz and Barney Childs, editors, *Contemporary Composers on Contemporary Music*, pp. 204–7. New York: Holt, Rinehart and Winston, 1967.

Vivier, Odile. *Varèse*. Paris: Editions du Seuil, 1973.

Vreedenburgh, C.G.J. "The Hyperbolic-Paraboloidal Shell and Its Mechanical Properties." *Philips Technical Review*, reprint of vol. 20, nos. 1 and 2/3 (1958), pp. 9–17.

Walden, Russell, editor. *The Open Hand: Essays on Le Corbusier*, Cambridge, Mass.: MIT Press, 1982.

Waldman, Frederic. "Edgard Varèse: An Appreciation." *Juilliard Review* 1/3 (1954).

White, Eric Walter. *Benjamin Britten*. London: Boosey & Hawkes, 1954.

Wollen, Peter. *Signs and Meaning in the Cinema*. 3d edition. Bloomington, Ind.: Indiana University Press, 1972.

Woolf, Virginia. *Orlando*. 1928. London: Virago Press, 1993.

Wörner, Karl H. *Stockhausen: Life and Work*. Translated by Bill Hopkins. Berkeley: University of California Press, 1976.

Xenakis, Iannis [Y]. "The Architectural Design of Le Corbusier and Xenakis." *Philips Technical Review*, reprint of vol. 20, nos. 1 and 2/3 (1958), pp. 2–8.

——[Janis]. "Corbusier's "Electronic Poem." *Gravesaner Blätter* 9 (1957), pp. 48–54.

——. "Notes sur un 'Geste Electronique.'" In *Musique Architecture*, pp. 143–50. Paris: Casterman, 1976.

——. "Le Pavillon Philips a l'aube d'une architecture." In *Musique Architecture*, pp. 45–59, 124–42. Paris: Casterman, 1976.

Zadkine Museum: General Guide, Paris: Paris Musées, 1991.

Zevi, Bruno. "The Closing at Brussels: Symbolism: Maggots of Architectonic Expression." *L'Architettura*, vol. 4, no. 6 (October 1958), pp. 364–65.

INDEX

Italic numbers refer to illustrations

Library of Congress Cataloging-in-Publication Data

Treib, Marc.
Space calculated in seconds : the Philips Pavilion, Le Corbusier, Edgard Varèse / Marc Treib.
p. cm.
Includes bibliographical references and index.
ISBN 0-691-02137-6 (cl. : alk. paper)

1. Philips Pavilion (Exposition universelle et internationale, 1958, Brussels, Belgium)
2. Le Corbusier, 1887–1965—Criticism and interpretation.
3. Varèse, Edgar, 1863–1965—Criticism and interpretation.
4. Group work in architecture —Belgium—Brussels.
5. Precast concrete construction —Belgium—Brussels.
6. Architecture—Technological innovations.
7. Music and architecture.
8. Brussels (Belgium)—Buildings, structures, etc.

I. Title.
NA6750.B7P497 1996
725′.91′0949332—dc20 96-4114
CIP

Marc Treib is Professor of Architecture at the University of California at Berkeley, a practicing designer, and frequent contributor to architecture, landscape and design publications. He is the author of *Sanctuaries of Spanish New Mexico* (University of California Press, 1993), co-author of *A Guide to the Gardens of Kyoto* (Shufunotomo, 1980), editor of *Modern Landscape Architecture: A Critical Review* (MIT Press, 1993) and *An Everyday Modernism: The Houses of William Wurster* (University of California Press, 1995), and co-editor of *Regional Garden Design in the United States* (Dumbarton Oaks, 1995).

Richard Felciano is Professor of Music at the University of California at Berkeley and founder of the Center for New Music and Audio Technologies, an interdisciplinary institute linking music, computer science, cognitive psychology and architecture. His interests in acoustics and space are central to a number of his musical compositions.